NATO ASI Series

Advanced Science Institutes Series

A series presenting the results of activities sponsored by the NATO Science Committee, which aims at the dissemination of advanced scientific and technological knowledge, with a view to strengthening links between scientific communities.

The Series is published by an international board of publishers in conjunction with the NATO Scientific Affairs Division

A Life Sciences B Physics	Plenum Publishing Corporation London and New York
C Mathematical and Physical Sciences D Behavioural and Social Sciences E Applied Sciences	Kluwer Academic Publishers Dordrecht, Boston and London
F Computer and Systems Sciences G Ecological Sciences H Cell Biology I Global Environmental Change	Springer-Verlag Berlin Heidelberg New York London Paris Tokyo Hong Kong Barcelona Budapest

PARTNERSHIP SUB-SERIES

1. Disarmament Technologies	Kluwer Academic Publishers
2. Environment	Springer-Verlag/Kluwer Acad. Publishers
3. High Technology	Kluwer Academic Publishers
4. Science and Technology Policy	Kluwer Academic Publishers
5. Computer Networking	Kluwer Academic Publishers

The Partnership Sub-Series incorporates activities undertaken in collaboration with NATO's Cooperation Partners, the countries of the CIS and Central and Eastern Europe, in Priority Areas of concern to those countries.

NATO-PCO DATABASE

The electronic index to the NATO ASI Series provides full bibliographical references (with keywords and/or abstracts) to about 50000 contributions from international scientists published in all sections of the NATO ASI Series. Access to the NATO-PCO DATABASE compiled by the NATO Publication Coordination Office is possible in two ways:

- via online FILE 128 (NATO-PCO DATABASE) hosted by ESRIN, Via Galileo Galilei, I-00044 Frascati, Italy.
- via CD-ROM "NATO Science & Technology Disk" with user-friendly retrieval software in English, French and German (© WTV GmbH and DATAWARE Technologies Inc. 1992).

The CD-ROM can be ordered through any member of the Board of Publishers or through NATO-PCO, Overijse, Belgium.

Series I: Global Environmental Change, Vol. 46

Springer
Berlin
Heidelberg
New York
Barcelona
Budapest
Hong Kong
London
Milan
Paris
Santa Clara
Singapore
Tokyo

Land Surface Processes in Hydrology

Trials and Tribulations of Modeling and Measuring

Edited by

Soroosh Sorooshian
Hoshin V. Gupta
Department of Hydrology and Water Resources
The University of Arizona
Tucson, AZ 85721, USA

John C. Rodda
Hydrology and Water Resources Department
World Meteorological Organization
CH-1211 Geneva, Switzerland
presently at:
International Association of Hydrological Sciences
Institute of Hydrology
Wallingford, Oxfordshire OX10 8BB, U.K.

With 117 Figures (6 colour plates) and 29 Tables

Springer
Published in cooperation with NATO Scientific Affairs Division

Proceedings of the NATO Advanced Research Workshop "Global Environmental Change and Land Surface Processes in Hydrology: The Trials and Tribulations of Modeling and Measuring", held in Tucson, Arizona, May 17–21, 1993.

Library of Congress Cataloging-in-Publication Data

Land surface processes in hydrology : trials and tribulations of modeling and measuring / edited by Soroosh Sorooshian, Hoshin V. Gupta, John C. Rodda.
p. cm. -- (NATO ASI series. Series I, Global environmental change ; vol. 46)
Proceedings of a workshop held in Tucson, Ariz., May 17-21, 1993.
"Published in cooperation with NATO Scientific Affairs Division."
Includes bibliographical references and index.
ISBN 978-3-642-64458-0 (hc) e-ISBN-13: 978-3-642-60567-3
DOI:10.1007/978-3-642-60567-3

1. Hydrology--Mathematical models--Congresses. 2. Hydrology--Remote sensing--Congresses. 3. Atmospheric circulation--Mathematical models--Congresses. I. Sorooshian, Soroosh. II. Gupta, Hoshin V. (Hoshin Vijai), 1957- . III. Rodda, J. C. IV. Series.
GB656.2.M33L36 1996
551.48'01'5118--DC20 96-43298
CIP

ISBN 978-3-642-64458-0 Springer-Verlag Berlin Heidelberg New York

Softcover reprint of the hardcover 1st edition 1997

Typesetting: Camera ready by the authors/editors
Printed on acid-free paper
SPIN: 10100909 31/3137 - 5 4 3 2 1 0

PREFACE

General Circulation Models (GCMs) predict certain changes in the amounts and distribution of precipitation, but the conversion of these predictions of impacts on water resources presents novel problems in hydrologic modeling, particularly with regard to the scale of the processes involved. In turn, the GCMs need a better hydrologic component because their results are dependent on land surface processes, especially evaporation. Conventional hydrologic models provide only a "lumped" representation of catchment processes, suited to the point nature of the generally available data. These models do not provide detailed representations of the variation of hydrologic processes over a large watershed, which are required to describe interactions within a GCM. Further improvements within GCMs depend, therefore, on the development of the distributed models now being studied by a number of research groups.

This interest in distributed models coincides with the availability of a number of new remotely sensed data sources for hydrology. These include radar and satellite imagery that provide data representing processes averaged over larger spatial and temporal scales, in contrast to the point data previously provided by conventional instruments. These spatially distributed data are essential for the development and verification of the distributed models mentioned above.

As is often true when there is sudden and rapid progress in one area of science, hydrologists now find themselves in the process of figuring out how to adjust to the changes. While the new technologies offer much potential, there are many attendant problems with the translation of remotely sensed signals into hydrologically relevant information. In some cases, it is not clear exactly how the remotely sensed signals are related to the hydrologic fluxes and variables of interest. In addition, the spatial and temporal scales of measurement are not necessarily consistent with the natural scales at which hydrologic processes operate.

There are also many difficulties in the development of hydrologic models that can take full advantage of the newly available data. Of course, the choice of modeling approach depends significantly on the objectives to be achieved. For example, a great deal of attention is being devoted to the representation of land surface hydrologic processes in General Circulation Models and regional and global scale climate studies. There is much debate about the level of complexity at which such processes should be represented in GCMs since the interests and needs of atmospheric modelers and those of watershed hydrologists are quite different. It is against this background that the North Atlantic Treaty Organization (NATO) Advanced Research Workshop (ARW) on "Global Environmental Change & Land Surface Processes in Hydrology: The Trials and Tribulations of Modeling and Measuring" was organized.

The idea for holding an Advanced Research Workshop on this topic grew out of conversations between *Professor Soroosh Sorooshian* and colleagues (primarily *Dr. John Rodda, Dr. Robert Clark, Dr. Alfred Becker*, and *Associate Professor Hoshin Gupta*) at various scientific meetings. Subsequently, NATO agreed to sponsor the ARW and an organizing committee was formed. The committee comprised the director *Professor Soroosh Sorooshian* (USA), co-director *Dr. John Rodda* (UK), and organizers *Associate Professor Hoshin V. Gupta* (USA), *Dr. Yann H. Kerr* (France), and *Professor James Shuttleworth* (USA).

The overall objective of the ARW was to bring together a number of prominent researchers from the areas of "modeling" (at both watershed hydrology and GCM scales) and "measurement" (including both traditional sources and the new remotely sensed, areally distributed data sources), to identify and discuss the problems that are hindering the development of hydrologic science in the area of land surface processes and to identify effective and efficient strategies and directions towards which current and future efforts should be directed. The ARW was held in Tucson, Arizona from May 17-21, 1993. It was attended by 52 scientists from 11 NATO countries and 8 non-NATO countries. In addition, from time to time, there were approximately 6 to 10 local graduate students and scientists in attendance, who provided valuable input to the discussions and assisted by taking notes and preparing written transcriptions.

The workshop was immensely successful. It was held at the Westward Look Resort in the foothills north of Tucson, and all the participants stayed at the hotel, and had their meals together on the patio outside the conference room. This arrangement enabled the participants to spend a great deal of time together, outside of conference times, resulting in a lot of camaraderie and discussion. There is no doubt that this contributed significantly to the success of the workshop. Keynote presentations were made by 19 speakers. In addition, many of the non-speaker participants were recruited to lead panel discussions and to prepare summaries of the discussions. This format produced a lot of active debate leading to several constructive recommendations regarding what kinds of data are needed by hydrologists and how such data should be collected. There was also a lot of communication between modelers and data measurement experts as to how to better coordinate their efforts. In particular, there was unanimous agreement that both modelers and measurers should coordinate their efforts to actively participate in and guide large-scale watershed experiments such as the Global Energy and Water Experiment (GEWEX) and observation efforts such as the Tropical Rainfall Measuring Mission (TRMM).

In addition to sponsorship by NATO and the Department of Hydrology and Water Resources of The University of Arizona (UA), financial support for the ARW was provided by The UA Foundation (*Richard Imwalle*), The UA Office of the Vice President for Research (*Michael A. Cusanovich)*, The UA College of Engineering (*Dean Ernest T. Smerdon*), the World Meteorological Organization in Switzerland (Permanent Representative of the U.S. to the WMO, *Elbert W. Friday, Jr.*), and the Salt River Project in Phoenix, Arizona (*Dan Phillips*). *Dr. Topozoglu* (of the Scientific and Technical Research Council of Turkey), the Turkish National Administrator for the NATO Science Fellowships, provided support for travel and living expenses for the Turkish participants. Numerous people worked in the background to ensure that the workshop proceeded smoothly. *Dr. da Cunha,* the Programme Director of the NATO Division des Affaires Scientifiques et de L'Environment, provided much needed guidance and advice. *Paul Baltes* and his staff in The UA Office of Engineering Professional Development did an outstanding job of handling all the logistical details. *Sue Hunter* of Century Travel helped the participants with their travel arrangements, booking most of the tickets and dealing gracefully with the complexities of travel from as far away as Russia. *Corrie Thies and Fran Janssen* helped keep track of the many organizational details and provided secretarial support.

This book consists of twenty-four chapters based on the invited lecturers, who gave presentations at the ARW, and on transcripts and summaries of the panel discussions. The introduction consists of broad discussions of issues related to hydrologic modeling and measuring. The remaining material is grouped under the chapter headings of Precipitation, Soil Moisture, Evapotranspiration, and Runoff.

We wish to thank each of the chapter authors (each of you a recognized leader in your field) for your contributions. We are very grateful to you for taking the time and effort from your busy schedules to prepare the manuscripts and participate in the ARW. Needless to say, we are indebted to the workshop participants for your contributions towards the immense success of the meeting. We can hardly begin to express our deep appreciation to *Ms. Corrie Thies* for your marathon efforts related to the preparation of this manuscript. Thank you also, *Bisher Imam, Emily Snider,* and *Mary Nett* for carrying out the non-trivial job of indexing, and *Keith Beven, Robert Dickinson, David Legates, David Goodrich, Barry Goodison, Roni Avissar, Jens Refsgaard, Bernard Seguin, Susan Moran, Jene Michaud, Francois De Troch, James Washburne, Xiaogang Gao, Paul Houser, Patrice Yapo, Jean Morrill, Xu Liang, Remigio Galarraga-Sanchez, Christa Peters, Helene Unland, Muhammud Shaikh, Chuck George, Mark Lynch-Steiglitz, Mariza Costa-Cabral,* and *Joseph Mas-Pla* for providing notes and summaries from the panel discussions. Finally, thanks to a very patient publisher.

Here then, is the record of this extremely productive and thoroughly valuable workshop. It is our sincere hope that this book will be as valuable to you, the reader, as the ARW was to those who attended it.

Soroosh Sorooshian
Hoshin Vijai Gupta
John Rodda

List of Contributors

A. Allen Bradley Iowa Institute of Hydraulic Research, University of Iowa, Iowa City, IA 52242, USA

Keith Beven University of Lancaster, Dept. of Environmental Science, Lancaster LA1 4Y2, UK

Ian R. Calder Institute of Hydrology, Maclean Building, Crowmarsh Gifford, Wallingford, Oxon OX10 8BB, UK

Bhaskar J. Choudhury NASA Goddard Space Flight Center, Hydrological Sciences Branch, Code 974, Greenbelt, MD 20771, USA

Chris G. Collier Department of Civil and Environmental Engineering, The University of Salford, Salford, M5 4WT, UK

Hoshin Vijai Gupta Department of Hydrology and Water Resources, The University of Arizona, Tucson, AZ 85721, USA

Reg W. Herschy CNS Scientific & Engineering Services, 2 Queensborough Drive, Reading RG4 7JA, UK

Katya Laval Lab. de Meteorologie Dynamique du CNRS, Ecole Normale Superieure, 24 Rue Lhomond, 75231 Paris Cedex 05, FRANCE

Sharon Nicholson Department of Meteorology B-161, The Florida State University, Tallahassee, FL 32306-3034, USA

Jens Refsgaard Chief Hydrologist, Water Resources Division, Danish Hydraulic Institute, Agern Alle 5DK-2970, Horsholm, DENMARK

John C. Rodda Hydrology & Water Resources Department, World Meteorological Organization, 41 Avenue Giuseppe Motta, Case Postale 2300CH-1211, Geneva 2, SWITZERLAND

Helmut Rott University of Innsbruck, Meteorology and Geophysics, Innrain 52, A-6020, Innsbruck, AUSTRIA

Thomas J. Schmugge USDA Agricultural Research Service, Hydrology Laboratory, Building 007, Room 104 BARC-WEST, Beltsville, MD 20705, USA

Bernard Seguin	Station de Bioclimatologie, Institut National Recherche Agronomique, Centre de Recherches d'Avignon, BP 9184140 Montfavet, FRANCE
Anatolij M. Shutko	Institute of Radio Engineering and Electronics, Russian Academy of Sciences, 1 Vvedensky Sq., Friazino, Moscow Region, 141120, RUSSIA
William J. Shuttleworth	Department of Hydrology and Water Resources, The University of Arizona, Tucson, AZ 85721, USA
James A. Smith	Department of Civil Engineering and Operations Resources, Princeton University, Princeton, NJ 08544, USA
Soroosh Sorooshian	Department of Hydrology and Water Resources, The University of Arizona, Tucson, AZ 85721, USA
Odon Starosolszky	Research Center for Water Resources Development, VITUKI1095 Budapest, Kvassay ut 1, HUNGARY

List of Workshop Participants

Maria D. Ananicheva	Institute of Geography, Russian Federation Academy of Sciences, Staromonetry 29, Moscow 109017, RUSSIA
Roni Avissar	Department of Meteorology, Rutgers University, Cook College, P.O. Box 231, New Brunswick, NJ 08903, USA
Alfred Becker	BAHC Core Project Office, Institute of Meteorology, Free University of Berlin, Car-Heinrich-Becker-Weg 6-10, D-1000 Berlin 41, GERMANY
Keith Beven	University of Lancaster, Dept. of Environmental Science, Lancaster LA1 4Y2, UK
Ian R. Calder	Institute of Hydrology, Maclean Building, Crowmarsh Gifford, Wallingford, Oxon OX10 8BB, UK
Abdelghani Chehbouni	Jet Propulsion Laboratories, Mail Stop 300-233, California Institute of Technology, 4800 Oak Grove Drive, Pasadena, CA 91109, USA
Bhaskar J. Choudhury	NASA Goddard Space Flight Center, Hydrological Sciences Branch, Code 974, Greenbelt, MD 20771, USA
Chris G. Collier	Meteorological Office, Room G6, Sutton House, London Road, Bracknell, Berkshire, RG12 2SY, UK
Mariza Costa-Cabral	Department of Civil Engineering, University of Washington, Seattle, WA 98185, USA
Henk deBruin	Department of Meteorology, University of Wageningen, Duivendaal 2, 6701 AP Wageningen, THE NETHERLANDS
Francois P. De Troch	Laboratory of Hydrology & Water Management, University of Ghent, Coupure Links 653, B9000 Ghent, BELGIUM
Robert Dickinson	Department of Atmospheric Sciences, The University of Arizona, Tucson, AZ 85721, USA
Jeff Dozier	Department of Geography, Center for Remote Sensing & Environmental Optics, Girvetz Hall, Univ. of California, Santa Barbara, CA 93106, USA
Edwin T. Engman	Goddard Space Flight Center, Code 974, NASA, Greenbelt, MD 20771, USA

Dara Entekhabi	Department of Civil and Environmental Engineering, 48-331, Ralph M. Parsons Laboratory, Massachusetts Institute of Technology, Cambridge, MA 02139, USA
Ozelkan C Ertunga	Department of Systems and Industrial Engineering, The University of Arizona, Tucson, AZ 85721, USA
James Famiglietti	Climate System Modeling Program, University Corporation for Atmospheric Research, P.O. Box 3000, Boulder, CO 80307-3000 USA
Jaime Garatuza	Instituto Technológico de Sonora, Dirección de Investigación y Estudios de Posgrado, 5 de Febrero No. 818 sur, Tel. 7-04-91 Apdo. Postal 541, 8500 Cd. Obregón, Sonora, MEXICO
Barry Goodison	Atmospheric Environment Service, Canadian Climate Centre, 4905 Dufferin St., Downsview, Ontario M3H 5T4 CANADA
David C. Goodrich	USDA-ARS, Southwest Watershed Research Center, 2000 E. Allen Rd., Tucson, AZ 85719, USA
Hoshin Vijai Gupta	Department of Hydrology and Water Resources, The University of Arizona, Tucson, AZ 85721, USA
Reg W. Herschy	CNS Scientific & Engineering Services, 2 Queensborough Drive, Reading RG4 7JA, UK
Katherine Hirschboeck	Climatology: Tree Ring Laboratory, West Stadium 105 B, The University of Arizona, Tucson, AZ 85721, USA
Pavel Kabat	Agricultural Research Department (DLO-NL), The Winand Staring Centre for Integrated Land, Soil and Water Research, Marijkeweg 11/22, PO Box 125, NL-6700 AC Wageningen, THE NETHERLANDS
Mehmet Karaca	ITU, Jeoloji Muhendisligi Bolumu, Genel Jeoloji Anabilim Dali, Ayazaga Kamposo 80626, Maslak, Istanbul, TURKEY
Yann H. Kerr	CESBIO, 18 Avenue Edouard, Belin F-31055, Toulouse, FRANCE
Helmut Kuhr	DLR /PT-USF, Sudstrasse 125, 5300 Bonn 2, GERMANY
Katya Laval	Lab. de Meteorologie Dynamique du CNRS, Ecole Normale Superieure, 24 Rue Lhomond, 75231 Paris Cedex 05, FRANCE
David R. Legates	Department of Geography, The University of Oklahoma, 100 East Boyd St., Room 684, Norman, OK 73019-0628, USA

Xu Liang	Department of Civil Engineering, Wilcox Hall, FX-10, University of Washington, Seattle, WA 98195, USA
Joseph Mas-Pla	Department of Hydrology and Water Resources, The University of Arizona, Tucson, AZ 85721, USA
Jené Michaud	Department of Hydrology and Water Resources, The University of Arizona, Tucson, AZ 85721, USA
Susan Moran	USDA Agricultural Research Service, Aridland Watershed Management Research Unit, 2000 E. Allen Rd., Tucson AZ 85719, USA
Sharon Nicholson	Department of Meteorology, Florida State University, Tallahassee, FL 32306, USA
Eni G. Njoku	Jet Propulsion Laboratory, Mail Stop 300-233, National Aeronautics and Space Administration, 4800 Oak Grove Drive, Pasadena, CA 91109, USA
Nazmi Oruc	Faculty of Engineering and Architecture, Anadolu University, Bademlik, Eskisehir, TURKEY
Christa Peters	Department of Civil Engineering, Princeton University, Princeton NJ 08544, USA
Jens Refsgaard	Chief Hydrologist, Water Resources Division, Danish Hydraulic Institute, Agern Alle 5DK-2970, Horsholm, DENMARK
John C. Rodda	Hydrology & Water Resources Department, WMO, World Meteorological Organization, 41 Avenue Giuseppe Motta, Case Postale 2300CH-1211, Geneva 2, SWITZERLAND
Helmut Rott	University of Innsbruck, Meteorology and Geophysics, Innrain 52, A-6020, Innsbruck, AUSTRIA
Osman Samsun	Faculty of Fisheries, Ondokuz Mayis University, Sinop Su Ürünleri Fakültesi, 57000 - Sinop, TURKEY
Thomas J. Schmugge	USDA Agricultural Research Service, Hydrology Laboratory, Building 007; Room 104 BARC-WEST, Beltsville, MD 20705, USA
Bernard Seguin	Station de Bioclimatologie, Institut National Recherche Agronomique, Centre de Recherches d'Avignon, BP 9184140 Montfavet, FRANCE

Andrei B. Shmakin	Institute of Geography, Russian Federation Academy of Sciences, Staromonetry 29, Moscow 109017, RUSSIA
Anatolij M. Shutko	Institute of Radio Engineering and Electronics, Russian Academy of Sciences, 1 Vvedensky Sq., Friazino, Moscow Region 141120, RUSSIA
William J. Shuttleworth	Department of Hydrology and Water Resources, The University of Arizona, Tucson, AZ 85721, USA
James A. Smith	Department of Civil Engineering and Operations Resources, Princeton University, Princeton, NJ 08544, USA
Odon Starosolszky	Research Center for Water Resources Development, VITUKI1095 Budapest, Kvassay ut 1, HUNGARY
Marc Lynch-Steiglitz	Goddard Institute for Space Studies, 2880 Broadway, New York, NY 10025, USA
Soroosh Sorooshian	Department of Hydrology and Water Resources, The University of Arizona, Tucson, AZ 85721, USA
Helene Unland	Department of Hydrology and Water Resources, The University of Arizona, Tucson, AZ 85721, USA
Chris Watts	CIDESON, AP G-47, Hermosillo, Sonora 83240, MEXICO

TABLE OF CONTENTS

Part V - Runoff

Part I - Models and Data

Data, Data Everywhere, Nor Any Drop to Drink

John C. Rodda, Director
Hydrology and Water Resources Department
World Meteorological Organization
Geneva, Switzerland

1. Water and Global Change

Water, particularly the fresh water that occupies the land phase of the hydrological cycle, is uniquely placed for characterizing global environmental change. Its movement powers the different geochemical cycles, and it moderates the global energy budget: at the same time, water is essential to virtually every human activity and is a key to development. As a consequence, virtually every human activity has an impact on water resources. Sometimes these are insignificant, but more often they result in changes to the volume and distribution of water in space and/or time and/or changes to its chemical, physical, and biological attributes. These changes are, of course, largely a consequence of the rapidly rising demand for water due to the population explosion and increasing affluence, coupled with the large volume of waste being discharged, often untreated, into the environment.

For these reasons and others, fresh water was identified as one of the nine major environmental issues for consideration at the United Nations Conference on Environment and Development (UNCED) which took place in Rio de Janeiro in June 1992. The International Conference on Water and the Environment (ICWE) held in Dublin in January 1992, as part of these preparations for UNCED, highlighted the main concerns and offered a strategy for overcoming them. This strategy was then incorporated into Chapter 18 of the AGENDA 21 Document agreed at UNCED as the blueprint for the future. In fact, some argue that water encapsulates the quintessential dilemma for UNCED and subsequent events: mankind demands more and more water, but virtually every move made to meet this demand leaves less and less water for the aquatic environment and for the ecosystems the water supports.

NATO ASI Series, Vol. I 46
Land Surface Processes in Hydrology
Trials and Tribulations of Modeling and Measuring
Edited by S. Sorooshian, H. V. Gupta, and S. C. Rodda

2. Improving Knowledge of Global Hydrology

Knowledge of the global hydrological cycle, of world water resources, and of their regional, national, and local subsets derives largely from the activities of the world's hydrological services and those allied to them. These bodies undertake regular and continuing observations of the hydrological and hydrometeorological variables through the national networks of instruments they operate. In addition, the many networks established by research institutes and academic bodies have contributed considerably to the sum total of knowledge, while over the recent past a large body of new observations has become available from space agencies to make a further contribution. Many of these data collection activities have, in the past, been harnessed to water resources project development and management, but increasingly they are being focused on environmental protection and on discerning global change.

On the global scale, information about the numbers of instruments of different types which make up the national networks that combine to form the world hydrological network is collected by WMO and published in the INFOHYDRO Manual (WMO, 1994a). Table 1 shows the summary of this information for the six regions into which WMO divides the world. Data from these networks are collected and archived nationally by hydrological services and agencies with allied missions, research bodies, etc. Some of these archives hold data for a number of variables with good coverage in space and time, but the majority lack these desirable attributes. Some services make their data freely available, some charge for data, others class hydrological data as a state secret. The latter attitude and increasing commercialization are making international exchange of data more difficult.

With regard to international data sets for water resources, there is no comprehensive global centre for their collection, but there are individual centres such as the World Glacier Monitoring Service at the Swiss Federal Institute of Hydrology in Zurich, the Global Runoff Data Centre at the Federal Institute of Hydrology in Koblenz, the Global Precipitation Climatology Centre at the German Weather Service in Offenbach, and the Collaborating Centre for Surface and Groundwater Quality at Burlington. However,

Table 1. Number of Hydrological Stations

Type of Stations	Africa	Asia	South America	North & Central America	S.W. Pacific	Europe	Total
Precipitation							
Nonrecording	16349	38385	16772	19629	15914	42962	150011
Recording	2452	17760	3539	5659	3634	7714	40758
Telemetry	0	1361	141	1137	60	73	2772
Radar	10	17	2	69	17	27	142
Evaporation							
Pans	1234	3274	1431	2645	833	1210	10627
Indirect Method	28	3	10	7	28	418	494
Discharge: Total	5217	10935	6125	15269	5137	18956	61639
Recording	1433	2435	1903	12105	3916	12867	34659
Nonrecording	2700	8500	4222	3166	833	6089	25510
Telemetry	10	1136	97	1703	69	749	3764
Water Level: Total	2781	5190	4165	16222	808	8726	37892
Recording	712	1022	519	611	284	3542	6690
Nonrecording	1698	4011	3646	611	540	5235	15741
Telemetry	3	1	0	1598	38	164	1804
Sediment Discharge							
Suspended	737	3657	1123	5103	662	2850	14132
Bedload	1	677	54	0	2	188	922
Water Quality	907	4200	1297	23006	2599	18761	50770
Groundwater Level							
Observation wells	1773	13232	71	18282	635	65793	99786
Production wells	12978	48524	10382	14067	4520	46623	137094

none of the data sets these centres hold provide a sufficient coverage either spatially or in terms of length of record to obtain a reliable appreciation of the world's water resources for any particular year, month, or day. For the global scenario over the long term, there are, of course a number of studies of the world water balance, the most exhaustive being conducted by Korzun (1978).

The absence of a reliable body of hydrological data for the globe has led WMO to propose a World Hydrological Cycle Observing System (WHYCOS), which would consist of about 1,000 stations world-wide, sited on the major rivers (Rodda et al., 1993). Each station would monitor about 15 variables which would be transmitted via a geostationary satellite such as Meteosat to national, regional, and global centres. At these centres, archives of data would be built up over the period of operation of WHYCOS (about 20 years) to create a tool for decision making and for science. WHYCOS would contribute to the Global Climate Observing System (GCOS), an initiative which WMO is sharing with a number of international bodies concerned with identifying global change.

3. Degrading Knowledge of Global Hydrology

Knowledge of the global hydrological cycle is getting worse. This is largely because government cutbacks in both developed and developing countries are hitting many different types of data collection but particularly the activities of hydrological services. Low priority is given to acquiring hydrological data by governments and by national and international funding agencies. These bodies appear to be willing to invest large sums of money in hydraulic structures, irrigation schemes, and developments of various types without concern for their failure or operation below design capacity. By way of contrast, they are unwilling to spend the much smaller sums of money, even to ensure that sufficient data are collected and archived to meet present and future design and operational needs. The WMO/UNESCO Water Resources Assessment Report (WMO/UNESCO 1991) identified declining instrument networks in many countries and decreasing capabilities to utilize the data from them. There are, of course, claims that the use of remote sensing from weather radar and satellites is counteracting the cutbacks in conventional networks. While this may be true in terms of research, it is not true globally for operational hydrology (WMO, 1994a). Few hydrological services are able to use remote sensing either to supplement existing conventional networks or to replace the data from networks which have deteriorated.

4. Hydrological Data

4.1 Data Properties

Dooge (1983) identified that hydrological processes operate over some 15 orders of magnitude in space, and that their range of time scales is even larger. Fortunately, restricted parts of these ranges concern the hydrological data relevant to global change, but this still presents considerable sampling and other problems. There are so many threats to the three intrinsic properties of hydrological data (Langbein, 1972), namely impartiality, relevance, and continuity. To obtain the ideal time series of high-quality data remains a dream for many hydrologists and others involved with the detection of global change. The committed international long-term effort aiming at rigorous standards of accuracy (Dozier, 1992) has been lacking.

4.2 Acquisition of Data

Regular and continuous hydrometeorological observations have been made in certain parts of the world for several hundred years and more. Today, many of these time series provide a basis for deciphering global change. Initially, much of these data were collected by individuals and voluntary bodies, but during the 19th Century, the responsibility for collecting them came more and more under the jurisdiction of various agencies of government (only rarely have these agencies been able to maintain their identity to the present).

Instruments installed in the field at representative sites have produced the bulk of these observations and, as procedures have become more formalized, instrument and observational sites have usually been selected to best sample the patterns of the variables in space and time, often to produce areal or basin-wide means.

Unfortunately, these sites frequently need to be located in hostile environments where the conditions make it most difficult to measure the variable concerned. In many cases, the conditions of the sites have changed with time, for example, because of urbanization or erosion. In others, the instruments may have been moved to another site on one or more occasions, while the instruments themselves may have been modified and/or upgraded while their means of deployment may have changed over a period of time (Figure 1). Then there is also the problem that a number of the most useful data sets are lacking because routine measurements are not practicable, for one reason or another. The most obvious example of this lack of records is for net radiation.

4.3 Standards and Accuracies

Designing measuring devices to satisfy these specialized requirements for hydrology is often only attractive to small companies. They attempt to meet the limited demand in short manufacturing runs, but at relatively high cost. Standards of performance are not always attached to these devices by the manufacturer, but frequently the customer may not know, or may not be aware of the standards he or she is endeavoring to attain. In any case, there are few absolute standards of measurements for the hydrological variables, most of the measurements being relative. One result of this is to cast doubt on the use of the word "accuracy". This is frequently employed in circumstances when it would be more correct to use the terms "precision" or "uncertainty". It also lends an air of respectability to hydrological measurements which they may not deserve. Many suffer from low precision because they are subject to a large number of errors in the observational process. These errors are derived from a variety of sources (Figure 2), some being random, some spurious, but more seriously some are systematic. Reducing these errors to maintain and, if possible, raise standards, is the aim of every hydrological service and the objective of much of the work of WMO, ISO, and similar bodies. For example, the Guide to Hydrological Practices (WMO, 1994b) (the 5th Edition is currently

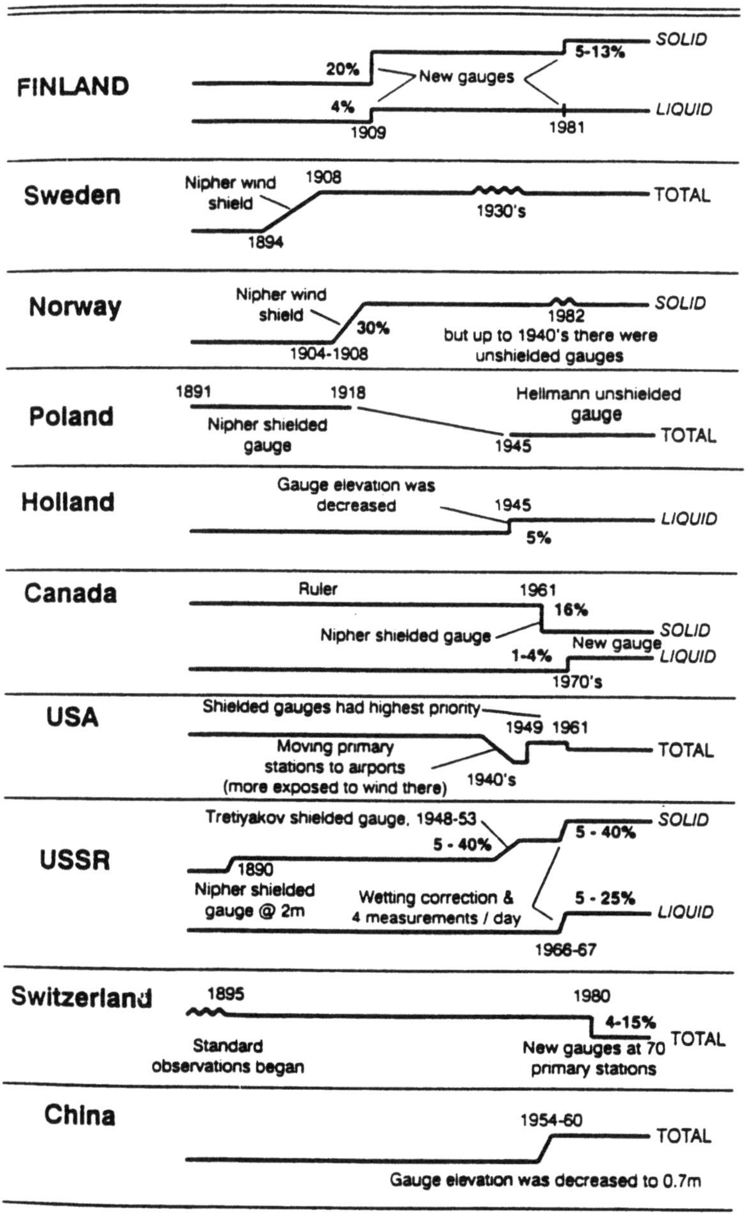

Figure 1. Discontinuuities in precipitation measurements (from Rasmussen et al., 1993)

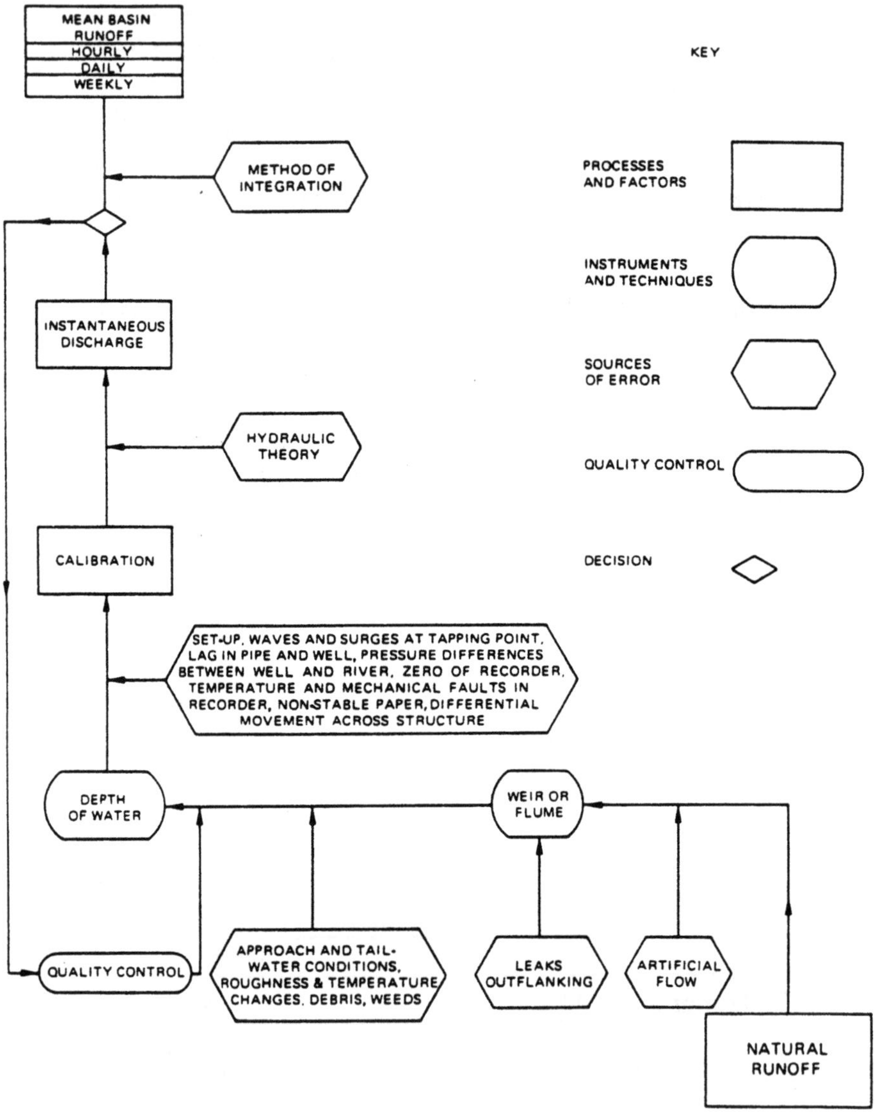

Figure 2. Sources of error in the measurement of natural runoff (from Edwards and Rodda, 1970)

in press) attempts to provide the hydrologist with an overview of the essential elements of practice in hydrology and the procedures for assuring their quality. The technology transfer system operated by WMO known as HOMS is another means for maintaining and raising standards. However, in the absence of detailed knowledge of the various accuracies, precisions, and errors associated with the measurements and observations of the different hydrological variables, the hydrologist is left in a quandary. To escape from this situation, surrogate values have had to be adopted namely "recommended accuracies". The question then is, do these accuracies have any relevance in global change studies?

4.4 Recommended Accuracies

For most of its 40-year history, the WMO Commission for Hydrology (CHy) has been foremost among those bodies and individuals grappling with the problems surrounding the acquisition and processing of data, including the errors involved. Table 2 (WMO, 1993) shows the "accuracy" recommendations that result from the deliberations of these representatives of the world's hydrological services, from their work together on instrument intercomparison projects, from the results of technical meetings, from HOMS activities, and from other cooperative ventures. Scientific theory, experimental evidence, prejudice, experience, compromise, and the like have all played a part in determining the values quoted. Some are only aspirations; others are realistic figures. Bedload transport is more likely to fall in the first category, while water level would come in the second. However, a number of the quoted levels of uncertainty invite more searching queries. In the case of suspended sediment transport, for example, it is not yet clear which is the best method of combining the spot measurements of sediment with the continuous measurement of flow, so the level of uncertainty quoted may be unrealistic. For precipitation there is, of course, no separation between the liquid and solid forms. Taking the latter, the errors involved are known to be well above the range of values listed, and the 3 to 7 percent may apply only to long period rainfall measurement at sheltered sites.

Table 2. Recommended Accuracy (Uncertainty Levels) Expressed at the 95% Confidence Interval*

Precipitation (amount and form)	3-7%
Rainfall intensity	1 mm/h
Snow depth (point)	1 cm below 20 cm or 10% above 20 cm
Water content of snow	2.5-10%
Evaporation (point)	2-5%, 0.5 mm
Wind speed	0.5 m/s
Water level	10-20 mm
Wave height	10%
Water depth	0.1 m, 2%
Width of water surface	0.5%
Velocity of flow	2-5%
Discharge	5%
Suspended sediment concentration	10%
Suspended sediment transport	10%
Bedload transport	25%
Water temperature	0.1-0.5°C
Dissolved oxygen (water temperature is more than 10°C)	3%
Turbidity	5-10%
Color	5%
pH	0.05-0.1 pH unit
Electrical conductivity	5%
Ice thickness	1-2 cm, 5%
Ice coverage	5% for $\geq$ 20 kg/m^3
Soil moisture	1 kg/m^3 $\geq$ 20 kg/m^3

* NOTE: When a range of levels is recommended, the lower value is applicable to measurements under relatively good conditions, and the higher value is applicable to measurements under difficult situations.

On behalf of the Commission for Hydrology, Kuittinen (1992) investigated the capabilities of remote sensing for determining the major hydrological variables. He detailed the observational requirements against the available and future satellite capabilities and identified (Table 3) the major remotely sensed hydrological variables that can or could be employed in hydrological models. It is suggested that research is concentrated on improving the measurements of those variables where the accuracy is low and on devising parameters for hydrological models which can be determined directly by remote sensing. Soil moisture is one of these, but perhaps more important is precipitation, where the possibilities for improved in situ and remote sensing methods may offer a bigger return in terms of reducing the errors involved.

Table 3. Remotely Sensed Hydrological Variables for Hydrological Models

Variable	Available Now		Available in Future	
	S	A	S	A
Precipitation	x			
Soil moisture		x		
Land use and vegetation	x	x		
Impervious area				
Frozen ground				x
Snow cover	x	x		
Snow water equivalent		x	x	
Evapotranspiration			x	x

(x) = air-based method; x = space-based method

5. Global Change

5.1 Problems of Detection

Global change, particularly climate change, has been the subject of an immense amount of scientific activity during recent years. It seems likely that this effort will be maintained and even grow in the future. However, the impression one gains is that science, including hydrology, is not very well equipped to establish whether or not change is occurring contemporarily, let alone to predict its future impact and the likely consequences. The biggest problem seems to be that, in many cases, the signal is at or below the noise level. In other cases, there is as yet no signal or the signal has not been received for a sufficient duration. Then there is the difficulty that natural changes are occurring which may mask those dependent on other sources.

5.2 Hydrological Change

Hydrologists have a history of involvement in detecting and cataloging change and its results, particularly land-use change. More recently, however, increasing attention has been given to climate change (Lins et al., 1990; Academy of Finland 1989) and to its monitoring such as through glacier mass balance studies (Haeberli and Herren, 1991). However, taking the example of precipitation measurement, there are so many problems associated with measuring this variable (Table 4) that it seems doubtful whether standard gauges in the existing global network can be used successfully to distinguish changes in total catch that might be the consequence of climate change. This is particularly unfortunate when the most important effects of climate change are likely to be felt through changes in hydrology and water resources. Would a universal gauge (Karl, 1993) be the answer to replace the plethora of national gauges (Figure 3) that now exist? Would improved precision in the measurement of precipitation and in the other variables be the answer to the uncertainties that now surround them?

6. Concluding Remarks

Chahine (1992) claimed that, within the short span of about 10 years, the hydrological cycle has emerged as the centrepiece of the study of climate, but that difficult mental adjustments are still to be undertaken to shape the nascent discipline of hydrology. Part of these adjustments have to be made to the difficult art of measurement, where that

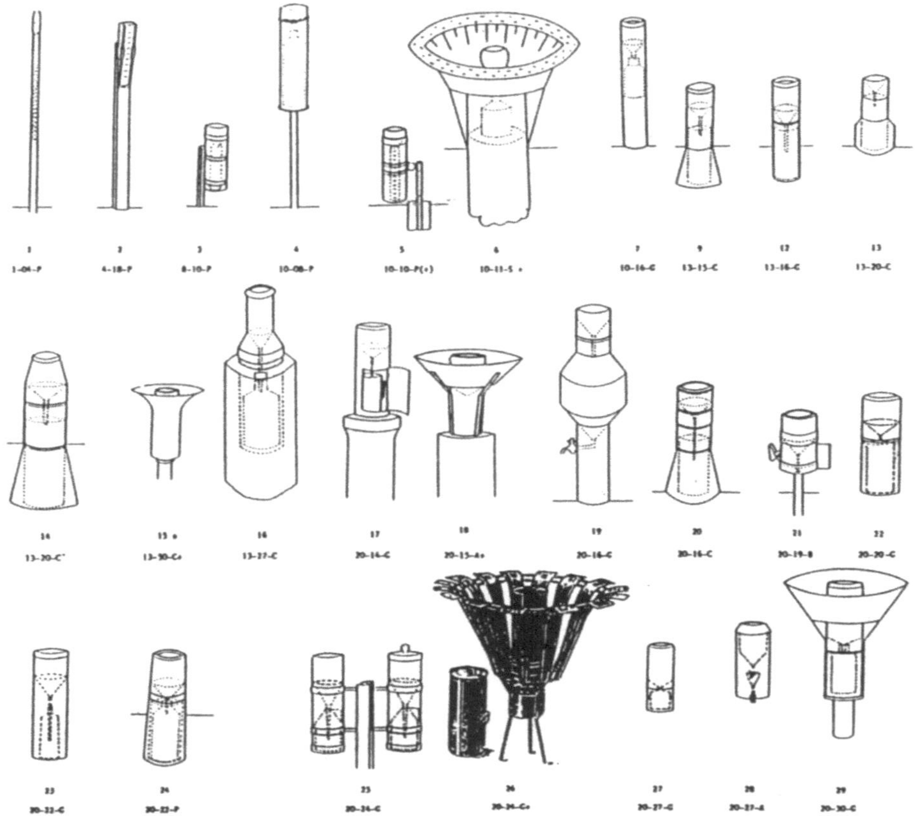

Figure 3. A small selection of national standard precipitation gauges (from Sevruk and Klemm, 1989)

and subsequent processes should be welded together as components of the hydrological information system (Figure 4). Indeed, all the signs point to the vital need for increasing discipline in the collection and compilation of hydrological data on a global basis. This discipline is demanded to meet the increasingly exact requirements within the field of hydrology and water resources as well as in those allied to it. WHYCOS and GCOS are both highly relevant initiatives which will employ this discipline. Without them and similar data management programs, the reductions in uncertainty surrounding future climate demanded by the 2nd World Climate Conference is likely to remain elusive.

Table 4. Problems of Assessing Global Change from Precipitation Measurements

1.	Spatial coverage is incomplete.
2.	Temporal coverage is incomplete.
3.	There are at least 54 different types of standard gauges in use in 136 countries covering about 90% of the land area of the globe.
4.	Errors of measurement have not been determined for each gauge.
5.	Changes have occurred in gauge exposure.
6.	Gauges have been moved.
7.	New types of gauges have been introduced.
8.	Observer practice has altered.
9.	Station histories are not documented.

References

Academy of Finland (1987) Conference on Climate and Water, Vols. 1 and 2, Helsinki, Finland.

Chahine MT (1992) The hydrological cycle and its influence on climate, *Nature*, Vol. 359, pp. 373-379.

Dooge JCI (1983) On the study of water, *Hydrological Sciences Journal*, 28, pp. 23-48.

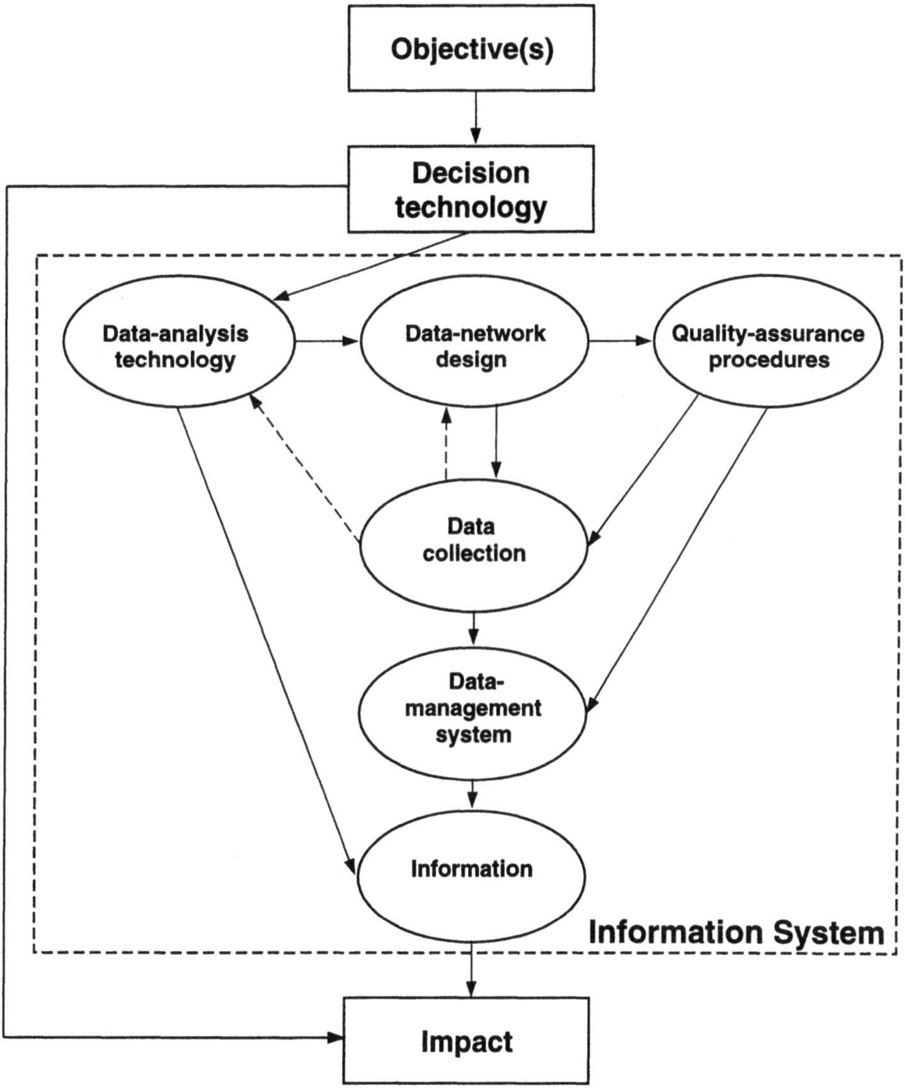

Figure 4. An idealized hydrological information system (from WHO, 1994)

Dozier J (1992) Opportunities to improve hydrologic data, *Reviews of Geophysics*, Vol. 30, pp. 315-331.

Edwards KA, JC Rodda (1970) A preliminary study of the water balance of a small clay catchment, Symposium of Wellington: Results of Research on Representative and Experimental Basins, IASH Publication No. 97, pp. 187-199.

Haeberli W, E Herren (1991) Glacier Mass Balance Bulletin, Bulletin No. 1, (1988-1989) IAHS/UNEP/UNESCO, 70 p.

ICWE (1992) International Conference on Water and the Environment--Development Issues for the 21st Century, World Meteorological Organization, 55 p.

Karl TR (1993) Detecting climate change: issues of special concern, Special Topics on Climate--Lectures presented at the 42nd Session of the WMO Executive Council, World Meteorological Organization, pp. 2-18.

Korzun VI (Ed.) (1978) World Water Balance and Water Resources of the Earth, Studies and Reports in Hydrology, No. 25, UNESCO.

Kuittinen R (1992) Remote sensing for hydrology, progress, and prospects, WMO Operational Hydrology Report, No. 36, 62 p.

Langbein WB (1972) Water Data Today and in Prospect, *Hydrological Sciences Bulletin*, Vol. 17, pp. 369-385.

Lins H, I Shiklomanov, E Stakhiv (1990) Hydrology and Water Resources, Chapter 4 in Climate Change, the IPPC Impacts Assessment Working Group (McG Tagart, OW Sheldon, DC Griffiths, Eds.), Commonwealth of Australia, pp. 4.1-4.42.

Rasmussen FM, RM Dickenson, JE Kutzback, MK Cleveland (1993) "Climatology" in: Handbook of Hydrology (Maidment DR, Ed. in Chief), McGraw-Hill, Inc.

Rodda JC, SA Pieyns, NS Sehmi, G Matthews (1993) Towards a world hydrological cycle observing system, *Hydrological Sciences Journal*, 38, pp. 373-378.

Sevruk B, S Klemm (1989) Types of Standard Precipitation Gauges, in "Precipitation Measurement", WMO/IAHS/ETH Workshop, St. Moritz, pp. 227-232.

WMO (1994a) Infohydro Manual, 2nd Edition, Operational Hydrology Report No. 28, World Meteorological Organization (in press).

WMO (1994b) Guide to Hydrological Practices, World Meteorological Organization, Fifth Edition (in press).

WMO/UNESCO (1991) Report on Water Resources Assessment, WMO/UNESCO, 64 p.

The Trials and Tribulations of Modeling and Measuring in Surface Water Hydrology

Soroosh Sorooshian
Dept. of Hydrology and Water Resources
The University of Arizona
Tucson, Arizona 85721

1. Introduction

Historically speaking, modeling of the precipitation runoff process got its start in the engineering side of hydrology. The development of these models was primarily driven by societal needs for water resources structural design purposes (i.e., dams and reservoirs, aqueducts, etc.), and operational requirements, such as flood forecasting. Most such models are "lumped", rather than "distributed", and their structures (i.e., process equations) are influenced by availability of observations and data constraints. Precipitation runoff models vary in their levels of complexity, ranging from the rational formula, the Soil Conservation Service (SCS) curve number model (SCS, 1964), the antecedent precipitation index (API) method in the "simple" category to the more complex Stanford-type conceptual rainfall-runoff (CRR) models and, finally to the physically based distributed models such as HEC-1 (Feldman, 1995) and KINEROS (Woolhiser et al., 1990; Smith et al., 1995). The spatial and temporal scale of their application varies from model to model, but ranges from tens to thousands of square miles and from minutes and hours to a few days, respectively.

The degree of parameterization varies from model to model. For instance, in the case of the rationale formula, there are two parameters involved. One is the surface area of the basin, and the other is termed the runoff coefficient, which must be user-specified. In this case, engineering experience with a lot of trial-and-error has resulted in values for the runoff coefficient that apply to different classes of catchments based on regions or soil type, etc. Information regarding these values can be obtained from most engineering handbooks and hydrology/hydraulic textbooks. At the other spectrum are the more complex hydrologic models that could well have a dozen or more parameters. The number of parameters is dictated by how elaborately the various processes involved in the conversion of precipitation to runoff in a catchment are represented. For

NATO ASI Series, Vol. I 46
Land Surface Processes in Hydrology
Trials and Tribulations of Modeling and Measuring
Edited by S. Sorooshian, H. V. Gupta, and S. C. Rodda

instance, the Sacramento Soil Moisture Accounting model (SAC-SMA; see Figure 1) of the U.S. National Weather Service River Forecast System (NWSRFS) has 16 or more adjustable parameters, excluding those associated with channel routing. While some of these parameters can be obtained from direct physical measurements, a good number of them must be obtained through some calibration process (Sorooshian, 1983). In a distributed model, parameter sets for each modeling element are required, thereby further increasing the number of required parameters.

The success (consistency and precision of their forecasts) in the application of these models is, therefore, a function of three inter-related activities: (1) the structural makeup of the model; (2) the success in calibration of the model; and (3) having the required data available for both the calibration and the application. This issue will be elaborated on later in this paper.

As recent as a decade ago, we have seen a shift in approaches to precipitation runoff modeling. There are two distinct--but still related--reasons for this shift. For one thing, there has been a surge of activities in development of "distributed" precipitation runoff models. Much of this has been fueled by an n-fold increase in computational power due to advances in computer technology, and by improved (and anticipated) observational capabilities which provide us with distributed data. Among this class of models, one might mention the Systeme Hydrologique Europeen (SHE) model (Bathurst and O'Connell, 1992) and KINEROS (Woolhiser et al., 1990; Smith et al., 1995). Nevertheless, the main purpose of these models is still for hydrologic and water resources applications which have the eventual operational forecast objective in mind.

Second, the other area where there has been a growing trend has been in development of "land-surface/atmosphere" modeling. The motivation for these developments has been fueled by the need for better representation of hydrology within the General Circulation Models (GCMs). Among this class of models, the most notable are the Biosphere Atmosphere Transfer Scheme (BATS) (Dickinson et al., 1986) and the Simple Biosphere model (SiB) (Sellers et al., 1986). The land-surface/atmosphere scheme models are also labeled as distributed models. This is due to the fact that these models represent the heterogeneity of the vertical processes governing the water-balance equations in a distributed fashion. However, from a spatial distribution point of view, they are typically "lumped" within each grid.

Theoretically, the main advantage of the distributed class of models is that they represent more accurately the heterogeneity (both in space and time) of the various hydrologic subprocesses. Of course, this comes at the expense of an even larger number of parameters, most of which are related to better representation of the physics involved. The practice to date has been that the parameters of the land-surface/atmosphere models are user-specified, and no calibration is attempted. The primary reason for this is that the observations which are required for calibration (as is commonly done in hydrology) are not available.

While the shift in modeling approaches in surface hydrology is welcomed and has the potential to make a difference, the distributed modeling approach is not without its share of problems. Hence, it is crucial that the lessons we've learned and the experiences we've gained with lumped modeling approaches guide us toward setting priorities for our modeling activities. In the following sections, a brief review of some of the lessons we've learned to date is presented. In the concluding section, some personal observations which highlight some critical problem areas are provided. Before proceeding with this, it is important to distinguish the two main classes of models discussed in this paper. The first main class of models--which are primarily the operational-type rainfall-runoff models--will be labeled as "Operational Catchment Models" (OCM). The second class of models are land-surface/atmosphere schemes which will be referred to as Land-Surface Models (LSM).

2. Operational Catchment Models (OCM)

Three elements for this class of models were identified, namely the development of model structure, the calibration, and the data requirements for both calibration and the application. A review of the literature shows that a great deal of effort has gone into the research of the first two above-mentioned elements. With regard to the third element, namely data requirements, there are high expectations with new observational technologies, such as radar and satellites, which are anticipated to result in improvements.

3. Model Structure

With respect to the model structure, there is a trend towards more sophisticated models (in terms of process representation, both vertically and horizontally). While more realistic representation--conceptually speaking--of the precipitation runoff processes has been achieved, the eventual result has been an increase in the number of parameters and the requirements for more data. Increased parameterization has translated into requirements for more rigorous calibration. As presented in a series of papers by Sorooshian et al. (1983) and Gupta and Sorooshian (1983), the identifiability and observability of model parameters are not only functions of the choice of the calibration procedure, but are highly dependent on the mathematical structure of process equations and the nature of data (in terms of variability, etc.). The experience of this author and his collaborators has been that the parameters of the percolation/infiltration equations in the SCA-SMA model which relate upper zone (surface storages) to the lower zone (soil storages) are the pivotal parameters, and their estimation poses the most difficulty in the calibration phase. The key to making further improvements in this area will depend on whether, with new technological advances in measurement techniques, a larger number of parameters could be perhaps obtained through direct or indirect measurements.

4. Calibration

Two approaches to calibration of hydrologic models are still in use. One is known as the "manual" approach, where the modeler attempts to adjust the values of the parameters based on observing the intended fit (i.e., the peaks of a hydrograph, the volume or, in general, the continuous hydrograph shape). The second approach, known as the "automatic" approach, is based on optimization theory, where an objective function (i.e., least squares, maximum likelihood criterion) would be implemented using an optimization algorithm.

The work with the automatic approach began in 1965 (Dawdy and O'Donnell), and many authors have made subsequent contributions to their work. This author and his collaborators have been heavily involved in research in this area, resulting in the publication of numerous articles in *Water Resources Research*. The most recent and perhaps the most significant quantum leap reported in the literature has been achieved with the use of random search techniques. In particular, Duan et al. (1992) presented a new algorithm called the Shuffled Complex Evolution algorithm method developed at The University of Arizona (SCE-UA). Subsequently, comparison results with other algorithms were presented in Sorooshian et al. (1993), Two outcomes from their paper are presented here to emphasize the effectiveness of the SCE-UA. The first outcome relates to some simulation studies they conducted to compare SCE-UA against the multi-start simplex (MSX). (The reason for selecting MSX was because of its superior performance over other conventional methods as reported in the literature). In this case, the model used was the SMA-NWSRFS. One hundred independent optimization trials for the MSX and ten trials for the SCE-UA were conducted. The convergence behaviors of these runs for four parameters of the SMA-NWSRFS model are displayed in Figure 2a-b for the MSX and SCE-UA, respectively. Given that these were simulation studies, the true values of the parameters were already pre-specified (shown by straight dotted lines). Therefore, each trial is considered as having an independent initial starting point in the parameter space, with the expectation that the convergence of the optimization run would lead to the true values. Comparing Figure 2a with Figure 2b clearly shows the convergence behaviors of the two algorithms for each of the parameters. For the MSX algorithm, the optimization seems to converge to almost 100 different points. For the SCE-UA method, all ten trials converged to the true parameter values.

The second set of results are obtained when the same two algorithms were used to calibrate the SMA-NWSRFS model using one year of historical data from the Leaf River watershed in Mississippi. An additional 7-year period of data was used to test and compare the performance of the calibrated model. The results (shown in Figure 3a-b) are for ten independent trials of each of the search algorithms using the maximum likelihood criterion. It is clear that the SCE-UA algorithm obtained more closely grouped parameter estimates while using 1/3 fewer function evaluations than the MSX algorithm (Duan, 1991; Sorooshian et al., 1993).

Due to recent developments in the optimization field, we have achieved great advances in our automatic calibration approaches. However, success in calibration has not necessarily improved the ability of models to perform better on a consistent basis. Among the reasons for this are peculiarities in the structural design of the modeling equations and the ensuing parameter interactions. The reader is referred to the article by Gupta and Sorooshian (1983). In that study, the non-identifiability of parameters of the percolation equation of the SMA-NWSRFS model was presented. While that study is almost a decade old, it typified the nature of parameter non-identifiability which plagues most models. This is an issue which is seldom considered in modeling studies and applications, but it's one that perhaps plays the most crucial role in the poor performance of models.

5. Data Requirements

The last element--data requirements--is perhaps one of the most crucial aspects of any modeling exercise. Numerous articles have been published discussing the role of data errors, information content, and availability for both calibration and application. Perhaps the biggest obstacle for application of precipitation runoff models has been the lack of available data for many basins, particularly headwater basins. Much of the discussion to date has been presented under the topic of "ungauged watershed". The implications are certainly more crucial for some nations, where data collection has not been a high priority. However, with the growth in population, there's now a degree of urgency in seeking ways to model precipitation- runoff for both water supply and flood prediction purposes. It is this author's expectation that this issue will generate a lot of challenges to the modeling community, particularly how well we can use remotely sensed data to overcome the difficulties for ungauged watersheds.

Another aspect of data requirements relates to procedures by which we aggregate or disaggregate data in order to meet the requirements of lumped and distributed modeling approaches, respectively. For example, let's assume that, through some remote-sensing technology, we are provided with soil moisture data which are spatially distributed. The main question then becomes: Are our distributed hydrologic models capable of using those data for initialization or parameter specification needs? An initial advance toward this goal was illustrated in Goodrich et al. (1994), who employed aircraft-derived soil moisture estimates from the push broom microwave radiometer (Schmugge et al., 1994) to define the prestorm initial soil conditions for the KINEROS (Woolhiser et al., 1990) rainfall-runoff model. This information was incorporated directly into the Smith-Parlange (1978) infiltration equations embedded in the model. It was found that the remotely derived estimates are as useful as scattered ground-based measures (gravimetric and time-domain reflectometry) for rainfall-runoff modeling in a semi-arid region. However, accurate estimation of basin response required reliable estimates of spatially distributed rainfall data, even at watershed scales as small as 4.4 ha (Goodrich et al., 1995; Faures et al., 1995).

In the case of lumped models, which are perhaps the most common type used for operational purposes, how would distributed information be used effectively? A good example is the case of distributed precipitation data that we expect from NEXRAD almost completely installed over the United States. The precipitation quantities are to be generated over 4 x 4 km grids. Even if we assume, for the moment, that such data will be reliable and that all the technical limitations of obtaining accurate precipitation estimates are overcome, are we truly ready to benefit from this important distributed precipitation product? In the case of flood forecasting, to this author's knowledge, most river forecast centers in the U.S. still rely on the use of the API method. As we all know, API uses one value for precipitation over the basin being modeled. Therefore, are agencies shifting their modeling approach to distributed types which can use distributed data effectively or, if they still intend to use lumped models, are they planning to use the radar data by applying some statistical averaging scheme to determine a lumped value? It is this author's view that there are many critical questions which need to be addressed to make best use of the distributed observational data.

6. Summary of Operational Catchment Modeling: Has Progress Been Made?

At this stage, this author will reference the work of some colleagues who have compared the performance of various models. As will be shown, results have generally indicated that simple models are approximately as accurate as complex models under certain conditions.

Naef (1981) compared six models on three watersheds, ranging in size from 1.7 to 120 km^2. The sophistication of models ranged from the rational formula to a conceptual-type model. The conclusions Naef presented were (1) even a simple model could explain a large part of the variance of streamflow, (2) some of the simple models were nearly as accurate as the complex ones, and (3) both simple and complex models will perform poorly in certain cases. In that study, all models were spatially lumped, calibrated, and validated on a split-sample approach.

Loague and Freeze (1985) compared two simple lumped models with a complex distributed model on three catchments ranging from 10 ha to 7.2 km^2. Using calibration (for the simple models) and split-sample verification, they determined that all models performed poorly under ordinary prediction mode. However, model performance improved in frequency prediction mode. In the case of one catchment, the complex model was applied with and without calibration of one of the key parameters. Split-sample verification showed that calibration had little impact on the eventual model performance. Later, Loague (1990) attempted to improve the performance of the complex distributed model by obtaining detailed field infiltration data. Even though model performance improved, it was still disappointing and, as Loague suggested, must have been due to a combination of data and model errors.

Wilcox et al. (1990) reported on daily and monthly runoff simulation studies for six small rangeland catchments (1.3-9.6 ha) using the simple SCS model and a complex model based on the Green and Ampt equation. Parameters for the Green and Ampt model were based on available soil and vegetation data, and no calibration was performed for either model. Here again, on some catchments, the simple model was more accurate than the complex one; on other catchments, the opposite was true.

One of the most recent studies was conducted by Michaud (1992). In her dissertation, which was subsequently published in *Water Resources Research* (Michaud and Sorooshian, 1994a), she compared the accuracy of simulation from a complex distributed model (KINEROS) (Woolhiser et al., 1990; Smith et al., 1995), a simple distributed model (based on the SCS method), and a simple lumped model (also based on the SCS method). The catchment was the 150 km^2 semi-arid Walnut Gulch experimental watershed of the USDA-ARS. Models were validated using 24 severe thunderstorm events and rain gauge densities similar to those at flash-flood warning sites (approximately 1 gauge per 20 km^2). The conclusions reported were that none of the models were able to accurately simulate peak flows or runoff volumes for individual events. Models showed somewhat more capability in matching time-to-peak and the ratio of peak-to-flow volume. When calibration was performed, the accuracy of the complex distributed model was practically similar to that of the simple distributed model (Figure 4). However, without calibration, the complex distributed model outperformed the simple distributed model (Figure 5). Woolhiser et al. (1996) noted important implications of the size of distributed model elements used by Michaud and Sorooshian (1994a) which caused significant incorporation of concentrated flow and channel processes into their large overland flow plane elements. Goodrich (1990) achieved exceptional runoff model validation results on four smaller subwatersheds within Walnut Gulch, ranging from 0.003 to 6.3 km^2. However, the distributed model element size employed by Goodrich (1990) was typically an order of magnitude smaller than that in Michaud and Sorooshian (1994a).

Woolhiser et al. (1996) also noted a dramatic interaction between distributed model element size and trends in infiltration properties, even after rainfall ceases. The point here is that, to achieve reliable model predictions, a very large number of model elements is required, which presents further barriers to operational implementation of these models.

While the study by Michaud and Sorooshian addressed the flash-flood forecasting capability of models over semi-arid watersheds, they re-emphasized the concern about the capabilities of our operational catchment models. It should be clear that forecasting in a semi-arid environment with predominantly localized convective thunderstorms subjects the model to a very rigorous test. Nevertheless, it is clear that the accuracy of rainfall-runoff forecasts would be poor, particularly in the case of other semi-arid watersheds, where the availability of data is likely to be more limited (Walnut Gulch has perhaps some of the best observational data because of its experimental nature).

In addition, Michaud (Michaud and Sorooshian, 1994b) looked into the effect of rainfall sampling errors on distributed hydrologic simulations using Walnut Gulch data. Rainfall fields based on observations from a very dense rain gauge network were compared to rainfall fields based on observations from a subset of the original gauges. The rationale for selecting the "sparse" network (one gauge per 20 km^2) was to simulate the condition of a typical gauge density recommended for the real-time density of an ALERT-type flash-flood warning system in the United States (note: ALERT stands for Automatic Local Evaluation in Real-Time). In the sparse network case, on the average, the errors in simulated peaks represented 58% of the observed peak flows. Half of that difference between observed and simulated peaks was attributed to rainfall sampling errors. Simulations were also conducted with rainfall which was similar to the NEXRAD precipitation coverage (4 km x 4 km pixels). Spatial averaging for the 4 x 4 km led to consistent reduction in simulated peaks that, on the average, represented 50% of the observed peak flow. The rainfall sampling studies highlight the very critical issue of anticipated difficulties for the distributed modeling of precipitation runoff processes. It is worth mentioning that, even if we have near-perfect distributed models, we will still have serious problems in dealing with sampling errors due to the heterogeneity of precipitation. This was emphasized further by Goodrich et al. (1994) and Goodrich (1990) who compared the impacts of using a single central rain gauge (uniform rainfall assumption) versus 10 well-distributed rain gauges in a 6.3 km^2 subwatershed in Walnut Gulch. The errors due to assuming uniform rainfall over this relatively small area were greater than those associated with simplifying the entire watershed to a single overland flow plane or a lumped watershed representation.

Significant spatial rainfall variability in this semi-arid environment persisted even to the catchment scales of 5 ha (Goodrich et al., 1995) and had significant impact on runoff model prediction (Faures et al., 1995), who concluded: "The results of this study indicate that, if distributed catchment modeling is to be conducted at the 5 hectare scale, knowledge of the rainfall variability on the same scale is required. A single rain gauge with the standard uniform rainfall assumption can lead to large uncertainties in runoff estimation." This, of course, will also lead to further difficulties in parameter optimization (Woolhiser, 1996).

Having provided a few examples from the recent literature, it is this author's belief that--while progress in specific areas (i.e., calibration, better model representation, etc.) has been made--when put together, there is little evidence that significant differences have been made at larger operational scales (> 100 km^2) to improve model performance in terms of accuracy and reliability of their forecasts. One could argue and attempt to provide justification why complex models did or did not perform as well as expected. It is fair to state that the results of any model comparison studies involving complex versus simple models could easily be debated, based on the type of initial conditions, boundary conditions, and type of data, etc. used in the studies. Almost all of these arguments are valid and justified because, as one attempts the use of any sophisticated model, there is a much greater sensitivity (i.e., the model output) to the factors listed above

(Woolhiser, 1996; Woolhiser et al., 1996; Smith et al., 1994). However, the bottom line is that, in most reported cases, there is a mixture of both simple and complex models performing similarly. Our arguments defending one model versus another might be more academic, and we could continue writing papers and discussing them in scientific meetings. Nonetheless, the fact remains that hydrologists on the operational side are likely to be reluctant to shift their current practices, because they do not see any convincing arguments to change their ongoing practices.

7. Land Surface Models (LSM)

Models classified under LSM were originally developed by the researchers involved in global-scale General Circulation Models (GCMs). The most frequently referenced in this class (as already mentioned) are BATS and SiB. A schematic representation of a model typifying the BATS class of models is displayed in Figure 6. These models provide an elaborate representation of the physics of the processes involved in the transfer of energy and water between the land and the atmosphere. Hence, the hydrology represented by them is the vertical movement of water, which takes place through evapotranspiration and infiltration and percolation processes in the soil. In terms of their spatial distribution, these models are lumped and represent an average value for the prespecified modeling grid. While they are highly parametric, their parameter values are user-specified, most of which are based on theoretical or experimental experiences at the point scale. Significant difficulty can arise for estimating these parameters at typical GCM grid scales. Because of their intended scale of application, their topographical representation is not as detailed as desired for most hydrologic simulations.

Much of the recent research activities, primarily promoted by the hydrologic community, have been to improve the hydrologic representation (in terms of improved spatial resolution) of LSMs. The most common classification given to these more detailed models is Soil Vegetation Atmosphere Transfer Schemes (SVATS). Notable among the work reported in recent literature are those of Famiglietti and Wood (1994). SVATS place a greater deal of emphasis on a finer representation of topography, which is critical for spatial generation of runoff at subgrid scale. Research in this area--being relatively recent--has not been subjected to close scrutiny. Big challenges for researchers in this area will be to achieve a balance in obtaining a reasonable spatial scale for which required data would be available. In this regard, there are high expectations that remotely sensed data will satisfy a great deal of the needs. A good number of field experiments (FIFE (First ISLSCP Field Experiment, 1992); Kustas and Goodrich, 1994; Monsoon '90; Washita, 1992; etc.) have been organized to test and improve SVATS. Other projects such as GCIP are intended to provide the interface between SVATS and GCMs in order to achieve the necessary improvements in water balance over continental scale. In all of these studies, the practicality of using remotely sensed data is also being explored.

Recent studies directed towards comparison of various land-surface schemes have undoubtedly identified the need for much more research in this important area. For instance, the Project for Intercomparison of Land-surface Parameterization Schemes (PILPS) is a notable intercomparison study (the report is WCRP/IGPO Publication #14, December 1994). Seventeen land-surface schemes were compared using the same data sets. Large differences in simulation results of partitioning of annual total precipitation into liquid water loss (runoff and drainage) and evaporation were discovered (Figure 7). In this study, each model was subjected to the same controlled-run conditions, and all models went through a "spinning up" period to avoid the effects of initial soil moisture setting. For the case exhibited in Figure 7, the annual total precipitation in the selected test site is about 870 mm. A ten-year average of observed ratio of runoff to precipitation was used to estimate the annual total liquid water loss (241-296 mm). The liquid water loss simulated by different land-surface schemes ranges from 100 to 300 mm. This error in predicting the amount of liquid water loss carries over to the error in the simulation of evaporation. Because these are equilibrium results, and no change in soil-water storage occurs, all models are expected to fall on a single line with the intercept representing the total precipitation of 870 mm. In an ideal world (assuming all models performed accurately and as expected), one should expect to see all 17 models clustered near each other. However, as shown in Figure 7, there is quite a wide range of performances in the way each model distributes runoff plus drainage and evaporation. The main message to be appreciated is that one can expect a wide range of performance results from these models, and that none of them could necessarily be trusted.

The latest intercomparison study was reported by Lau et al. (1995) where the ability of 29 global climate models (GCMs) in simulating various aspects of regional and hydrologic processes were presented. Specific to the hydrologic aspects, two conclusions from Lau et al.'s work are most relevant to this report. Quoting from their work:

> "The models' depiction of the hydrologic cycle tends to be more realistic with least intermodel variability where there is a strong annual cycle and where long-term local moisture balance is maintained, i.e., (P-E) = 0, i.e., over large interior land regions in the extratropics. In regions of strong dynamic control, i.e., (P-E) >> 0, such as the tropical western Pacific, monsoon regions, and ITCZ, the model differences are very large. The GCM runoffs are consistent with the climatological observed runoffs for the Amazon and Mississippi basins. However, the intermodel variability is very large. Moreover, the models yield rather divergent results over other world river basins. The results suggest that, while some GCMs are better than others in simulating the land hydrology cycles, it is premature to use them for climate studies related to continental scale water balance."

From the conclusions of the two above-mentioned studies, which are the most recent, it is clear that the hydrologic components of LSMs will provide a great deal of challenge and opportunity to the research community. There are many ongoing investigations in which the distributed approach is viewed as a solution to achieving satisfactory representation of the hydrology within LSMs. In one case, it is rumored that the entire U.S. is being represented by 40,000 hydrologic units in order to simulate the water balance needed for GCMs. However, it is this author's belief that, while computer technology and Geographic Information Systems (GIS) capabilities may make this possible, we should not get carried away if there are not enough data to do an adequate hydrologic balance within each of the model subunits.

8. Conclusions

The above discussions have highlighted the fact that, while incremental progress in the various aspects of modeling of the hydrologic processes has been achieved, the overall performance has not necessarily improved. More sophisticated modeling (i.e., more distributed approaches, better calibration methodologies, etc.) has not produced the expected results at larger scales. It is difficult to decide where to invest in order to obtain the most progress in our modeling efforts.

The scientific community most likely is divided with regard to the opinion of where the most progress could be obtained for resources invested. For what it's worth, this author's opinions are as follows.

- We still have a long way to go to integrate our knowledge within the hydrologic sciences. As an example, most hydrologic models of the OCM class rely on the traditional "engineering approaches" of separating baseflow from the total hydrograph in their routing schemes. Much of the advances made in the field of subsurface hydrology in groundwater modeling have not been utilized or extended to represent the subsurface streamflow interactions. The majority of research conducted by scientists in the vadose zone and infiltration fields, again, has not been utilized to represent the surface/subsurface interaction in most hydrologic models. It is this author's belief that a greater deal of integration to incorporate such knowledge into OCMs (which are viewed to be more of a surface water hydrologist's tool) is necessary.

- The handling of the evapotranspiration in OCMs is perhaps one area (from a process study point-of-view) which is most primitive. It is encouraging to see a high degree of research activities in this area. Furthermore, it is hoped that the resulting knowledge will be incorporated into OCMs.

- Because precipitation is the primary input (in input/output sets) to a catchment, it is crucial that problems associated with the heterogeneity, sampling error, and other considerations be given the attention they deserve.

- It is encouraging to see a great deal of research activities in the development and definition of SVATS. However, it is important that the knowledge and the experience with OCMs not be ignored. In other words, with advances in computer and graphics technologies, it is relatively simple to create new models as sophisticated as one's imagination allows. However, the limitations of data availability and required sophistications should be addressed. It's easy to re-invent the wheel, but it's difficult to ensure that the invention continually has something substantial to offer.

The author wishes to acknowledge the support of NSF, NASA, and NOAA over the years under many different projects, which have resulted in the basis for the discussions in this paper. Collaboration and input from several colleagues and students are also acknowledged, particularly Hoshin Gupta, James Shuttleworth, Xiaogang Gao, James Washburne, Jene Michaud, and David Goodrich. The typing and formatting of this paper by Corrie Thies is greatly appreciated.

References

Bathurst JC, PE O'Connell (1992) Future of distributed modeling: the Systeme Hydrologique Europeen, *Hydrologic Processes*, 6, 3:265-278.

Dawdy DR, T O'Donnell (1965) Mathematical models of catchment behavior, *Journal of Hydraulic Division, American Society of Civil Engineering*, 91(HY4):113-137.

Dickinson RE, A Henderson-Sellers, PJ Kennedy, MF Wilson (1986) Biosphere-Atmosphere Transfer Scheme (BATS) for the NCAR Community Climate Model, NCAR Technical Note, NCAR/TN-275+STR.

Duan Q, (1991) A global optimization strategy for efficient and effective calibration of hydrologic models, Ph.D. Dissertation, Dept. of Hydrology and Water Resources, The University of Arizona, Tucson, Arizona, 371 p.

Duan Q, S Sorooshian, VK Gupta (1992) Effective and efficient global optimization for conceptual rainfall-runoff models, *Water Resources Research*, 28, 4:1015-1031.

Famiglietti JS, EF Wood (1994) Multiscale modeling of spatially variable water and energy balance processes, *Water Resources Research*, 30, 11:3061-3078.

Faures JM, DC Goodrich, DA Woolhiser, S Sorooshian (1995) Impact of small-scale spatial rainfall variability on runoff simulation, *Journal of Hydrology*, 173:309-326.

Feldman AD (1995) HEC-1 Flood Hydrograph Package, Chapter 4 in Computer Models of Watershed Hydrology, Edited by VP Singh, pp. 119-150.

FIFE (First ISLSCP Field Experiment, Special issue reprinted from *Journal of Geophysical Research*, 97, D17:18343-19109.

Goodrich DC (1990) Geometric simplification of a distributed rainfall-runoff model over a range of basin scales, Ph.D. Dissertation, Dept. of Hydrology and Water Resources, The University of Arizona, Tucson, Arizona, 361 p.

Goodrich DC, DA Woolhiser (1994) Comment on "Physically based hydrologic modeling 1: a terrain-based model for investigative purposes," by RB Grayson, ID Moore, TA McMahon, *Water Resources Research*, 30, 3:8485-847.

Goodrich DC, TJ Schmugge, TJ Jackson, CL Unkrich, TO Keefer, R Parry, LB Bach, SA Amer (1994) Runoff simulation sensitivity to remotely sensed initial soil water content, *Water Resources Research*, 30, 5:1393-1405.

Goodrich DC, JM Faures, DA Woolhiser, LJ Lane, S Sorooshian (1995) Measurement and analysis of small-scale convective storm rainfall variability, *Journal of Hydrology*, 173:283-308.

Gupta VK, S Sorooshian (1983) Uniqueness and observability of conceptual rainfall-runoff model parameters: the percolation process examined, *Water Resources Research*, 19, 1:269-276.

Kustas WP, DC Goodrich (1994) Preface to the special session on Monsoon '90, *Water Resources Research*, 30, 5:1211-1225.

Lau WK-M, YC Sud, J-H Kim (1995) Intercomparison of hydrologic processes in global climate models, NASA Technical Memorandum 104617, 161 p.

Loague K (1990) R-5 revisited, 2: re-evaluation of a quasi-physically based rainfall-runoff model with supplemental information, *Water Resources Research*, 26, 5:973-987.

Loague K, RA Freeze (1985) A comparison of rainfall-runoff modeling techniques on small upland catchments, *Water Resources Research*, 21, 2:229-248.

Michaud J (1992) Rainfall-runoff modeling of flash floods in semi-arid watersheds, Ph.D. Dissertation, Dept. of Hydrology and Water Resources, The University of Arizona, Tucson, Arizona, 318 p.

Michaud J, S Sorooshian (1994a) Comparison of simple versus complex distributed runoff models on a midsized semiarid watershed, *Water Resources Research*, 30, 3:593-605.

Michaud J, S Sorooshian (1994b) Effect of rainfall-sampling errors on simulations of desert flash floods, *Water Resources Research*, 30, 10:2765-2775.

Naef F (1981) Can we model the rainfall-runoff process today? *Hydrological Sciences Bulletin*, 26, 3:281-289.

Schmugge T, TJ Jackson, WP Kustas, R Roberts, R Parry, DC Goodrich, SA Amer, MA Weltz, Push broom microwave observations of surface soil moisture in Monsoon '90, *Water Resources Research*, 30, 5:1321-1327.

Sellers PJ, Y Mintz, YC Sud, A Dalcher (1986) A simple biosphere model (SiB) for use within general circulation models, *Journal of Atmospheric Sciences*, 43:505-531.

Smith RE, DC Goodrich, DA Woolhiser, JR Simmons (1994) Comment on "Physically based hydrologic modeling, 2: is the concept realistic?" by RB Grayson, ID Moore, TA McMahon, *Water Resources Research*, 30, 3:851-854.

Smith, RE, DC Goodrich, DA Woolhiser, CL Unkrich (1995) KINEROS--A Kinematic Runoff and Erosion Model, Chapter 2 in Computer Models of Watershed Hydrology, Edited by VP Singh, pp. 697-732.

Soil Conservation Service (1964) Hydrology: part 1--watershed planning, SCS National Engineering Handbook, Sec. 4, U.S. Dept. of Agriculture, Soil Conservation Service, Washington, D.C.

Sorooshian S (1983) Surface water hydrology: on-line estimation, *Reviews of Geophysics and Space Physics*, 21, 3:706-721.

Sorooshian S, VK Gupta, JL Fulton (1983) Evaluation of maximum likelihood parameter estimation techniques for conceptual rainfall-runoff models: influence of calibration data variability and length on model credibility, *Water Resources Research*, 19, 1:251-259.

Sorooshian S, Q Duan, VK Gupta (1993) Calibration of rainfall-runoff models: application of global optimization to the Sacramento soil moisture accounting model, *Water Resources Research*, 29, 4:1185-1194.

WCRP/IGPO (1994) Project for Intercomparison of Land-Surface Parameterization Schemes (PILPS), in Soil Moisture Simulation: A Report of the RICE and PILPS Workshop, IGPO Publication Series #14, December.

Wilcox BP, WJ Rawls, DL Brakensiek, JR Wight (1990) Predicting runoff from rangeland catchments: a comparison of two models, *Water Resources Research*, 26, 10:2401-2410.

Woolhiser DA, RE Smith, DC Goodrich (1990) A kinematic runoff and erosion manual, Documentation and User Manual, *Rep. ARS 77*, U.S. Dept. of Agriculture, Agricultural Research Service, Washington, D.C., 130 p.

Woolhiser DA, RE Smith, J-V Giraldez (1996) Effects of spatial variability of saturated hydraulic conductivity on Hortonian overland flow, *Water Resources Research*, 32, 3:671-678.

Woolhiser DA (1996) Search for physically based runoff model--a hydrologic El Dorado?, *Journal of Hydraulic Engineering*, 122, 3:122-129.

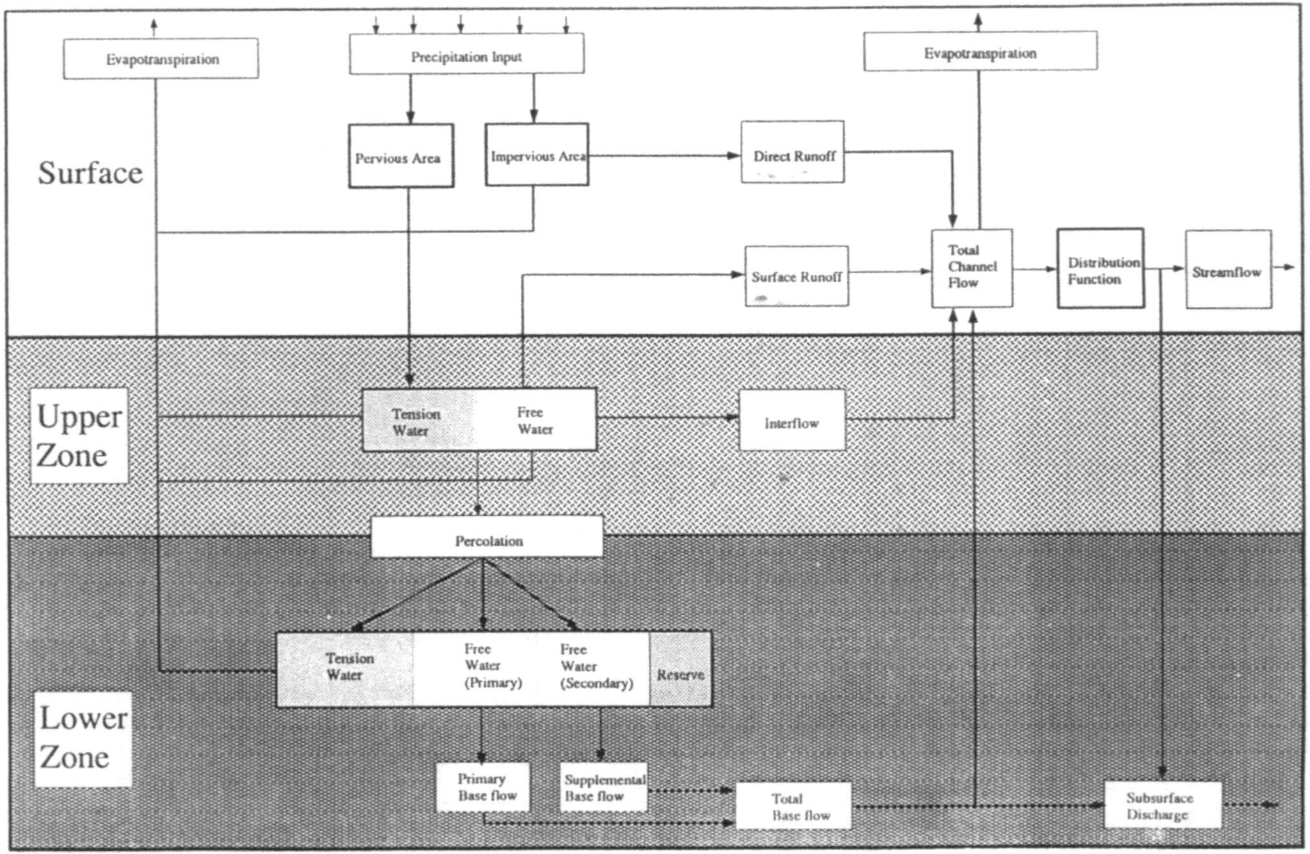

Figure 1. Schematic representation of the Sacramento Soil Moisture Accounting model (SAC-SMA).

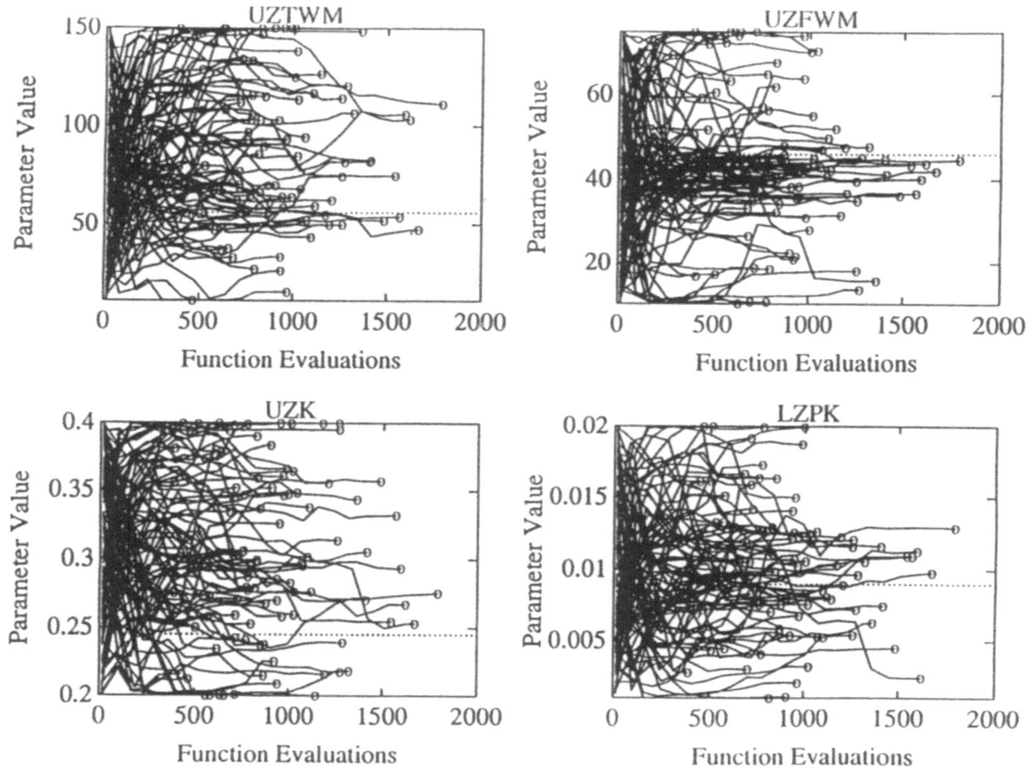

Figure 2a. Convergence behavior of four of the parameters for 100 independent trials of the simplex algorithm (synthetic data study); dotted lines represent the "true" parameter values. (Source: Sorooshian et al., 1993).

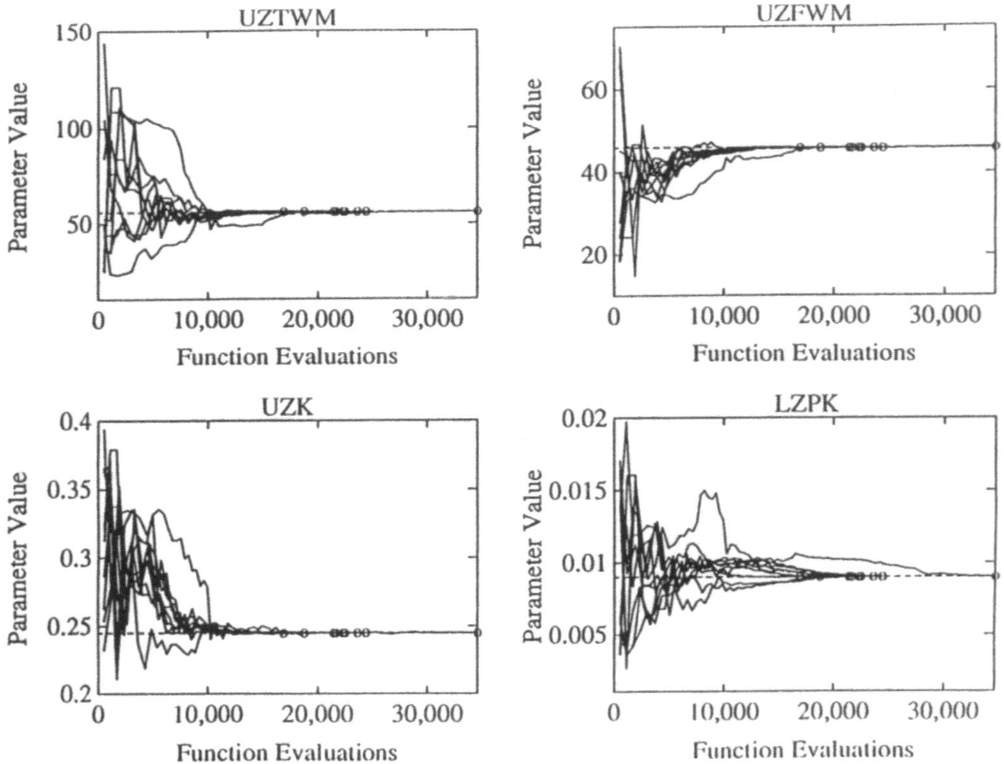

Figure 2b. Convergence behavior of four of the SAC-SMA model parameters for 10 independent trials of the SCE-UA algorithm (synthetic data study); dashed lines represent the "true"parameter values (Source: Sorooshian et al., 1993).

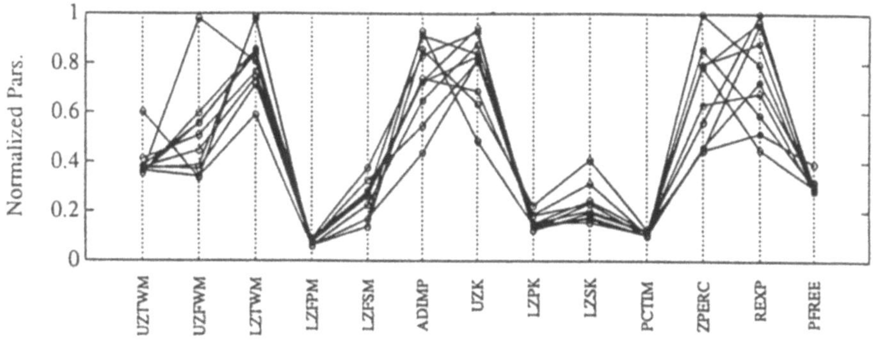

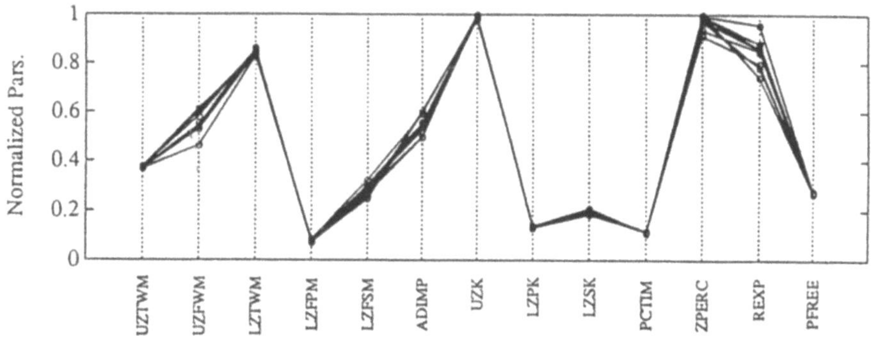

Figure 3. Comparison of the best parameter sets obtained for 10 independent trials of the search algorithms using the HMLE criterion: (a) MSX algorithm (average of 18,266 function evaluations); and (b) SCE-UA algorithm (average of 10,510 function evaluations). (Source: Sorooshian et al., 1993).

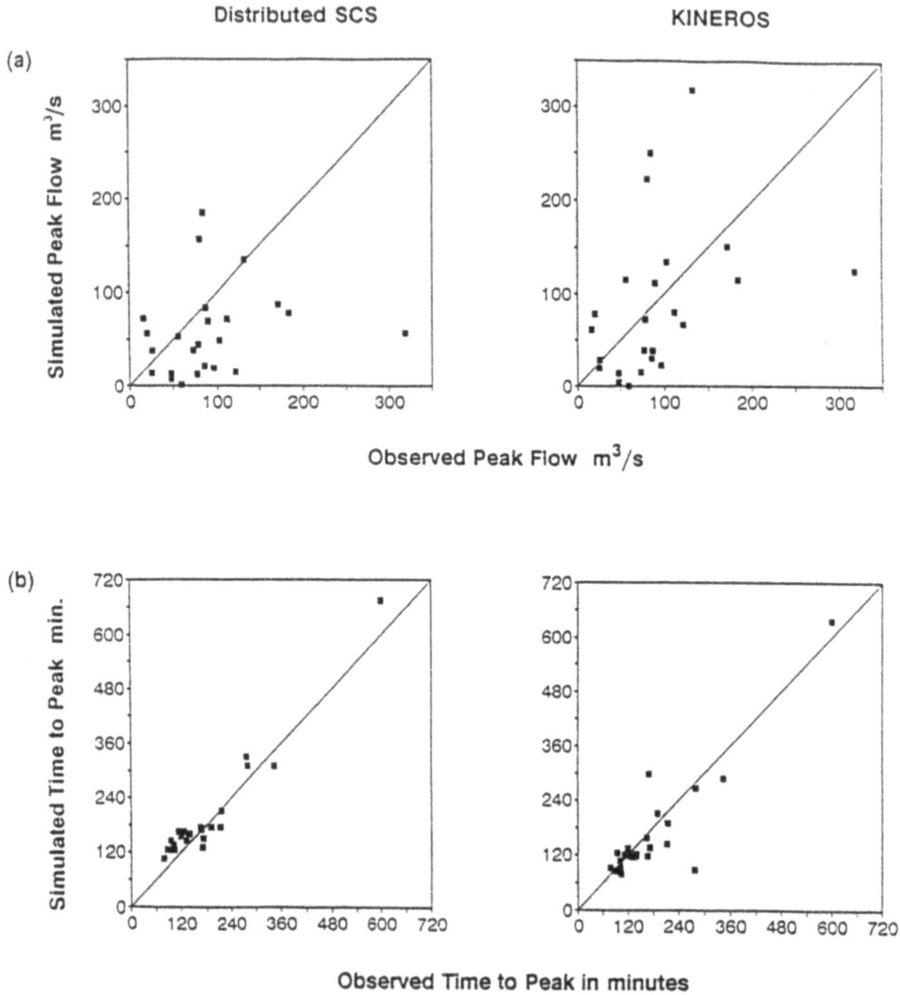

Figure 4. Simulated versus observed values for the calibrated distributed models: (a) peak flow; and (b) time to peak. Data represent individual events at the basin outlet. The 1:1 diagonal represents the line of perfect fit. (Source: Michaud and Sorooshian, 1994a).

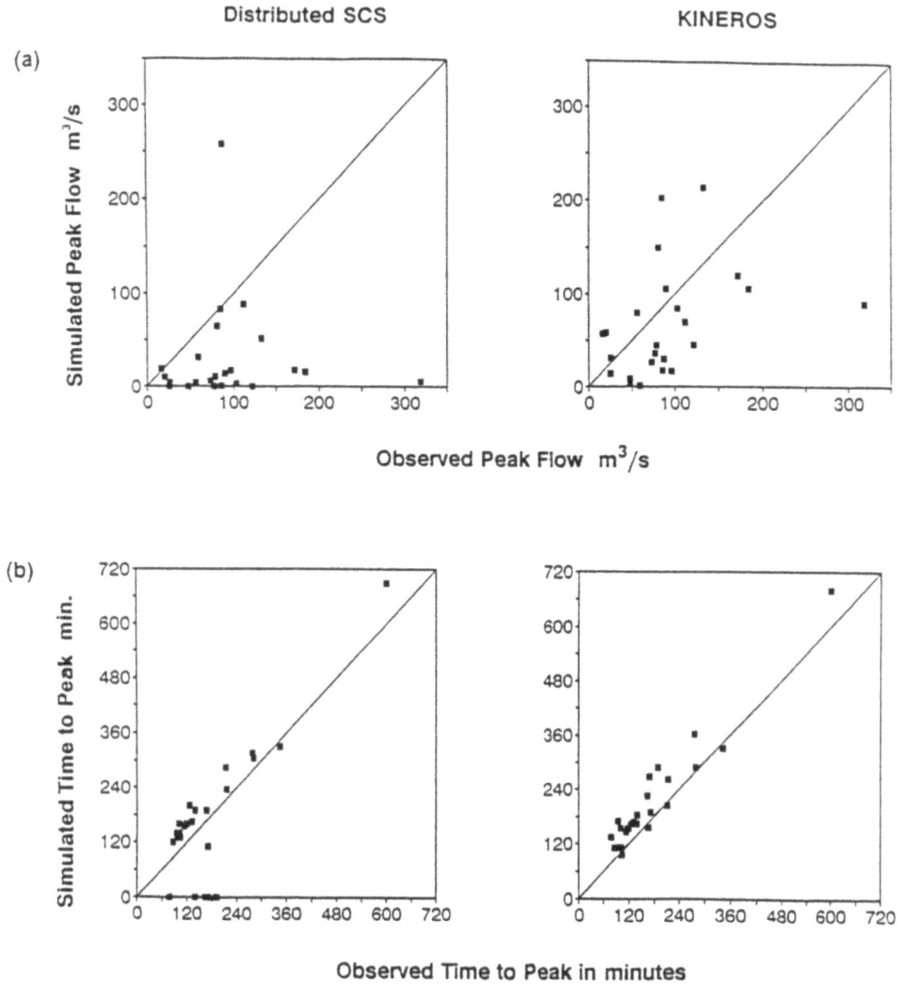

Figure 5. Simulated versus observed values for the uncalibrated distributed models: (a) peak flow; and (b) time to peak. (Source: Michaud and Sorooshian, 1994a).

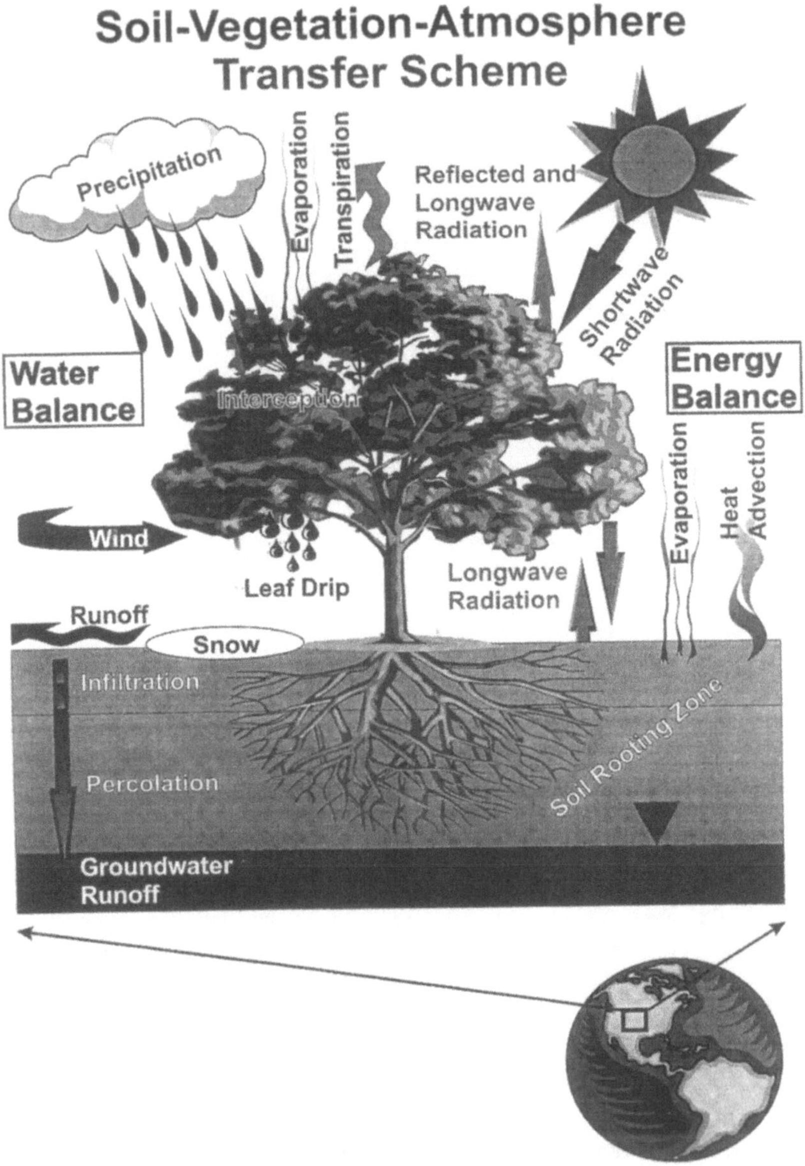

Figure 6. A schematic of the processes involved in the Soil-Vegetation-Atmospheric Transfer Scheme (SVATS).

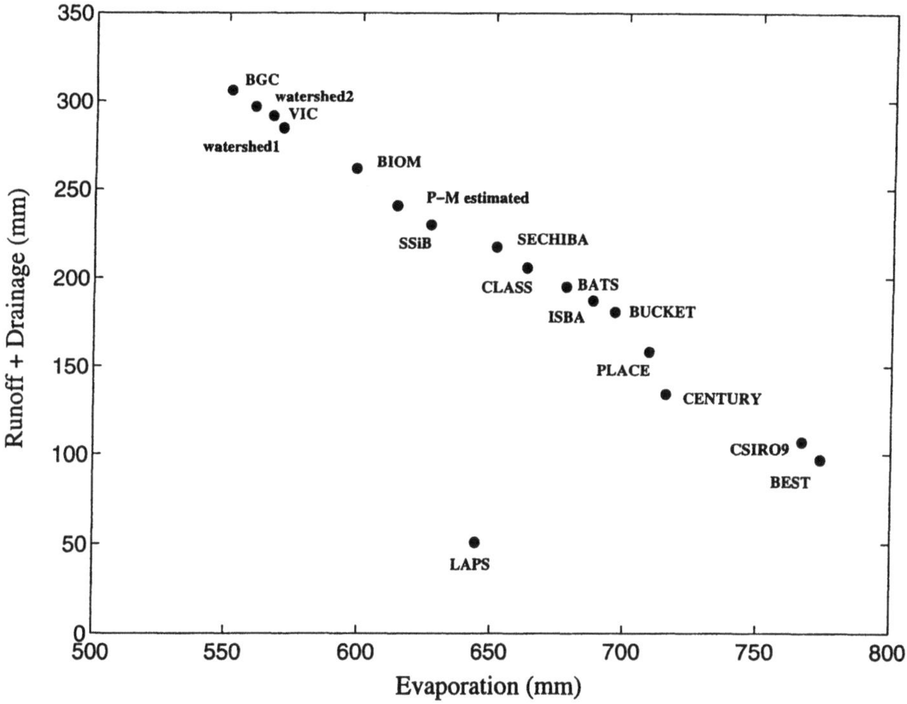

Figure 7. Comparison of annual total runoff plus drainage against evaporation (mm) for different schemes from Experiment 13. For the data point represented by symbol '●',the evaporation is estimated from the Penman-Monteith equation. (Source: WCRP/-IGPO #14, 1994).

Hydrological Processes in GCMs

K. Laval
Laboratoire de Meteorologie Dynamique du CNRS
Paris, FRANCE

1. Introduction

The first attempt to represent hydrological processes in General Circulation Models (GCMs) was defined by Manabe (1969), who introduced into these models a parameterization of land evaporation depending on a soil moisture content simulated by the model. Since then, many sensitivity experiments have shown the importance of hydrologic land-surface processes in defining the simulated climate by GCMs (Mintz, 1984). The possible climatic changes are even more dependent on the hydrologic budget of atmosphere and land. Furthermore, the possible impact of climatic changes on water resources has induced the scientific community to consider small- to large-scale hydrological processes as major issues in future climate studies.

In this paper, the different parameterizations of hydrological processes over land used in GCMS are discussed.

2. Latent Heat Flux Between the Earth's Surface and Atmosphere

The latent heat flux is made of evaporation from bare soil and evapotranspiration from plants. Vegetation can store liquid water over the foliage by intercepting rainfall or snow. This water can contribute significantly to the evaporation rate over forests (Rutter, 1975). To transpire, plants can also extract water from the soil by roots.

The sum of precipitation which falls directly to ground and water which drips from the canopy is the throughfall. This water--which reaches the surface--can partly infiltrate in the soil; the excess is surface runoff.

NATO ASI Series, Vol. I 46
Land Surface Processes in Hydrology
Trials and Tribulations of Modeling and Measuring
Edited by S. Sorooshian, H. V. Gupta, and S. C. Rodda

2.1 Interception Loss and Storage

The importance of interception loss and storage over a vegetal cover was emphasized by Monteith (1965). When rainfall occurs, foliage is capable of storing liquid water that evaporates without any stress. For example, over the Amazonian forest, Lloyd et al. (1988) showed that this evaporation may be up to 0.2 mm/h. As long as vegetal cover was not represented in GCMs, these models fail to simulate this important process. After the pioneering work of Deardorff (1978), many modelers introduced the computation of the interception loss and storage in their GCMs.

A fraction of the rainfall P can be intercepted by the foliage which holds a canopy water store. The governing equation for this reservoir is:

$$\frac{\delta w_c}{\delta t} = P_i - E_i \tag{1}$$

where w_c is the canopy water content which cannot exceed a maximum value w_{cmax}, P_i is the intercepted rainfall, and E_i is the evaporation rate from wetted leaves. According to Deardorff (1978), P_i is the product of the precipitation rate by the fractional area of the vegetal cover.

According to Rutter (1975), the storage capacity of a canopy lies within the range of 0.4 to 2.0 mm. He mentioned that "there is no indication of a separation of range between forests and herbaceous communities." However, in GCMs, w_c is often taken as a function of the leaf-stem area index (Dickinson, 1984).

This parameter may be tuned to obtain a realistic interception loss in a model, as was done by Sellers et al. (1989) over the Amazonian forest.

2.2 Bare-Soil Evaporation

Bare soil is a porous medium, and the evaporation rate depends on soil moisture availability. One defines a potential evaporation rate, EP, and lowers this potential value to real evaporation rate when the soil dries up. The potential evaporation rate is computed by a bulk aerodynamic formula:

$$EP = \frac{\rho}{r_a}\left[q_s(T_s) - q_a\right] \tag{2}$$

where ρ is the air density, r_a is the aerodynamic resistance, $Q_s(T_s)$ is the specific humidity of the surface at its saturation value, and q_a is the specific humidity of the air.

The simple way to compute the real rate is to introduce an aridity parameter β which connects the real to potential evaporation and which depends on soil moisture content. This parameterization was first introduced by Manabe (1969), who defined β as a simple function of total soil moisture content. Another way to deal with the reduction of the rate is to introduce a relative humidity at the Earth's surface to decrease the humidity gradient between air and soil (Noilhan and Planton, 1989). This is also used in the ECMWF model (Blondin, 1988). Another choice is to introduce a surface resistance that slows down the evaporation rate. This surface resistance has been introduced in the SECHIBA scheme (Ducoudre et al., 1993) as the product of the dry soil depth by a constant which represents the resistance of one meter of dry soil.

2.3 Transpiration

The water that transpires originates within the leaves which exert a stomatal control. This effect is taken into account by adding a stomatal resistance to aerodynamic resistance. Generalizing this concept to a bulk canopy, one can introduce a surface resistance r_s (Monteith, 1965) as:

$$E = \frac{\rho}{r_a + r_s}\left(q_s(T_s) - q_a\right) \tag{3}$$

The differences between this surface resistance and a "mean" value of a physiological stomatal resistance are discussed by Thom (1975).

Let us note the significant "leap" that modelers have made by extending the concept of surface resistance to the resistance of a vegetation canopy covering few 10^4 km^2. However, it is clear that the introduction of this term reduces the evaporation rate in a GCM (Sato et al., 1989) compared to the bulk aerodynamic formula. This is not the only way to reduce evaporation, and an attempt has been made to use the Priestley and Taylor (1972) formulation to compute the potential evaporation in a GCM (Laval et al., 1984). These authors showed that the evaporation simulated by the model was significantly reduced and more realistic with this formulation compared to the bulk aerodynamic formula used in the bucket model.

To define the surface resistance, one considers a characteristic value associated with a particular type of vegetation together with a dependence upon leaf area index, photosynthetically active radiation, temperature, soil moisture stress, or hydric air deficit. Thus, for each point of the model, one must prescribe the particular vegetation which covers the area. As an example, in the SECHIBA model (Ducoudre et al., 1993), the surface resistance is given by:

$$r_s = \frac{1}{LAI} \frac{R_s + R_{so}}{R_s} \frac{a + \lambda \delta c}{k_o} \qquad (4)$$

where R_{so} is a half-light saturation factor of radiation R_s, set equal to 125 Wm^{-2}, the parameters a and λ are, respectively, equal to 23 10^{-3} Kg m^{-2} and 1.5, and δc is the hydric air deficit. However, the parameters LAI and k_o are assigned to each land cover class. Note that this equation does not contain explicitly the dependence on temperature, but exists through the hydric air deficit. It is also interesting to note that some agreement with reality can be obtained when one uses a simple model, depending only on time for the Amazonian forest (Dolman et al., 1991).

3. Soil Moisture Content

3.1 One Layer

Manabe (1969) introduced the first method to simulate soil moisture in GCMs by the equation:

$$\frac{\delta w}{\delta t} = P - E + M - R \qquad (5)$$

where w is the soil moisture content of 1 meter of soil, M is the snow melt, R is runoff rate, P is the precipitation rate, and E is the evaporation rate.

3.2 Two Layers

In order to simulate a high evaporation rate over a bare soil after a rainfall event, Deardorff (1977) proposed writing an equation for the soil-surface moisture content and defined the aridity coefficient as dependent on this surface value. The prediction equations for soil surface moisture fraction ω_g and total soil moisture fraction ω_b are:

$$\frac{\delta \omega_g}{\delta t} = c_1 \frac{P - E}{\rho_1 D_1} - c_2 \frac{\omega_g - \omega_b}{\tau} \qquad (6)$$

$$\frac{\delta \omega_b}{\delta t} = \frac{P - E}{\rho_1 D_2} \qquad (7)$$

$$w_b = \frac{\omega_b}{\rho_1 D_2} \qquad (8)$$

where the soil moisture fractions are volume of water divided by volume of soil, ρ_1 is the density of liquid water, D_1 is the depth to which diurnal soil moisture cycle extends, D_2 is the depth of the layer of total soil moisture, and τ is one day.

These equations can simulate a rapid wetting of a surface after precipitation and the consequent increase of evaporation rate. The coefficients c_1 and c_2 were defined as:

$$c_2 = 0.9$$

$$c_1 = 0.5 \qquad \text{if} \quad \frac{\omega_g}{\omega_{max}} \geq 0.75 \tag{9}$$

$$c_1 = 14 - 22.5\left(\frac{\omega_g}{\omega_{max}} - 0.15\right) \quad \text{if} \quad 0.15 < \frac{\omega_g}{\omega_{max}} < 0.75$$

$$c_1 = 14 \qquad \text{if} \quad \frac{\omega_g}{\omega_{max}} \leq 0.15$$

The three equations which define c_1 correspond to the different stages of the drying of soil. ω_{max} is field capacity for the surface layer.

Another method, with two layers, was introduced by Laval (1988) in the LMD GCM. This method was based on the ideas of Choisnel (1984). In this scheme, one also considers deep soil moisture content w_2 and soil surface moisture content w_1.

After the dry season, there is only moisture content in deep soil (w_2) which is less than field capacity (ω_{max}). When the precipitation rate becomes sufficiently high to be larger than the evaporation rate, the excess water is used to create a surface layer of moisture, saturated, and with a depth d_1 directly related to this excess. As long as the water convergence is positive, this water supply is used to increase the depth of the soil moisture layer which remains saturated.

When soil begins to dry because the evaporation rate exceeds the precipitation rate, the latent heat flux is supplied by this surface layer which dries. Additionally, it is only when this surface layer is sufficiently dry that the water loss is provided by deep soil moisture content.

This scheme has the advantage of defining a surface layer with a depth that varies geographically as it depends on water convergence during the wet period. There are regions where the maximum of the depth d_1 remains small during wetting, and the soil surface moisture content w_1 varies on a time scale much shorter than the bulk moisture content. However, in the regions where water convergence is important, the depth d_1 may become large. Then, the surface layer will have the same behavior of a deep layer, and the interest of this scheme for long integrations diminishes. This is why the concept of a maximum value for this depth d_1, which remains variable, was introduced by Laval et al. (1993).

In the recent schemes used in GCMs, soil moisture content is defined by two or three layers of depth D_1, D_2, ... and the diffusion of water between layers is computed (Rind, 1982; Dickinson, 1984; Sellers et al., 1986). For simplicity, let us consider two layers. The equations for soil moisture wetness ω_1 are:

$$\theta_s D_1 \frac{\delta \omega_1}{\delta t} = P_r - E_1 - Q_{12} - R_s \tag{10}$$

$$\theta_s D_2 \frac{\delta \omega_2}{\delta t} = E_2 + Q_{12} - Q_2 \tag{11}$$

where $\omega_i = \theta_i/\theta_s$, θ_i is volumetric soil moisture, θ_s its saturation value, D_1 and D_2 are the layer depths, P_r is the precipitation rate reaching the soil, E_1 and E_2 are the transpiration of water resulting from transfer from layer 1 or 2 through plants, Q_{12} is diffusion and/or drainage between the two layers, Q_2 is subsurface drainage, and R_s is surface runoff.

Water diffusion between layers depends on the humidity gradient in the soil and is computed by Darcy's law:

$$Q = K \left(\frac{\delta \Psi}{\delta z} + 1 \right) \tag{12}$$

where the first term represents capillary forces, and the second term represents gravity force. K and Ψ are, respectively, the unsaturated hydraulic conductivity and soil moisture potential related to the saturated values by the relations discussed by Clapp and Hornberger (1978):

$$K = K_s \omega^{2B+3} \tag{13}$$

$$\Psi = \Psi_s \omega^{-B} \tag{14}$$

where B is a coefficient which characterizes the soil type.

In the bucket model, surface runoff was introduced only when soil moisture content reaches field capacity. In these recent models, surface runoff occurs when the water convergence is larger than the diffusion rate through the two layers.

4. Subgrid Scale Spatial Variability (SSV) in GCMs

In a grid box of a GCM that covers a domain of the order of 10^4 km^2, the assumption of homogeneity is highly questionable. For the land surface, the spatial inhomogeneities may concern the different types of vegetation or bare soil which cover the area, the spatial distribution of rainfall over the grid area, or the distribution of soil moisture capacities. All of these inhomogeneities may have an impact on the exchanges between the Earth's surface and the atmosphere. The different vegetal covers evaporate at different rates; if rainfall is unevenly distributed, surface runoff may occur over a subarea of the grid for which precipitation is sufficiently high to produce saturation. For a uniform rate of precipitation, subareas with low field capacity may produce surface runoff more often than subareas with higher field capacity.

We don't know the importance of taking into account these spatial inhomogeneities in the numerical experiments of many years that are conducted with GCMs. Abramopoulos et al. (1988) compared hydrological quantities obtained within a grid box of 8° x 10° where parameters are area weighted means, with the average of computations performed on a 1° x 1° cells inside the grid box, but this comparison was carried out as offline tests. This comparison was done for four regions: Brazil, Sahel, Sahara, and India. They concluded that the composite procedure works well except in the Sahel. However, it is necessary to do more studies, defined with different models and conditions to obtain a definite answer. It will also be instructive to carry out comparisons with GCMs where interactions between the Earth's surface and the atmosphere are allowed.

4.1 Heterogeneities of Rainfall

There was an early attempt in the UKMO GCM to take into account the fact that convective rainfall P is not homogeneous over the grid area of a GCM (Warrilow et al., 1986). In the case of convective rain, it was assumed that only a proportion μ of the grid area receives rain, and that the probability density distribution for local rain p over this proportion is an exponential function defined as:

$$F(p) = \frac{\mu}{P} \exp\left(\frac{\mu p}{P}\right) \tag{15}$$

An obvious consequence of this assumption is the increase of the surface runoff because a part of the grid area receives more rainfall than the average and becomes saturated easier. The surface runoff is related to P by the equation:

$$R_s = P \exp - \left(\frac{\mu I}{P}\right) \tag{16}$$

where I is the maximum infiltration rate, assumed constant over the area.

Because this process was introduced in the UKMO simultaneously with many changes of the land-surface scheme, one is not able to evaluate the improvement that this precise parameterization has induced. However, this parameterization introduces a physical process, and it is certainly interesting to use it. It may be useful also to define a parameter μ which depends on latitude or which is geographically distributed, as this parameter may be different in middle latitudes and semi-arid zones.

One can also consider the impact of the spatial heterogeneity of rainfall on the amount of water stored over the canopy. The interception loss and storage depend on the proportion μ of the grid area which receives rainfall (Shuttleworth, 1988). This effect was tested, with the BATS model, in a stand-alone mode by Pitman et al. (1990). They found that, when atmospheric forcing is prescribed, the seasonal evolution of the evaporation rate and runoff are changed significantly when one considers the effect of rainfall heterogeneities on interception. However, this was not exactly the conclusion that was obtained by Dolman and Gregory (1992) who studied the same effect with a one-dimensional model. Although the transpiration and interception loss were dependent on the heterogeneity of rainfall, total evaporation was not modified much by this assumption.

4.2 Heterogeneities of Vegetal Cover

The importance of heterogeneities of vegetal cover has been emphasized by Verstraete (1989) and Avissar and Verstraete (1990). In the LMD GCM, the choice was to define within a grid box eight subareas (Ducoudre, 1990; Ducoudre et al., 1993). Bare soil and seven various vegetal types may exist simultaneously and are defined by their fractional area. For each vegetal cover, the leaf area index (LAI) is prescribed. Each vegetal cover has a canopy store, depending on LAI and, consequently, has its particular interception loss and storage. Two parameters, necessary to compute the total resistance to transpiration, are prescribed also. The parameter k_o is related to the minimum of the surface resistance defined for each canopy (see Section 2.3). The other term is the architectural resistance introduced by Perrier (1975). This last term, important only for tall vegetation, is added to the surface resistance in order to represent the aerodynamic resistance between the different horizontal layers of the canopy and the canopy top. This term takes into account that the different layers of the canopy do not evaporate at the same rate because ventilation of the foliage is not perfect and variations of humidity exist within the canopy (McNaughton, 1987).

One may consider that this effect is introduced in SiB in another way. In this scheme, one considers an effective LAI (Sellers et al., 1989) for a tropical forest which is lower than the usual physiological parameter considered in the literature. As a result, canopy resistance is enhanced and canopy water store is lowered, which produces a reduction of transpiration and interception loss, as in our scheme.

In a SECHIBA scheme, it is easy to compute the transpiration and interception loss and storage for each subarea of vegetal cover because each cover has its own parameters. In this scheme, it is not necessary to define the morphological parameters of a total vegetation of a grid box that consists of many distinct types.

4.3 Subgrid Scale Spatial Variability of Water Field Capacities

Another important cause of SSV is related to different water field capacities associated to orography or to different types of soil. As a result, surface runoff may occur over a portion of the grid area, even though the entire grid is not saturated. Hydrologists have developed river discharge models for catchment areas which take into account this feature (Moore, 1985). One parameterization has been introduced in the Max Planck Institute Model (ECHAM-GCM) by L. Dumenil, who has used the Arno model (Dumenil and Todini (1992). This model defines a statistical distribution of water field capacity inside the grid box that follows the equation:

$$F(w) = 1 - \left(\frac{w}{w_{max}}\right)^b \tag{17}$$

where w_{max} is, in this context, the maximum of the water field capacity which exists inside the area, and b is a characteristic parameter.

When it rains, part of the area of the grid box is saturated and produces surface runoff, whereas the remaining area stores all the water that it has received.

Because of the particular formulation of the statistical distribution, it is easy to compute the apportionment between the storage and the runoff.

This Arno model was slightly modified to be introduced in the LMD-GCM in order to take into account the fact that soil moisture content was simulated by two layers.

The effect of this SSV was studied in a dimensionless model by analyzing the results of a control run which assumed homogeneity of water field capacity and an SSV run where the Arno model was introduced. The inputs were the precipitation rate, radiative fluxes, air temperature, wind, and humidity for one year. The temporal variation of the runoff, obtained by the O-D model, showed that the season during which runoff occurs lasts longer when SSV is assumed and the maximum of the surface runoff reached during the wet season is lower. Another consequence of the early occurrence of the surface runoff in the SSV run was that the soil moisture reservoir fills at a slower rate and saturation occurs less often. When this SSV was introduced in longer simulations carried out with the LMD GCM, we were unable to obtain significant changes in the hydrological budget simulated by the model (Savoye, 1992). However, with a slight change of the Choisnel model, the sensitivity of the GCM became significant (Laval et al., 1993).

5. Perspective

In Section 4, we described a number of processes that are related to subgrid scale spatial variability. There are other processes, as described by Li and Avissar (1994) and Mahrt and Ek (1993), not mentioned here. Modelers have here a large area of investigation with various mathematical tools.

The first problem is to choose between geographical and statistical distributions. The first can be chosen to represent a particular spatial variability (such as vegetal cover), and the second can be chosen to represent another variability (such as rainfall). How can we deal with the problem of associating two processes with SSV? Entekhabi and Eagleson (1989) defined a scheme which takes into account statistical distributions of rainfall and of soil moisture content. Do we need to consider that two distributions are independent of each other? For example, if we assume a geographical distribution of vegetation inside the grid area, do we need to differentiate the variability of rainfall associated with the various types of vegetal cover? This problem is related to the discussion between Eltahir and Bras (1994) and Dolman and Gregory (1994) or Shuttleworth (1994). These authors discussed the parameterization of the interception loss. The most important process is attributed either to the precipitation variability (Shuttleworth, 1988) or to the inhomogeneity of the canopy water content (Eltahir and Bras, 1993).

The second problem is to relate as much as possible these studies with data. All of these mathematical exercises are interesting only if we find that some real atmospheric state is better simulated with SSV introduced in a GCM. Moreover, it will be very fruitful if, with a given GCM, simulation with various subgrid scale variability schemes are compared with data.

The last problem concerns the impact of SSV on the climate simulated by a GCM. We usually agree that, if a process occurs in reality, we need to represent it in GCMs. However, the issue is also to know how sensitive the model is to this process. We have mentioned already that, although there was an influence with the O-D scheme, the GCM simulations of five years were not sensitive to SSV of soil moisture. We had to slightly change the SECHIBA scheme to obtain an impact on simulations (Laval et al., 1993). Furthermore, we have mentioned that, when the SSV of rainfall was taken into account, the impact on total evaporation was more important with a stand-alone model (Pitman et al., 1990) than with a one-dimensional model (Dolman and Gregory, 1992).

It may be that GCMs are not "realistic" enough and that we need to improve other physics of the model. Then, we must check not only the sensitivity of a GCM to one process (such as land-surface exchanges), but we must also study how this sensitivity is changed when we change other parameterizations of physical processes (boundary layer, clouds, etc.). This was addressed by Laval (1988) with the LMD GCM, where the sensitivity to initial soil moisture was considered using alternatively three different schemes of soil moisture.

We must consider also that, possibly, a small-scale process, even "real", has no impact on long-term simulations because, in reality, it has no influence on the mean climate. For instance, McNaughton and Jarvis (1991) showed that the "sensitivity of transpiration to a change in stomatal resistance decreases as the spatial scale increases from leaf to region" and that "the importance of stomatal control of transpiration depends on the amount of environmental feedbacks."

The scientific community needs to answer basic questions about the future of the climate. International projects such as PILPS (in GEWEX-GCIP) or AMIP, with the collaboration of scientists of many fields, are necessary in order to make definite progress with models.

Acknowledgments

I would like to thank M.C. Roos for typing this manuscript.

References

Abramopoulos F, C Rosenzweig, B Choudhury (1988) Improved ground hydrology calculations for global climate models (GCMs): soil water movement and evapotranspiration. *Journal of Climate*, 1:921-941.

Avissar R, MM Verstraete (1990) The representation of continental surface processes in mesoscale atmospheric models. *Reviews of Geophysics*, 28:35-52.

Blondin C (1988) Research on land surface parameterization schemes at ECMWF, Parameterization of Fluxes over Land Surfaces, ECMWF, 285-330.

Choisnel E (1984) Un modele agrometeorologique operationnel de bilan hydrique utilisant des donnees climatiques. Les besoins en eau des Cultures, *Conference Internationale CIID*, Paris, France.

Clapp RB, GM Hornberger (1978) Empirical equations for some hydraulic properties. *Water Resources Research*, 14(4):601-604.

Deardorff JW (1977) A parameterization of ground surface moisture content for use in atmospheric prediction models. *Journal of Applied Meteorology*, 16:1182-1185.

Deardorff JW (1978) Efficient prediction of ground surface temperature and moisture, with inclusion of a layer of vegetation. *Journal of Geophysical Research*, 83:1889-1903.

Dickinson RE (1984) Modeling evapotranspiration for three-dimensional global climate models. Climate Processes and Climate Sensitivity, JE Hansen, T. Takahashi (Editors), 58-72.

Dolman AJ, JHC Gash, J. Roberts, WJ Shuttleworth (1991) Stomatal and surface conductance of tropical rain forests. *Agricultural and Forest Meteorology*, 54:303-318.

Dolman AJ, D Gregory (1992) The parameterization of rainfall interception in GCMs. *Quarterly Journal of Royal Meteorological Society*, 118:445-467.

Dolman AJ, D Gregory (1994) Reply to Comments by EAB Eltahir, RL Bras on "The parameterization of rainfall interception in GCMs," by AJ Dolman, D Gregory. *Quarterly Journal of Royal Meteorological Society*, 120:735-736.

Ducoudre NI (1990) Sensibilite du climat simule a la parametrisation des echanges de vapeur d'eau entre la Biosphere et l'Atmosphere. These de Doctorat, Universite de Paris VI.

Ducoudre N, K Laval, A Perrier (1993) SECHIBA: a new set of parameterizations of the hydrologic exchanges at the land/atmosphere interface within the LMD atmospheric general circulation model. *Journal of Climate*, 2:248-273.

Dumenil L, E Todini (1992) A rainfall-runoff scheme for use in the Hamburg climate model, In Advances in Theoretical Hydrology: A Tribute to James Dooge, European Geophysical Society Series on Hydrological Sciences, 1, Elsevier, 129-157.

Eltahir EAB, RL Bras (1993) A description of rainfall interception over large areas. *Journal of Climate*, 6:1002-1008.

Eltahir EAB, RL Bras (1994) Comments on "The parameterization of rainfall interception in GCMs, by AJ Dolman, D Gregory, *Quarterly Journal of Royal Meteorological Society*, 120:733-734.

Entekhabi D, PS Eagleson (1989) Land surface hydrology parameterization for atmospheric General Circulation Models including subgrid scale spatial variability. *Journal of Climate*, 2:816-831.

Laval K, C Ottle, A Perrier, Y Serafini (1984) Effect of parameterization of evapotranspiration on climate simulated by a GCM, In New Perspectives in Climate Modeling, AL Berger, C Nicolis (Editors), Elsevier, Amsterdam, The Netherlands, 223-247.

Laval K (1988) Experience with surface processes at LMD. Parameterization of Fluxes over Land Surfaces, ECMWF, 171-205.

Laval K, A Ducharne, J Polcher (1993) Sensitivity of climate simulations to hydrological processes with the LMD GCM. *Proceedings of IAMAP-IAHS*, Yokohama, Japan.

Li B, R Avissar (1994) The impact of spatial variability of land-surface characteristics on land-surface heat fluxes. *Journal of Climate*, 7:527-537.

Lloyd CR, JHC Gash, WJ Shuttleworth, A de O Marques (1988) The measurements and modeling of rainfall interception by Amazonian rain forest. *Agricultural and Forest Meteorology*, 43:277-294.

Manabe S (1969) Climate and the ocean circulation: the atmospheric circulation and the hydrology of the Earth's surface. *Monthly Weather Review*, 97:739-774.

Mahrt L, M Ek (1993) Spatial variability of turbulent fluxes and roughness lengths in Hapex-Mobilhy. *Boundary Layer Meteorology*, 65:381-400.

McNaughton KG (1987) Comments on "Modeling the effects of vegetation on climate," Geophysiology of Amazonia, RE Dickinson (Editor), John Wiley and Sons, 526 p.

Mintz Y (1984) The sensitivity of numerically simulated climates to land-surface boundary conditions. The Global Climate, JT Houghton (Editor), Cambridge University Press, 79-105.

Monteith JL (1965) Evaporation and environment, Symp. Soc. Exp. Biol. XIX, 205-234.

Moore RJ (1985) The probability distributed principle and runoff production at point and basin scales. *Hydrological Sciences Journal*, 30(2):273-297.

Noilhan J, S Planton (1989) A simple parameterization of land surface processes for meteorological models. *Monthly Weather Review*, 117:536-549.

Perrier A (1975) Etude physique de l'evapotranspiration dans les conditions naturelles. *Annales Agronomiques*, 26:1-18, 105-123, 229-243.

Pitman AJ, A Henderson-Sellers, ZL Yang (1990) Sensitivity of regional climates to localized precipitation in global models. *Nature*, 346:734-737.

Priestley CHB, RJ Taylor (1972) On the assessment of surface heat flux and evaporation using large-scale parameters. *Monthly Weather Review*, 100:81-92.

Rind D (1982) The influence of ground moisture conditions in North America on summer climate as modeled in the GISS GCM. *Monthly Weather Review*, 110:1487-1494.

Rutter AJ (1975) The hydrological cycle in vegetation, Vegetation and the Atmosphere, JL Monteith (Editor), Academic Press, 1, 111-154.

Sato N, PJ Sellers, DA Randall, EK Schneider, J Shukla, JL Kinter III, Y-T Hou, E Albertazzi (1989) Effects of implementing the Simple Biosphere model in a General Circulation Model. *Journal of Atmospheric Sciences*, 46:2757-2782.

Savoye S (1992) Utilisation d'un modele de climat pour etudier l'influence de la variabilite spatiale de l'eau dans le sol sur les bilans hydriques d'une region. Rapport de stage. Universite Paris 6.

Sellers PJ, Y Mintz, YC Sud, A Dalcher (1986) A Simple Biosphere (SiB) model for use within General Circulation Models. *Journal of Atmospheric Sciences*, 43:505-531.

Sellers PJ, WJ Shuttleworth, JL Dorman, A Dalcher, JM Roberts (1989) Calibrating the Simple Biosphere model for Amazonian tropical forest using field and remote sensing data: 1. average calibration with field data. *Journal of Applied Meteorology*, 28:727-759.

Shuttleworth WJ (1988) Macrohydrology: the new challenge for process hydrology. *Journal of Hydrology*, 100:31-56.

Shuttleworth WJ (1994) Reply to Comments by EAB Eltahir, RL Bras on "The parameterization of rainfall interception in GCMs," by AJ Dolman, D Gregory, *Quarterly Journal of Royal Meteorological Society*, 120:737.

Thom AS (1975) Momentum, mass, and heat exchange of plant communities, Vegetation and the Atmosphere, JL Monteith (Editor), Academic Press, 1, 57-109.

Verstraete MM (1989) Land surface processes in climate models: status and prospects, in Climate and Geosciences: A Challenge for Science and Society in the 21st Century, A Berger, S Schneider, J Cl Duplessy (Editors), Kluwer Academic Publishers, Dordrecht, The Netherlands, 321-340.

Warrilow DA, AB Sangster, A Slingo (1986) Modeling of land surface processes and their influence on European climate. Dynamical Climatology Technical Note, 38, Meteorological Office, Bracknell.

The Challenges We Face: Panel Discussion on Data and Models

Edited by Hoshin Gupta and Soroosh Sorooshian

1. Introduction

This lively and engaging panel discussion began with introductory remarks by the moderators *Keith Beven* and *Robert Dickinson*. Based on this preamble, an extensive discussion ensued. The following is a summary of the major points of discussion, based primarily on notes prepared by the moderators *Keith Beven* and *Robert Dickinson*, and by additional written contributions from participants *Jim Washburne, Xiaogang Gao, Paul Houser and Patrice Yapo*. Their assistance in documenting the discussions is greatly appreciated.

2. Our models are invalid and not validatable!

Beven began by reminding the participants that both data and models have been the recent focus of many critical discussions among hydrologic scientists. A major reason for this has been the rapid proliferation of variants of hydrological and meteorological models. The problem is, however, that there is no clear way of distinguishing which models are more realistic or are likely to be more successful. *Beven* suggested that:

(i) All our process models are invalid, and

(ii) None of our models can be validated by the available data (see also discussion in *Beven, 1993*)

He stated that what we are doing is not, in fact, prediction, but rather "prophesy or divination...", and suggested that if there were real personal consequences for false prophesy, modelers would be more circumspect of their results. As stated in a popular Chinese proverb, "Prophesy is very difficult, especially with respect to the future"!

NATO ASI Series, Vol. I 46
Land Surface Processes in Hydrology
Trials and Tribulations of Modeling and Measuring
Edited by S. Sorooshian, H. V. Gupta, and S. C. Rodda

3. We must evaluate the risk of false prophesy.

If we accept that all models are "invalid" and " not validatable", we are faced with two possible responses. The first is to suspend our disbelief in the validity and predictions of our models despite the considerable evidence to the contrary. This, in fact, is our usual mode of operation and examples of this response abound in the hydrological, meteorological and soil-vegetation-atmosphere-transfer-scheme (SVAT) model literatures. The second response is to estimate the risk of false prophesy or prediction through uncertainty estimation and risk analysis. *Beven* pointed out that an important, immediate and desirable result of producing realistic uncertainty estimates is an increased focus on care in data collection and on the value of data in constraining levels of predictive uncertainty.

4. Model development is driven by the availability of data.

In counterpoint to the above, *Dickinson* emphasized that much of the interest in the processes of climate change has been driven by the availability of long term data sets, such as of CO_2 and ozone concentrations. Certainly in the realm of GCMs, model development has been at least partly driven by a real and recognized need to predict climate variability into the future. He stressed the following issues:

(i) The atmosphere is an important component in the hydrological cycle and that there is a critical need for more interaction between the hydrological and meteorological modeling communities

(ii) Studies on how to aggregate and disaggregate processes so as to make better use of model predictions and data at different scales are in their infancy and are severely data limited.

(iii) There is a need for better techniques for the intercomparison of models, as these can help to identify modeling errors.

5. Questions posed.

Beven then posed the following questions for the participants to address as part of the discussion.

(1) Is the available data useful?

What is useful and for what purpose? What belief can we place in a measurement?

(2) Can models be validated by the available data?

Is the data clean? What is validation?

(3) Is there a better way to model?

What are adequate inputs? How do we improve understanding? How do we choose between simple and complex models?

(4) How do we achieve parsimony?

The ensuing discussions then focused largely on the following topics.

6. We need to improve communication between modelers and data collectors.

It is very clear that there needs to be more interaction between the people collecting the data and the modelers who use it. The primary problem is that at present, many of the variables and parameters required by hydrological and meteorological models cannot be measured or estimated at the required scales. On the one hand, ground based measurements tend to be at much smaller scales. On the other, remotely sensed data while allowing the evaluation of spatial patterns, often give only a relatively inaccurate indication of the variables involved.

The participants felt that it is very important to establish the *quality* of the data, but in many situations *any* data would be welcome. Data shortcomings are particularly evident at global scales. *Nicholson* and *Dickinson* pointed out that information about soil types and characteristics such as albedo is very limited for 80% of the land surface. *Nicholson* also remarked that there may be very significant errors in the processes of data analysis and transmission. *Smith* commented on the issue of trade-off between quantity and quality of data. With increasing availability of remotely sensed data and correspondingly little ground based data for validation, it may become more difficult to know the quality of the available data.

There is also very little known about the *value* of different types of data in different situations. *Choudhury* pointed out that there is an inevitable trade-off or association between the appropriate level of model complexity and the consequent data requirements. Therefcre, there is an important need for modelers and data collectors to work together to establish compromises regarding what type and scale of data is required for improving land surface parameterizations at different predictive scales and in different environments. *Beven* highlighted the current mismatch of data and model scales (eg. point versus grid scales) and *Goodison* remarked that the onus is on modelers to provide clear guidelines for their data needs.

7. Model validation is still a desirable goal.

In any modeling exercise we are faced with the problem of validation in the presence of boundary condition uncertainties, data uncertainties and errors in model structure. *Beven* pointed out that success in hydrologic modeling relies on calibration of the model against observed data. *Gupta* remarked that validation must be approached in the context of our modeling objectives and the scales, both temporal and spatial, at which the validation is required. In other words, a model may seem to provide acceptable performance at one temporal-spatial scale while failing miserably at another. *Sorooshian* agreed and suggested that in one sense, no model can ever be validated because no model can ever achieve a perfect representation of the actual system. *Avissar* and *Sorooshian* reminded the group that from a pragmatic point of view, the practical use of hydrologic models has and will continue to require the use of an engineering "factor of safety", which as *Beven* remarked is often a factor of two or more. *Kabat* and *Beven* mentioned that the development of stochastic approaches to the estimation of uncertainty would be very useful in this regard.

It seemed to be the consensus, however, that the concept of model validation is still a desirable goal and that it may be necessary to define the appropriate set of conditions for different levels of validation of the model predictions, particularly for distributed predictions of the internal hydrological variables within a watershed.

8. The method of intercomparison of models needs attention.

The importance of representing land surface processes in GCMs was emphasized, particularly for arid regions (such as Arizona) where 98% or more of the precipitation may be rapidly returned to the atmosphere. *Dickinson* stressed that there is still much to be done in improving the hydrological and SVAT components of land surface parameterizations for atmospheric models.

The role of intercomparisons as a way of detecting modeling errors received extensive discussion. *Collier* raised a concern that the use of intercomparisons of GCMs to understand physical processes may not be a proper approach. He suggested, instead, that comparisons of small scale models should be used to correct the algorithms incorporated into GCMs. *Dickinson* agreed, mentioning that GCMs are cumbersome to work with and involve many interdependent mechanisms whose effects that are difficult to isolate (for example, changes in GCM ocean temperature cause radiation changes that are dependent on cloud cover). Further, because model intercomparisons are often done with portions of GCMs that do not incorporate feedback processes, the answers obtained can be wrong.

Kabat suggested that intercomparisons of models may not be valid if only the data are compared and that model intercomparisons should also include consideration of the concepts implemented in the models. *Sorooshian* pointed out that the intercomparisons of watershed models using historical data have been able to reveal model deficiencies and serious weaknesses in parameterizations, and asked if GCM modelers expect to detect similar information from their intercomparison studies. *Dickinson* answered in the affirmative, but remarked that it is more likely that the outcome will be information regarding the sensitivities of GCM simulations to model parameters.

Becker voiced a concern that various basic and known parameters whose values should be constant appear to assigned values over quite a large range. *Dickinson* commented that the problem is that modelers sometimes "make up the data" without going to the literature to obtain the correct values. However, the major problem is that it is difficult to get the proper data, and as a result many key parameters often have to be estimated (for example Manabe estimated global average soil moisture capacity at 15 cm)! A number of the participants supported suggestions by *Sorooshian* and *Becker* that some standard data sets of hydrological variables and parameters should be collated to aid in model intercomparisons. It was pointed out, however, that different data sets might be required for different environments. Another issue raised by *Lynch-Steiglitz* is that because of differences in model structures and differences in conceptualizations of parameters, it may be very difficult to establish a set of common parameter values.

9. We need to improve land surface parameterizations at different scales.

The appropriate scale for model intercomparisons was discussed. The GCM scale is often cumbersome to work with. Further, because of the inclusion or exclusion of feedback mechanisms, the results may be difficult to interpret. *Dickinson* pointed out that sensitivity analyses conducted using only parts of GCMs may be biased by leaving out important feedback mechanisms. *Washburne* suggested that better intercomparisons may be achieved by using smaller scale models for which the necessary data to describe temporal and spatial variations may be available. In support of this point of view, *Collier* suggested that these small scale model intercomparisons can be used to correct the algorithms used in GCMs. This met with agreement from *Dickinson*.

10. The type and scale of data required is still not understood.

The type and scale of data required for improving land surface parameterizations is still not well understood. *Shuttleworth* and *Choudhury* raised the critical questions "Just how much detail is necessary?" and "What should the level of model complexity be?". The issue is, whether it is necessary to quantify every patch of soil, or is it possible to work at a somewhat larger scale, and if the latter is true, how do we determine what scale is appropriate? *Dickinson* offered, that while he had no precise answer to this question, it is clear that a proper representation of the "land" and associated land-atmosphere feedback processes is important. For example, in arid Southwestern US, one would expect 98% of the rainfall to be returned to the atmosphere from the land. *Calder* stated that a recent UK study has indicated that the representation of interception in GCMs needs to be made more complex, and that we should consider incorporating drop size into interception modeling. Again, there needs to be more interaction between data collection and modeling strategies with a view to improving confidence in land surface parameterizations at various scales.

11. The appropriateness of a model depends on many factors.

It is important to keep in mind that there is no such thing as a universal model. *Gupta* pointed out that the appropriate model geometry and process complexity in any model must depend on the temporal and spatial scales of interest, the modeled environment, the availability of data, and the modeling objectives. For example, a model may prove to function very well at one temporal and spatial scale while being quite inadequate at another. *Dickinson* and *Beven* commented on the need for modelers and data collectors to better understand the importance of identifiability and parsimony in the use of parameters in various situations. It was envisioned by *Goodrich* that one "Hydrologic Holy Grail" for the future might be to define a set of rules for distinguishing the dominant hydrologic processes in different environments, that would help guide modeling and data collection efforts.

12. We need to better understand how to aggregate and disaggregate variables and processes.

Associated with the above is the issue of aggregation and disaggregation of variables and model predictions between different scales (global, continental, GCM, mesoscale, watershed etc..). Work in this area has begun but, as indicated by *Dickinson*, is severely limited by the availability of data. The participants expressed support for the WMO global measurement site initiative, but it was stressed that to maximize the value of data the data sets should be as complete as possible and include subsurface and flow variables.

13. Challenges for the hydrological community.

The excitement generated during this panel discussion set the tone for the rest of the interactions during the workshop. Two questions, posed as challenges to the hydrologic community, stimulated much discussion. These are:

a) Can you demonstrate that your data is 'clean', 'documented' and 'available' (*John Rodda*).

b) Can you demonstrate that your model is 'validated' (*Keith Beven* offers a prize for any suitably convincing demonstration!).

Participant Opinion Survey: How Satisfied Are We with Hydrologic Data?

Compiled by Jene Michaud

During the workshop, *Michaud* conducted an ad-hoc questionnaire based survey among the participants to determine what the general perception of the community was of its satisfaction with existing hydrologic flux data. The survey produced the following result. The numbers should not be interpreted to have any more significance than a relative measure of satisfaction; positive numbers indicating relative satisfaction and negative numbers indicating relative dissatisfaction.

Hydrologic Flux	Satisfaction Score
Runoff	+ 14
Rainfall	+ 10
Potential evapotranspiration	+ 2
Snow	- 7
Soil moisture	- 11
Actual evapotranspiration	- 22

The survey results indicated that the workshop participants felt most satisfied with runoff data and were seriously dissatisfied with estimates of actual evapotranspiration (AET). However, each of these data components have their own "pluses" and "minuses". Some of the issues that were commented on were:

a) Runoff

(+) This is the hydrologic flux we have the most experience measuring, and indeed the measurements and methods are deeply institutionalized. Runoff has the advantage of being a strong integrator of watershed behavioral components

(-) Most major flood flows are poorly gaged. Data for the equatorial and Arctic regions are extremely scarce. Baseflow is poorly understood.

NATO ASI Series, Vol. I 46
Land Surface Processes in Hydrology
Trials and Tribulations of Modeling and Measuring
Edited by S. Sorooshian, H. V. Gupta, and S. C. Rodda

b) Rainfall

(+) This is the most common point measurement.

(-) Global coverage is poor. Distributed areal coverage is poor. Remote sensing sensors so far provide only limited improvements.

c) Potential evapotranspiration

(-) The few pan measurements that are routinely available suffer from serious calibration errors. It is important to distinguish between potential evaporation and potential transpiration.

d) Snow

(+) Snow is critical for water supply

(-) Snow is highly distributed over complex terrain. At present, the most useful remote sensors have very large footprints.

e) Soil moisture

(+) Remotely sensed data can provide clear patterns of the wetness of the ground surface.

(-) We cannot remotely sense soil moisture at depths below 5 cms or so. Also the passive microwave signal is masked by vegetation.

f) Actual evapotranspiration

(-) Virtually no systematic measurements exist.

Based on this survey, *Michaud* posed the following question: "If this (survey) reflects the consensus of the larger community of researchers and scientists, and taking into account issues such as cost-effectiveness and scientific necessity, etc., which data should we place our effort on improving?"

Part II - Precipitation

The Measurement of Rainfall

C G Collier
METSTAR Consultants, Meteorological Office
Bracknell, UNITED KINGDOM

1. Introduction

Over 97% of the water in our world resides in the oceans, and only 0.001% resides in the atmosphere. Indeed, only about 0.005% of the total world water supply is in motion at any one time. However, this very small amount of water is associated with all weather and river flow and has a profound impact upon the activities of man. The measurement of rainfall is of primary importance for hydrological calculations.

A wide range of techniques which have been developed worldwide for measuring rainfall will be discussed in this paper. These techniques range from point gauge measurements to methods based upon the interpretation of sometimes very indirect data obtained from space-based instrumentation. However, all of the techniques are complementary. Often, one technique is used to calibrate or adjust another, and it is the purpose of this paper to place the sometimes quite different performance characteristics of different methods in perspective.

2. Point Measurements Using Gauges

The commonly accepted standard method of rainfall measurement is the use of a can-type raingauge. However, such raingauges exposed above ground-level are subject to appreciable systematic error (-3 to -30% or even more) (Sevruk, 1982). This error is due to wind-field deformation above the gauge rim, to wetting, evaporation, splashing of raindrops, and blowing of snow. The components of the systematic error are related to the meteorological and instrumental factors listed in Table 1 and can be statistically analyzed. The general model for corrections takes the following form:

NATO ASI Series, Vol. I 46
Land Surface Processes in Hydrology
Trials and Tribulations of Modeling and Measuring
Edited by S. Sorooshian, H. V. Gupta, and S. C. Rodda

Table 1. Main Component of the Systematic Error in Precipitation Measurements and Their Meteorological and Instrumental Factors Listed in Order of General Importance (after Sevruk, 1982)

Symbol	Component of error	Magnitude	Meteorological factors	Instrumental factors
k	Loss due to wind-field deformation above the gauge orifice	2-10% *10-50%	Wind speed at the gauge rim during precipitation and the structure of precipitation	The shape, orifice area and depth of both the gauge rim and collector
$\Delta P_1 +$ ΔP_2	Losses from wetting on internal walls of the collector and in the container when it is emptied	2-10%	Frequency, type and amount of precipitation, the drying time of the gauge and the frequency of emptying the container	The same as above, and, in addition, the material, colour and age of both the gauge collector and container
ΔP_3	Loss due to evaporation from the container	0-4%	Type of precipitation, saturation deficit and wind speed at the level of the gauge rim during the interval between the end of precipitation and its measurement	The orifice area and the isolation of the container, the colour and in some cases the age of the collector, or the type of funnel (rigid or removable)
ΔP_4	Splash-out and splash-in	1-2%	Rainfall intensity and wind speed	The shape and depth of the gauge collector and the kind of gauge installation
ΔP_5	Blowing and drifting snow		Intensity and duration of snow-storm, wind speed and the state of snow cover	The shape, orifice area and depth of both the gauge rim and collector
* Snow				

$$P_K = kP_C = k(P_g + \triangle P_1 + \triangle P_2 + \triangle P_3 \pm \triangle P_4 - \triangle P_5) \pm \triangle P_r \qquad (1)$$

where:

P_K = corrected amount of precipitation;

k = conversion factor due to wind-field deformation;

P_c = the amount of precipitation caught by the gauge;

P_g = the amount of precipitation measured in the gauge;

$\triangle P_1$-$\triangle P_5$ = corrections for various components of systematic error according to Table 1; and

$\triangle P_r$ = random observational and instrumental errors.

The conversion factor k, as well as the corrections $\triangle P_{1\text{-}5}$ for a particular gauge, can be estimated experimentally by field comparisons or laboratory tests. The meteorological factors needed can be estimated using standard meteorological observations at the gauge site or in its vicinity.

Although the components of systematic error vary with the instrument and meteorological factors from very small to significant values, not all of them must be taken into consideration for certain gauge types, seasons, and regions. For example, the use of pit gauges prevents loss from wind-field deformation. Evaporation losses can be reduced by placing oil in the container or by designing the gauge so that only a small part of the water surface is exposed, ventilation is minimal, and the inside temperature of the gauge is not allowed to become excessive. In addition, the wetting loss can be reduced by making the internal wall of the collector and container as smooth as possible, so that the water does not adhere to it. Splash-in and -out can be reduced by the proper design of the collector, its rim, and in the case of the pit gauge, the construction of the non-splash surface.

It can be seen from Table 1 that errors due to deformation of the wind field can be very large if gauges are not installed in pits. Therefore, considerable effort has been put into aerodynamic raingauges. In the United Kingdom, Folland (1986) used computer models of airflow over gauges to explore possible new shapes for collectors, and Strangeways

(1984) reported comparable field experiments. This work led to the development of "champagne-glass" gauges as shown in Figure 1, and recent field trials reported by Hughes et al. (1993) have demonstrated significant improvements of such gauges over more conventional designs.

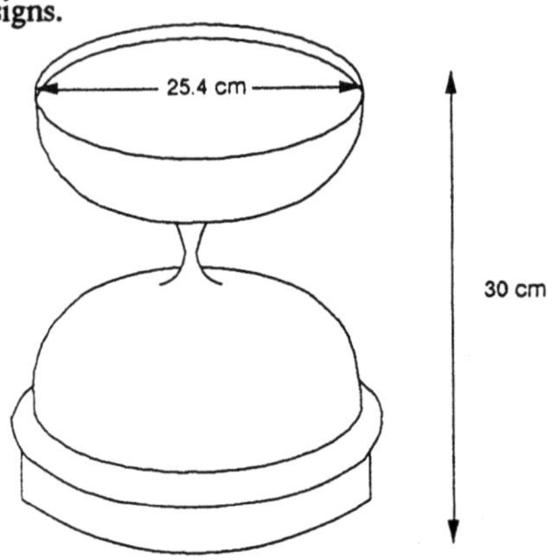

Figure 1. Prototype or 'champagne-glass' gauge

3. Areal Measurements of Rainfall Derived from Raingauge Networks

Weisner (1970) summarized the techniques which have been developed to extrapolate point measurements of rainfall to areal measurements. Empirical formulae have been derived having the form:

$$\bar{P} = a - bA^{1/2} \qquad (2)$$

where a and b are constants which represent the maximum rainfall at the focus and the mean rainfall gradient, and A is the area over which the measurement is required. However, such formulae are only applicable if carefully tuned for particular climate and orographic regimes. It is much more reliable to use a dense raingauge network using one of the following techniques:

- Arithmetic mean $\bar{P} = \sum_{i}^{n} \frac{Pn}{n}$

· Areal weighting or Thiessen , where $\bar{P} = \sum_{i}^{n} \frac{WnPn}{100}$

W_n are percentage weighting factors derived from the areas of polygons formed by the gauge locations.

- Triangulation
- Isohyetal with linear interpolation
- Isohyetal with subjective interpolation
- Weights based on isohyetal pattern or other data

Clearly, all these techniques depend upon the density of the raingauge network being employed. Considerable work over many years has been carried out to investigate the effect of raingauge network density upon areal measurement of rainfall. A recently reported example is that shown in Figure 2 from Seed and Austin (1990a,b) derived for networks in Florida and South Africa. Krstanovic and Singh (1992a,b) described an approach based upon an evaluation of entropy for gauge networks in Louisiana and showed that, for larger sampling intervals (greater than one day), the number of necessary raingauges to produce acceptable areal measurements could be reduced significantly. Nevertheless, care is needed when areal measurements are used to estimate streamflow. Duncan et al. (1993) showed that gauge density has a very strong effect on the estimation accuracy of hydrograph parameters, with the standard error generally falling off as a power law with increasing gauge density.

4. Measurements of Rainfall by Ground-Based Radar

Ground-based radar offers areal measurements of rainfall from a single location, over large areas in near real time. Techniques are based on four types of measurements: (1) the intensity of the backscattered radiation (radar reflectivity), (2) the difference in reflectivity between vertically and horizontally polarized radiation, (3) the attenuation of radar energy, and (4) attenuation and reflectivity determined simultaneously at two wavelengths. Of these measurements, only the first has been used widely and implemented operationally. Hence, we confine our discussion to this method, although technically more advanced techniques offer potential for operational development, and some discussion of the dual polarization technique is given in what follows.

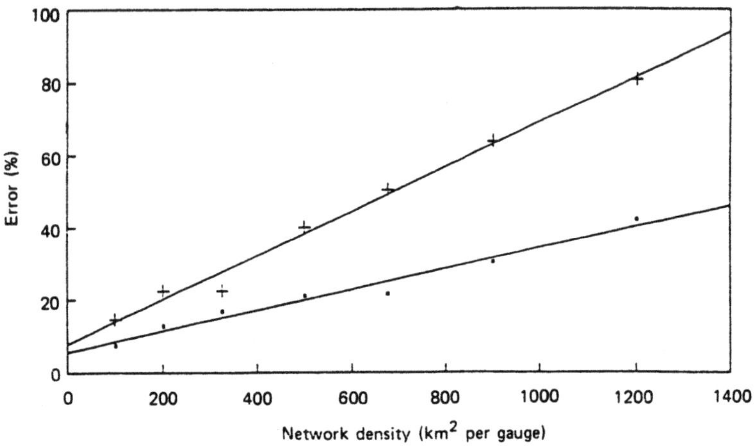

Figure 2. Errors in the estimation of mean areal daily rainfall over 45,000 km^2 using distance-weighting interpolation: •Florida + Nelspruit (Seed and Austin, 1990a,b)

Given a radar wavelength λ, and considering a spherical raindrop with diameter D, we may define a backscattering cross section $\sigma_\alpha(D)$ and a total attenuating cross section $\sigma_\alpha(D)$ is proportional to $D^6\lambda^4$. This is the justification for introducing a physical parameter called 'radar reflectivity factor' Z defined as:

$$Z = \int N(D) d^6 dD \tag{3}$$

where N(D) is the drop size distribution (DSD) within the resolution cell (Z in $mm^6 \cdot m^{-3}$, D in mm, N(D) in $mm^{-1}m^{-3}$). In the absence of attenuation along the radar path, and as long as the Rayleigh theory holds, the backscattered power from a resolution cell is proportional to Z. However, when the ratio $\pi D/\lambda$ becomes larger than 0.1, then the Mie theory should be used in place of the Rayleigh theory. To take account of this effect, an 'equivalent' radar reflectivity factor Z_E is generally considered. Z_E is the same as Z for light rain (mainly composed of small rain drops) but departs from it as the rainfall rate increases; the departure increases more rapidly as the radar wavelength decreases.

Similarly, $\sigma_a(D)$ may be modeled as σ_a = cD, where the a and c coefficients are tabulated as functions of λ and ambient temperature. The linear attenuation, K, of the incoming microwave radiation within a volume characterized by the mean drop size distribution N(D) is given by:

$$K = c \int_0^\infty N(D)\, D^{a} dD \tag{4}$$

Similarly, the rainfall rate R may be expressed as a function of N(D) as:

$$R = \pi c_1/\sigma \int_0^\infty N(D) D^3 + a_1 dD \tag{5}$$

where c_1 and a_1 are the coefficients defining the power law relationship for the terminal fall velocity v_t of a raindrop as a function of its diameter D:

$$vt(D) = c_1 D_{a_1}$$

If now we assume that N(D) can be described by an exponential function of the form $N(D)=N_o \exp -\Lambda D$, the integrals in Equations (3), (4), and (5) may be integrated analytically. Subsequent elimination of Λ produces the following two relations between Z, R, and K:

$$Z = EN_O^{(1-B)} R^B \tag{6}$$

$$K = CN_O^{(1-F)} R^F \tag{7}$$

where coefficients E, B, C, and F may be expressed analytically as functions of c, a, c_1, and a_1, respectively.

Table 2 specifies the numerical values of coefficients E, B, C, and F at 3 cm, 2 cm, and 0.86 cm, assuming an ambient temperature of 0° C. It should be remembered that Equation (3) assumes the Rayleigh approximation. The introduction of the Mie correction (i.e., of the Z_E parameter) complicates the relation (without changing it qualitatively) and reveals a Z_C/λ dependence (not apparent in Table 2).

Table 2. Coefficients Used in the K-R and Zr-R Models

(cm)	E	B	C	F
0.86	0.66 x 106	1.50	0.209	1.0
2.0	0.66 x 106	1.50	0.181	1.11
3.2	0.66 x 106	1.50	0.139	1.169

When using a radar for estimating rainfall rate, the reason for measuring two independent parameters, like Z_E and K, simultaneously is to minimize the dependence of estimates of rainfall rate, R, on the variability of the droplet size distribution (the N_O parameter may be eliminated between Equations (6) and (7), allowing R to be expressed as a function of Z and K).

As a radar beam rotates about a vertical axis, measurements are made at many ranges out to 100 km or more, and at different azimuths, of the energy backscattered from precipitation particles in volumes above the ground (Figure 3a). It may be shown that, if liquid precipitation uniformly fills the pulse volume, then the average power returned from precipitation at range r is proportional to Z/r^2, where Z is the so-called radar reflectivity factor given by the sum of the precipitation particle diameters raised to the sixth power. Equation (6) may be written as $Z = AR^B$, where A and B depend on the type of rainfall. Many values have been suggested, but values often used for rain at 5.6 cm (C-band) and 10 cm (S-band) wavelengths are $A = 200$ and $B = 1.6$ (Marshall and Palmer, 1948). An example of a radar image comprising data from several radars in the United Kingdom is shown in Figure 4. Modern weather radars may generate a wide range of products. Crum and Alberty (1993) described the products produced by the NEXRAD radars in the USA.

Use of R:Z relationships to measure rain, modifying A and B as appropriate, would appear to be straightforward. There are a number of problems, however, arising from the characteristics of both the radar and the precipitation. The importance of these problems will depend upon the particular radar configuration in use and the meteorology of particular situations. Hence, they are not listed in order of importance.

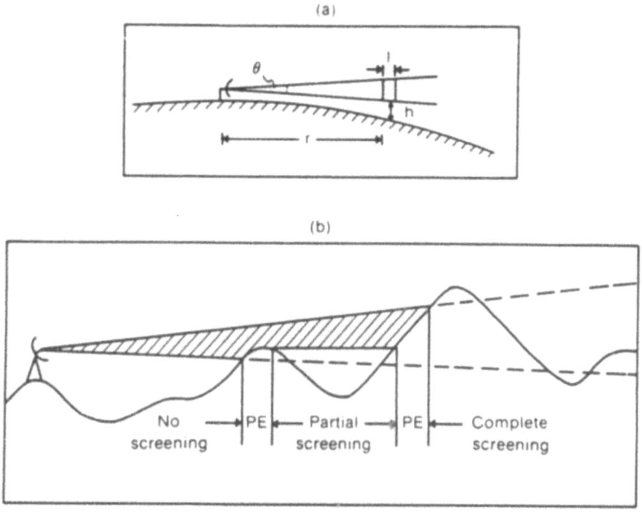

Figure 3. (a) The geometry of the radar beam relative to the curvature of the Earth. A radar pulse having length l is located at a range r from the radar. The height of the base of the beam at range r having an angular width of θ is h. (b) Permanent echoes (PE) and screening caused by hills close to the radar site are also shown (from Browning and Collier, 1989).

4.1. Problems Arising from the Characteristics of the Radar Site

Inadequate Radar Hardware Stability and Calibration

Maintenance of a stable system is extremely important. Joss and Waldvogel (1990) stated that, by using up-to-date solid state electronics, it is possible to keep the cumulative error of transmitted power, antenna parameters, noise figure, amplification in the receiver chain, and analogue to digital conversion well within 2 dB. This corresponds to a 36% error in precipitation rate.

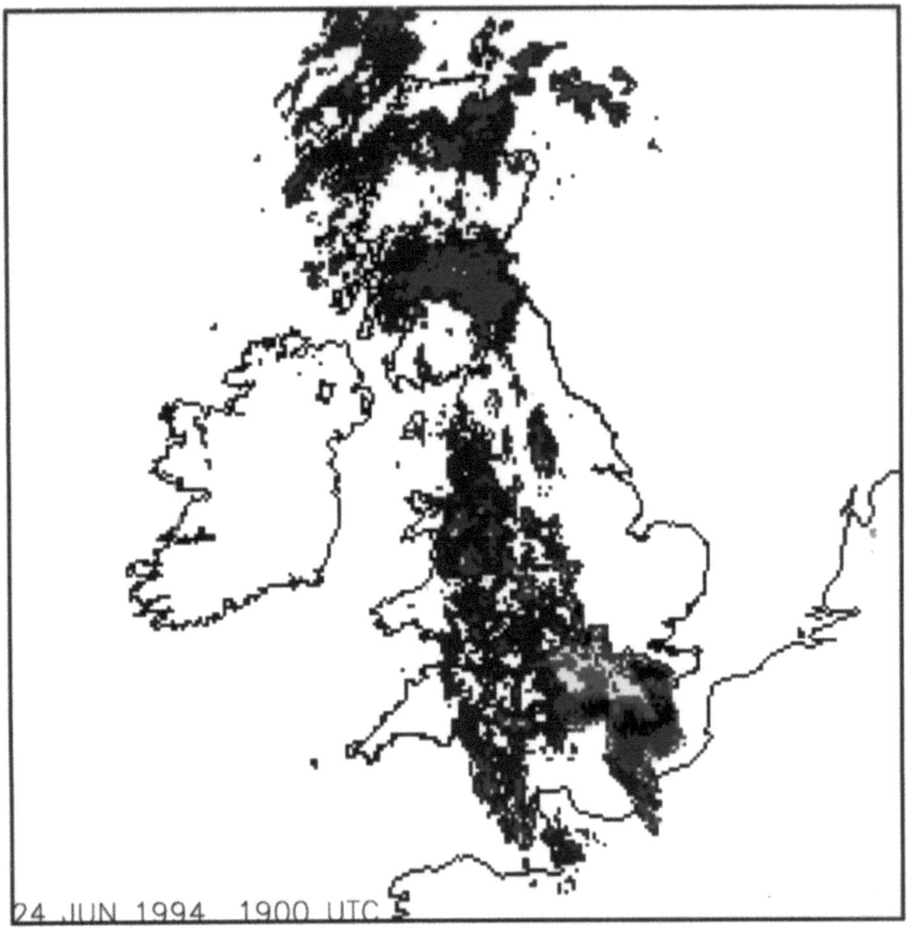

Figure 4. Radar composite image over the United Kingdom at 1900 UTC on 24 June 1994. Data from 15 radars are used. Each grid square is 5 km x 5 km. Blue is less than 1 mm h^{-1}, yellow 16-32 mm h^{-1}, and black greater than 64 mm h^{-1}.

Ground Clutter and Occultation

Many aspects of a radar site affect the quality of radar measurements. In general, to obtain good estimates of surface precipitation, a low (less than 0.5°) horizon is required to enable the beam to be kept low. Figure 3a shows an idealized radar beam from a well-sited radar under normal conditions of atmospheric propagation. However, Figure 3b, which is for less ideal situations, shows that the main radar beam (or subsidiary beams outside the main beam known as side lobes which contain much less, but still significant, amounts of microwave power), may encounter ground targets, causing strong persistent echoes (ground clutter) which may be misinterpreted as rainfall. Interception of the beam by the ground also causes occultation or screening of part of the beam, such that only a fraction of the power illuminates the rain at longer ranges. Anomalous refraction of the radar beam, caused by warm, dry air overlaying cooler, moister air, may cause additional ground echoes known as anomalous propagation echoes (ANAPROP). Use of Doppler radar (for a review, see Doviak and Zrnic, 1993) provides a method for removing most of these spurious radar echoes.

Attenuation or Poor Resolution Depending on Radar Wavelength

Use of long wavelengths (10 cm or more) avoids errors caused by attenuation due to precipitation. However, for the same antenna size, the beam width at 10 cm is twice as large as at 5 cm. This causes reduced spatial resolution at long ranges and leads to more ground clutter (Harrold, 1974). Moreover, Ulaby et al. (1986) pointed out that precipitation echoes observed at 10 cm are 12 dB weaker than those observed at 5 cm for the same output power, whereas the ground clutter echo strength is almost independent of wavelength. Although a 3-cm radar would be even more effective than a 5-cm radar having the same beam width in reducing clutter, it would suffer unacceptably from attenuation in only moderately heavy rain. In practice, the choice of wavelength is between 5 and 10 cm; it is influenced by cost and the frequency of occurrence of very heavy rain and large hail which cause significant attenuation of a 5-cm radar beam. Operational radars in the United States, including the Next

Generation Radar Network (NEXRAD) radars, will be 10-cm radars if used for long range surveillance, as hail and exceedingly heavy rain are very frequent in much of the United States (Crum and Alberty, 1993). Attenuation by moderate to heavy rainfall can be corrected by using a 5-cm radar beam, although when the beam intersects the melting level, such corrections are somewhat unreliable. Hence, radars used for airport surveillance in the USA and for much of northern Europe operate at C-band.

Radome Attenuation

The antenna of a radar are often housed in a radome made from rubberized material or fiberglass and either inflated or supported by aluminium strips. Such structures protect the antenna from precipitation and pollutants. By eliminating wind loading, they also allow less powerful drive motors to be used. Unfortunately, precipitation on the radome attenuates the radar beam. The degree of wetting and attenuation depends on the radar wavelength, the nature of the surface, and the size of the radome. Wilson (1978) found that a rainfall rate of 40 mm h^{-1} produced a two-way attenuation of a 5-cm radar of 3 db for a radome having a diameter of 5.5 m. This type of attenuation can be reduced somewhat by the use of Teflon® coatings on the radome, although regular cleaning is also essential. Snow can produce a much bigger effect, and it is necessary to heat the radome to prevent a buildup of snow or ice.

4.2. Problems Arising from the Characteristics of the Precipitation

Errors Due to the Fluctuating Nature of the Signal from Precipitation and Averaging Over Areas of Nonuniform Precipitation

Scattering of microwave radiation from a population of precipitation particles within the radar pulse volume produces a fluctuating signal at the radar receiver. To produce a measurement of precipitation intensity, it is necessary to average the received power over a number of independent samples either in time or in space. Rainfall rate may vary by a factor of 10 within a 10-minute period or within a distance of 2 km, so the averaging must be done on smaller scales. If such variations occur within the radar pulse volume (for example, at long ranges where the beam width is several kilometres wide), this may lead to a large error of several decibels (Mueller, 1977; Zawadski, 1984). These effects are often noted when comparing radar and raingauge measurements of rainfall; for example, Joss and Waldvogel (1990).

Variability of the R:Z Relationship

The values of A and B in the R:Z relationship depend upon the nature of the raindrop size distribution which differs greatly between drizzle at one extreme and thunderstorms at the other. Battan (1973) listed over 60 R:Z relationships. The presence of updrafts or downdrafts also affects the R:Z relationship. Austin (1987) calculated that, in a strong downdraft of 8 m s^{-1}, the reflectivity value for a given rainfall rate would be about 3 dB less than in still air, producing an underestimate of the rainfall rate of 40%. Another important problem is associated with the detection of hail. Reflectivity is overestimated when hail is present in the radar beam. Recently Auer (1994) (see also Hardaker and Auer, 1994) demonstrated a useful algorithm to separate rain and hail using single polarization radar and IR cloud top temperatures.

Variability of Reflectivity in the Vertical

The radar beam at far ranges from the radar site is at a considerable height above the surface of the Earth. For a beam elevation of 0.5°, the axis of the beam is at a height of 2 km at 130 km range and 4 km at 200 km range. Reflectivity varies in the vertical because of growth or evaporation of precipitation, vertical air motion, and melting (where melting of snowflakes occurs, there is a layer of enhanced reflectivity known as the bright band). These factors can lead to large differences between the apparent precipitation intensity measured by radar and the intensity of precipitation reaching the ground. Average vertical reflectivity profiles have been derived for different precipitation conditions. The top three profiles shown in Figure 5 are for convective rain

(Figure 5a), widespread rain with bright band (Figure 5b), and snow or shallow rain (Figure 5c). Although there may be large deviations from these profiles on particular occasions, it is instructive to use them as an indication of the percentage of the precipitation observed by radar at various ranges allowing for Earth curvature. In particular, the profiles in Figures 5b and 5c draw attention to the difficulties of making measurements in the presence of bright band or shallow rain. Considerable effort is being put into developing correction procedures (for example Hardaker et al., 1994; Kitchen et al., 1994). Another problem in some parts of the world is orographic growth of precipitation at low levels over hills exposed to strong moist maritime air flows. To some extent, orographic low-level growth can be estimated by applying climatological correction factors, but where possible the radar should be sited so that it can observe the low-level precipitation directly. As shown by Figure 5d, considerable orographic enhancement may occur within 1/2 km of the ground, and so it is observable only at very close ranges.

4.3 Adjustment of Radar Measurements

Considerable work has been carried out over the past 20 years or so to address all of the above problems. This has led to the development of procedures for adjusting radar measurements of precipitation using data from rain gauges (heated gauges for snowfall). Wilson and Brandes (1979) (see also Collier, 1989) summarized much of this work. Techniques have ranged from the application of an adjustment factor derived by simple averaging or statistical procedures for the whole area of interest to the use of a large number of rain gauges to define the detailed spatial variations in the adjustment factor (see for example Hudlow et al., 1991; Krajewski, 1987).

Many of the reported estimates of the accuracy of rainfall measurements by radar refer to the results of carefully controlled research assessments rather than to operational systems. An exception is the study described by Collier (1986a,b). This was carried out using an operational system in northwest England for a 12-month period during 1982/1983. A raingauge adjustment procedure (see Collier et al., 1983) was used

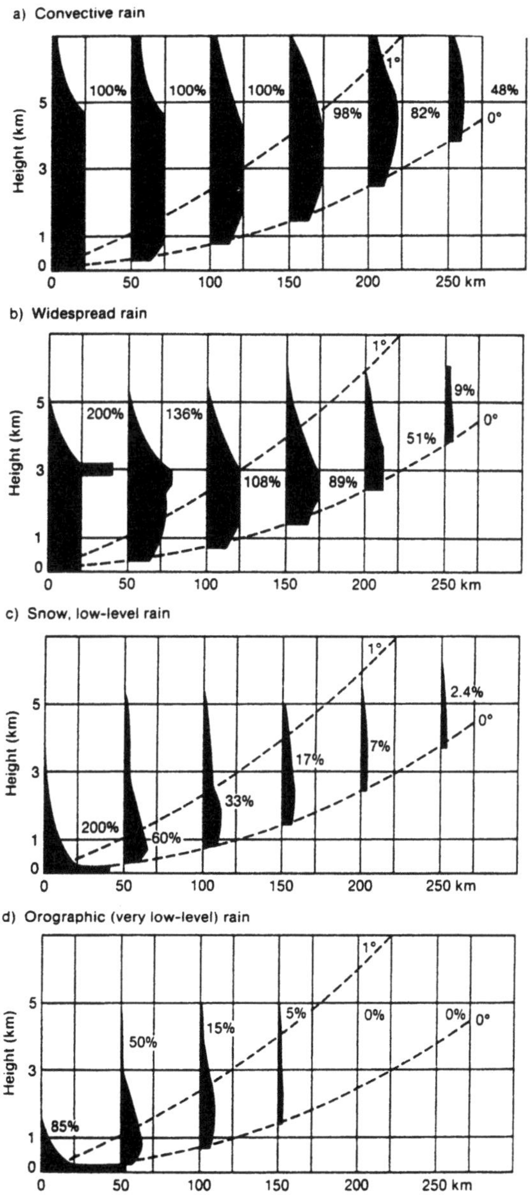

Figure 5. Vertical profiles seen by the radar at various ranges in convective and widespread rain, in low-level rain or snow, and in orographic rain. The numbers in each figure give the percentage (referred to the true melted water value which would be measured at ground level using a raingauge) of rain rate deduced from the maximum reflectivity of the profile. A radar with a 1° beam width is assumed, in a flat country. If the radar is lower than nearby obstacles, then much less precipitation can be observed at far ranges. The top and bottom of the main part of a 1° beam elevated at 0.5° are shown as dashed lines labeled 0° and 1° (partly after Joss and Waldvogel, 1990), from Browning and Collier (1989).

consistently throughout the period, and all data were included regardless of their quality. Figure 6 shows the variation of the monthly mean adjustment factors (radar/gauge) over the 12-month period for four raingauge sites at different ranges from the radar. Three of these sites were within the area covered by the real-time adjustment system (up to 75 km range); the fourth, Nottingham, is included to illustrate the performance at greater range. During the summer, when deep convective rainfall was common, the radar estimated rainfall well at all ranges. The average performance was not as good during the winter. The record for Nottingham shows the underestimation which can occur at far ranges in winter because of the intensification of reflectivity below the radar beam. The other graphs indicate a tendency to overestimate at near ranges, probably as a result of uncorrected bright-band effects within the radar beam.

At any time of the year, the adjustment factors at individual raingauge sites may vary considerably from the mean monthly values. When these variations are examined, the bright-band effect is seen to be very significant. If all observations influenced by bright band are excluded, the average percentage difference between hourly unadjusted radar estimates and raingauge readings in conditions of widespread frontal rainfall is 60% (Collier, 1986a). Real-time adjustment reduced this average difference to 45%. In convective rainfall, without bright band, adjustment improves the figures from 37% to 21%. When bright-band effects are present, the average difference is 100%, and adjustment reduced this to 75%. The probability of getting a particular error, and the improvement achieved by using real-time raingauge adjustment, are shown in Figure 7 for three categories of rainfall, with and without bright band.

4.4 Area-Average Method

Chiu (1988) found a high correlation between the instantaneous area-average rainfall rate and the fractional area covered by radar echoes above a predefined rainfall-rate threshold, using data from the Global Atmospheric Research Programme (GARP) Tropical Atlantic Experiment (GATE). Rosenfeld et al. (1990) also found the same high correlation using GATE data and data from South Africa, Texas, and Darwin.

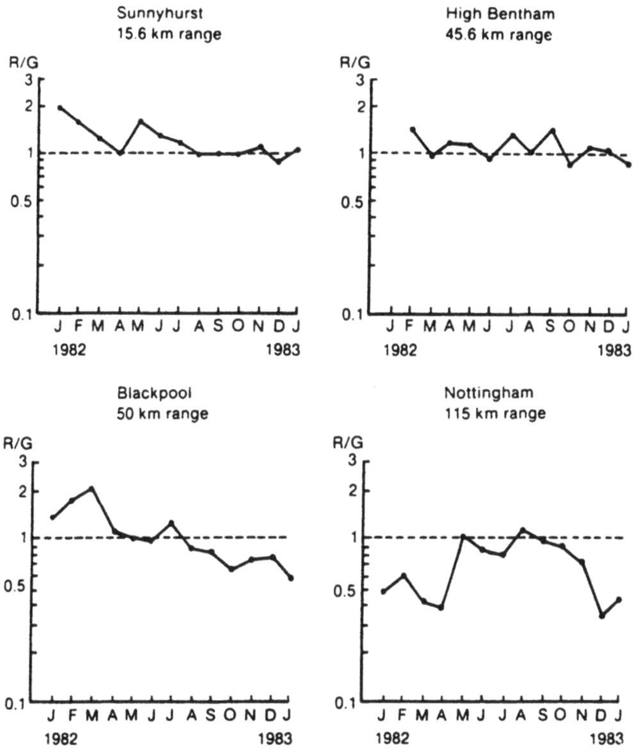

Figure 6. Variation of monthly mean hourly assessment factors (radar divided by gauge) for four raingauge sites at different ranges from the Hameldon Hill radar (from Collier, 1986a).

Atlas et al. (1990) showed that the relationship depends on the existence of a well-behaved probability density function (pdf) of rainfall rate. A theoretical explanation of the correlation between the instantaneous area-average rainfall rate and fractional area occupied by rain of intensity greater than a set threshold was also provided by Kedem et al. (1990), based on the assumption of a homogeneous pdf. Recently Cheng and Brown (1993) demonstrated that this technique could be used to estimate instantaneous frontal rainfall over areas of about 10^5 Km^2 with an accuracy of around 14%.

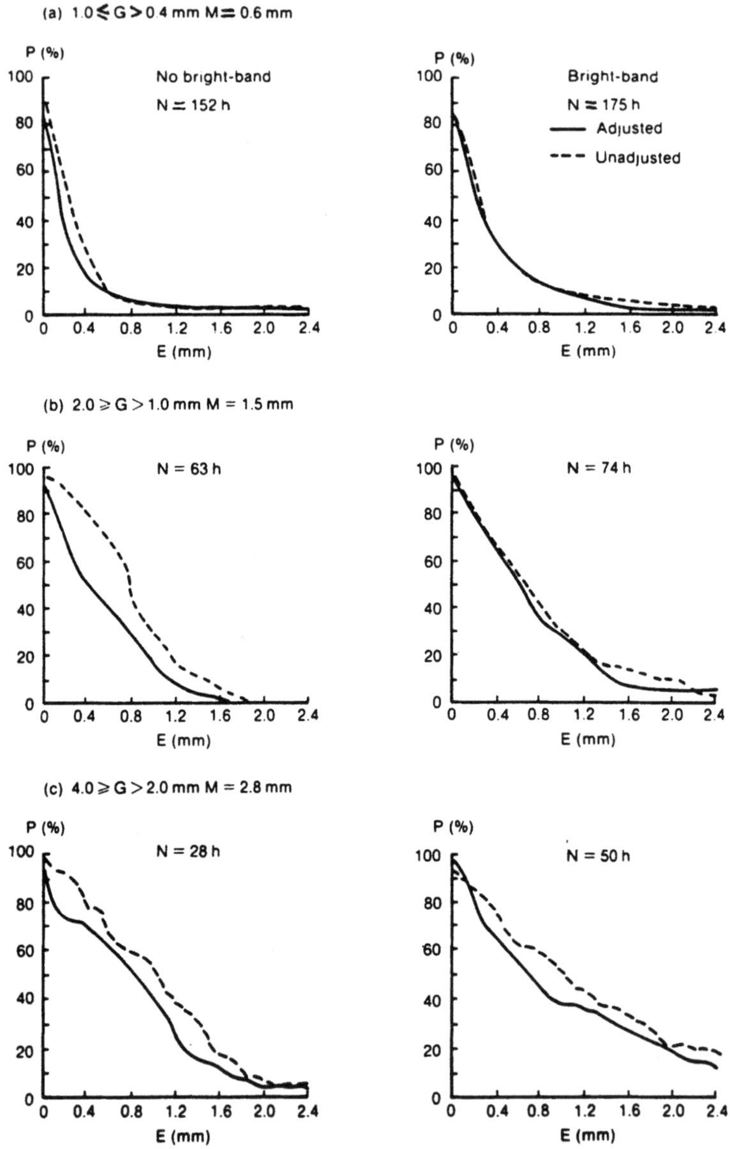

Figure 7. Probability (P, percent) of particular errors (E, millimeters) occurring in measurements of hourly point rainfall for three classes of rainfall amount (G) and for cases (left) with no bright band and (right) with bright band present. Both real-time-adjusted (solid lines) and unadjusted (dashed lines) plots using raingauge data analyses are distinguished. The number of hours in each category (M) is also given. The data are for the Hameldon Hill radar in northwest England during 1982 (after Collier, 1986a,b).

4.5 Dual Polarization Radar

Another 'multiparameter' technique currently used from the ground is the dual polarization radar, measuring the differential reflectivity ZDR, between vertically and horizontally polarized waves. Because the raindrops are not spheres, but oblate spheroids with axes aligned vertically, the equivalent reflectivity at vertical polarization Z_V is smaller than that at horizontal polarization Z_H. ZDR is defined as:

$$ZDR = 10 \log_{10} (Z_H/Z_V) \tag{8}$$

The eccentricity of the spheroid increases with the rain drop diameter D. This relationship is known. ZDR provides an estimate of the median volume diameter D_o, which is related to the Λ parameter of the distribution. It can be shown that the knowledge of the two parameters (Z, Λ) provides a unique determination of R. However, although this technique is attractive from the ground and has been used successfully (see, for example, Joss and Waldvogel, 1990), it cannot be used from space (Section 5.2) because the angle of incidence is then too close to the vertical, making the determination of the ratio Z_H/Z_V very inaccurate (or even impossible at vertical incidence).

5. Measurement of Rainfall Using Space-Based Techniques

There is a lot of literature on the measurement of rainfall from space. In what follows, we make no attempt to review this literature, but refer the reader to existing reviews such as Lovejoy and Austin (1979), Atlas and Thiele (1981), and Collier (1989).

5.1 Passive Methods

Visible/Infrared Techniques

Visible/infrared techniques derive qualitative or quantitative estimates of rainfall from satellite imagery through indirect relationships between solar radiance reflected by clouds (or cloud brightness temperatures) and precipitation. A number of methods have been developed and tested during the past 15 years with a measured degree of success (see, for example, Arkin and Ardanwy, 1989).

There are two basic approaches, namely the 'life-history' and the 'cloud-indexing' techniques. The first type makes use of data from geostationary satellites which produce images usually every half hour. It has been mostly applied to convective systems. The second type, also based on cloud classification, does not require a series of consecutive observations of the same cloud system.

It must be noted, however, that up to now none of these techniques has been shown to be 'transportable'. In other words, relationships derived for a given region and a given time period may not be valid for a different location and/or season.

Other problems include difficulties in defining rain/no rain boundaries and inability to cope with the rainfall patterns at the meso or local scales. An example of a cloud not producing rainfall in close proximity to cloud producing rainfall is shown in Figure 8. Scientists working in this field are perfectly aware of these problems, and this is why it is current practice to speak of the derivation of 'precipitation indices' rather than rain rates. Nevertheless, the Climate Analysis Centre, NOAA, provides operational estimates of accumulated tropical rainfall for pentads and months on a 2.5° grid in the latitude belt from 40° S to 40° N (eight images per day) using the algorithm described by Janowiak and Arkin (1991).

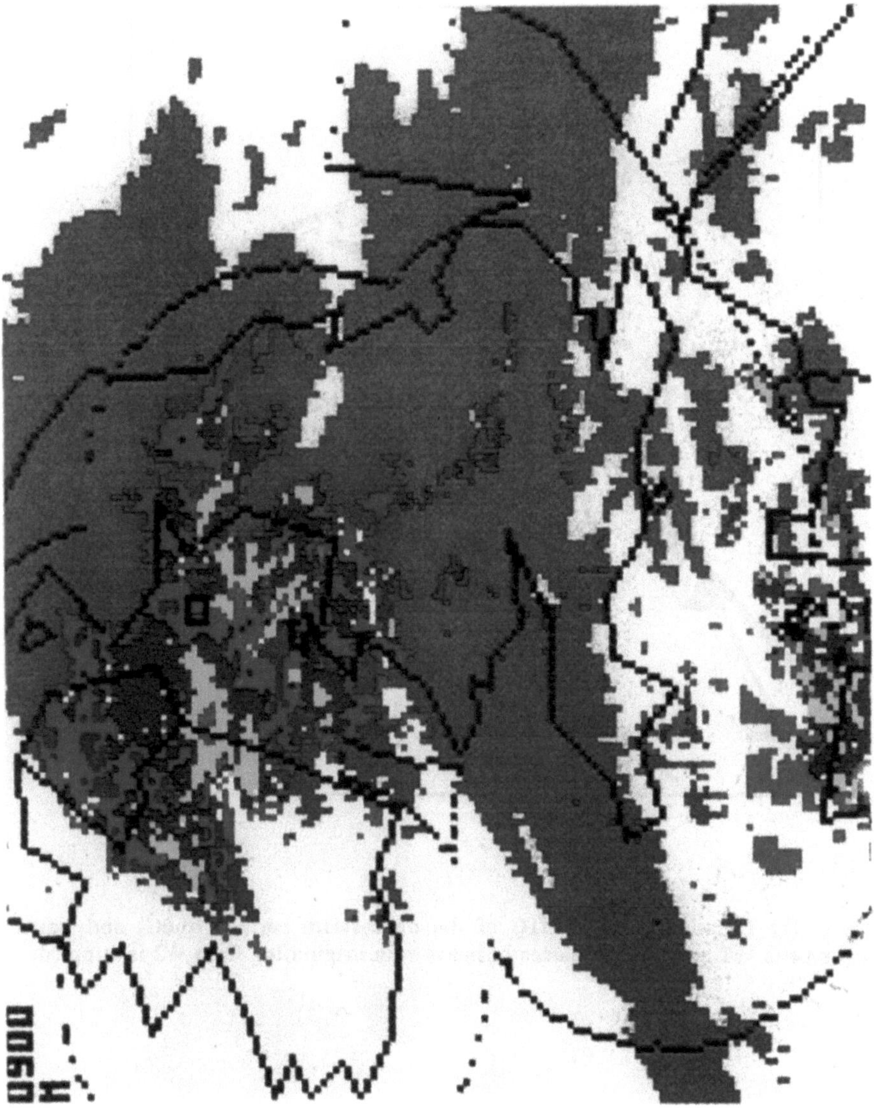

Figure 8. (a) Meteosat-derived infrared cloud-top temperatures for 0900 UTC, light blue colder than -15°C. Radar rainfall intensity (mm h^{-1}) and pink < 1, green 1-3, yellow 3-10, red > 10. Coastlines of northwest Europe, national boundaries, and radar network boundaries are shown in black.

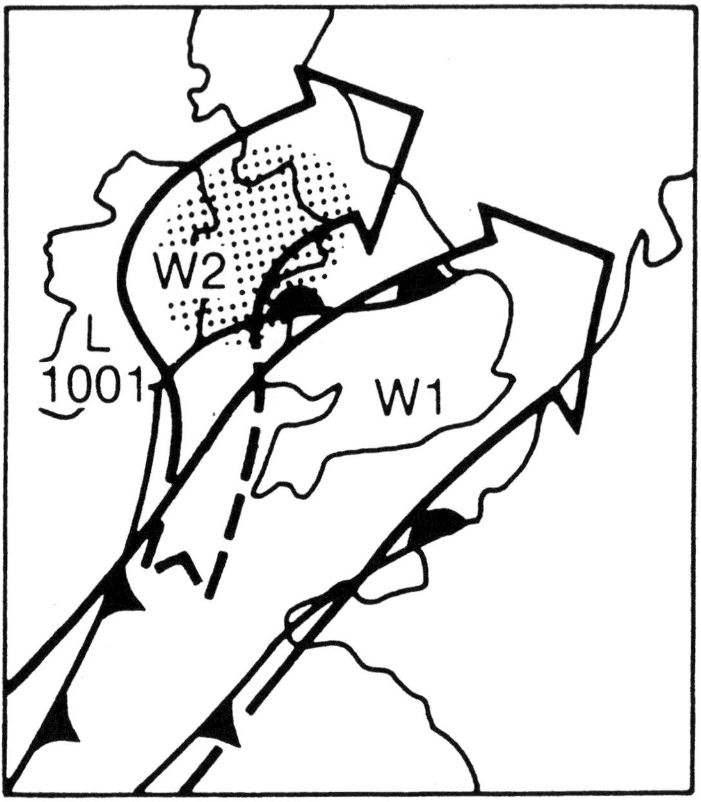

Figure 8. (b) Location at 0900 UTC of the depression, surface fronts, and warm conveyor belts W1 and W2. The area of heavy rain originating from W2 is stippled.

Cloud Indexing Methods

Cloud indexing was the first technique developed to estimate precipitation from space. It is based on the assumption that the probability of rainfall over a given area is related to the amount and type of cloudiness present over this area. Hence, one may postulate that precipitation can be characterized by the structure of the upper surface of the associated cloudiness. In addition, in the case of convective precipitation, one may also postulate that a relationship exists between the capacity of a cumuliform cloud to produce rain and its vertical as well as horizontal dimensions. The vertical extent of a convective cloud is related to the cloud top brightness temperature (higher cloud tops are associated with colder brightness temperatures).

Therefore, the approach is to perform a cloud structure analysis (objective or subjective) based on the definition of a criterium relating cloudiness to a coefficient (or index) or precipitation. This characteristic may be, for instance, the number of image pixels above a given threshold level.

The general approach for cloud indexing methods involving infrared observations is to derive a relationship between a Precipitation Index (PI) and a function of the cloud surface area, S[TBB], associated with the brightness temperature (TBB) colder than a given threshold T_o. This relationship can be generally expressed as follows:

$$PI = A_0 + \sum_i A_i * S_i[TBB_i] \tag{8}$$

for $TBB_i < T_o$

If desired, an additional term related to the visible image can be included on the righthand side of Equation (8). The next step is to relate PI to a physical quantity related in some way to rain. This is done by adjusting the coefficients A and the threshold level T_o by comparison with independent observations such as raingauge or radar data.

One of the problems inherent to this technique is the bias created by the potential presence of high-level nonprecipitating clouds such as cirrus. Another limitation resides in the fact that the satellite measurement represents an instantaneous observation integrated over space, while raingauge observations are integrated over time at a given site.

Life-History Methods

Life-history methods, as indicated by their name, are based on the observation of a series of consecutive images obtained from a geostationary satellite. It has been observed that the amount of precipitation associated with a given cloud is also related to its stage of development; therefore, two clouds presenting the same aspect (from the VIS-IR images point of view) may, produce different quantities of rain depending on whether they are growing or decaying.

As with the cloud indexing technique, a relationship is derived between a Precipitation Index (PI) and a function of the cloud surface area, S[TBB], associated with a given brightness temperature (TBB) lying above a given threshold level. In addition, the cloud evolution is taken into account and expressed in terms of rate of change of S[TBB] between two consecutive observations.

An equation, as complex as desired, may be derived between PI and functions of S[TBB] and its derivative with respect to time:

$$PI = A + A * S[TBB] + A' \frac{d}{dt} S[TBB] \qquad (9)$$

for TBB $< T_o$

Here, again, another step is necessary in order to relate the Precipitation Index defined by the equation to a physical quantity related to rain.

Many such relationships have already been published. These publications have been extensively discussed and it was demonstrated, at least for one instance, that taking into account the cloud evolution with time added unnecessary complexity and that comparable success could be obtained with a simple cloud indexing technique.

Recently, more physics have been introduced to the various schemes. Improvements include:

- use of cloud models to take into account the stratiform precipitation often associated with convective rainfall and to help with the cloud classification;
- introduction of simultaneous upper tropospheric water vapor observations; and
- introduction of a time lag between the satellite observations and ground-based measurements.

It has also become evident that satellite data could be used in conjunction with radar observations, not only to validate a method, but as a complementary tool. The FRONTIERS (Forecasting Rain Optimized using New Techniques of Interactively Enhanced Radar and Satellite) developed by the UK Meteorological Office represents an example of combined use of satellite imagery and radar observations (see Browning and Collier, 1989).

Quite a few comparisons between different methods over the same test cases have now been performed and published, but one must remain extremely cautious about any final statement concerning the success (or lack of it) of VIS-IR methods. The degree of success is very strongly related to the space-time scales considered, and one cannot expect that a regression developed and tested for use in climate studies will be also valid for estimation of mesoscale precipitation. One must also keep in mind that it is always easy to adjust regression coefficients for a particular case and claim that the method has been validated.

Microwave Techniques

VIS-IR measurements represent observations of the upper surface of clouds only. In contrast, it is often believed that microwave radiation is not affected by the presence of clouds. This statement is not generally true. Its degree of validity varies with the microwave frequency used as well as with the type of cloud being observed.

One major difference between infrared and microwave radiation is the fact that, while the ocean surface emissivity is nearly equal to one in the infrared, its value (although variable) is much smaller in the microwave region (taken from 5 to 200 GHz). Therefore, the background brightness temperature (TBB) of the ocean surface appears much colder in the microwave. Over land, the emissivity is close to one, but varies greatly depending on the soil moisture.

As far as microwaves are concerned, several different effects are associated with the presence of clouds over the ocean. They are highly frequency-dependent.

- Absorption by cloud droplets is roughly proportional to the square of the frequency. A 2 km thick (nonprecipitating) cumulus cloud transparent at 18 GHz may be totally opaque (through absorption) at 160 GHz, resulting in an increase of brightness temperature (through emission) over the cold background.
- The increase of cloud water content associated with the presence of raindrops in a cloud produces an enhanced absorption-emission effect. The 2 km thick precipitation (a few mm/hr^{-1}) cumulus cloud will this time be detected over the cold surface background at 18 GHz, and the observed TBB will increase with the amount of 'vertically integrated' raindrops contained in the cloud. This is the dominating effect below about 22 GHz and the basis for estimation of rainfall at these frequencies.
- When the amount of rain present in the cloud becomes sufficient to totally absorb incoming radiation, the value of TBB reaches a value close to the

cloud's infrared (or thermal) temperature. This is referred to as the 'saturation' effect.

- Scattering by raindrops increases with drop size. The effect of scattering is to reduce the apparent TBB and is also frequency-dependent.
- The presence of ice in a precipitation cloud results in a decrease of TBB through scattering (ice particles scatter but do not absorb microwave radiation). This effect increases with frequency and particle size and is used to indirectly delineate rain over the Earth's surface.

Measurements of rainfall using passive microwaves fall into one of two classes, depending on the particular effect used to detect precipitation, namely absorption or scattering. To exploit absorption, a cold background is necessary. Thus, techniques based on this effect can be applied only over the ocean using frequencies below 19 GHz. Scattering may be exploited over land as well as over the ocean surface. The brightness temperature depends on the number and size of scattering particles, and the major scattering effect results from the presence of frozen hydrometeors at the top of convective clouds. The scattering approach utilizes frequencies above 60 GHz.

5.2 Active Method: Rain Radar

Measuring rain from a space radar is very attractive because this instrument is potentially able to provide the most thorough investigation of the precipitation field one can presently imagine from space. Two strong arguments mitigate in favor of the radar: its ability to discriminate in range (i.e., in altitude when operating from space) and the fact that the measured parameters like radar reflectivity, or radar attenuation, are directly related to the rainfall rate.

In this section, we will review the main techniques that have been envisaged up to now for measuring rain using a space-borne radar. The basic theory upon which all these techniques are based was discussed in Section 4. As with passive instruments, sampling problems are important for satellites in any orbits other than geostationary orbits.

None of the techniques that have been proposed up to now for use from space has envisaged the use of dual polarisation measurements (Section 4.5). All of the techniques are based on measurements of Z_E, K, or both:

- Direct method: rainfall rate R is estimated using standard Z_E-R relationships empirically determined.
- Surface reference technique (SRT): the scattering properties of the ocean surface vary on a much larger horizontal scale than the rain rate field does. Hence, the level of the signal received from the ocean surface through the rain cell relative to that outside the rain cell provides a means of monitoring the two way attenuation along the ray path. Using an attenuation to rainfall relationship, rainfall can be estimated.
- Dual wavelength technique: the apparent reflectivities at the separate wavelength are compared within the same resolution cell, and the differential attenuation is estimated.
- Dual-beam technique (or stereoradar): stereoscopic observations are made from the same space vehicle using two antennae operating at the same frequency.

None of these techniques is yet used operationally, although a rain radar is planned for launch in the late 1990s as part of the joint U.S.-Japan Tropical Rainfall Measurement Mission (TRMM) (Simpson et al., 1988). This system will fly in a tropical orbit (inclination 35°) and includes both active (14GHz) and passive (10, 19, 17, 90 GHz) microwave instruments plus visible and infrared radiometers.

6. Combining Satellite with Ground-Based Techniques

It is evident that there are limitations with both existing and planned techniques for precipitation measurements from space based upon direct sensing of the visible, infrared, and microwave properties of cloud and precipitation. This has prompted investigation of the contribution that ground-based measurements of precipitation and independent measurements of cloud characteristics might make to these methods.

To date, the blending of data from different sources has received restricted application as effort has concentrated on the estimation of precipitation over oceans and other data-sparse areas. One exception has been in satisfying the requirements of nowcasting. Here, the need to extend the area of ground-based radar coverage has encouraged attempts to combine satellite data with surface observations and ground-based radar data. More recently, as part of the World Climate Program, the Global Energy and Water Cycle Experiment (GEWEX) has been proposed in which the importance of combining ground-based and space-based measurements of precipitation is recognized.

Use of Ground-Based Weather Observations. Although ground-based manual and automatic observations of cloud and precipitation have been used for some time to assess the performance of space-based techniques, there have been few attempts to devise objective procedures for combining these data with the precipitation estimates produced from satellite data. This is probably because of the limited coverage provided by conventional observations from raingauges and from observers.

While there are clearly problems in many parts of the world, there are areas, for example, the USA and Europe, which do have extensive coverage (even in real time). Work is underway in Europe, as elsewhere, to make better use of these observations, including both manual and automatic measurements of present weather on land and from merchant ships and buoys. Areas of nonprecipitating clouds are identified from the surface observations and then removed. The removal process is usually subjective, although objective methods may be possible in areas where the observation density is high. This possibility has been exploited most extensively in attempts to blend surface-based weather radar data with satellite data.

Within the last few years, encouraging developments have taken place in the use of microphones and acoustic techniques generally to detect and measure rainfall at the surface of the oceans (see, for example, Nystuen and Farmer, 1989). Considerably more work remains to be done in this area before a fully operational system can be deployed.

Use of Ground-Based Radar Observations. Radar systems can make measurements of precipitation over wide areas from a single location in real time and are easier and more cost-effective to operate than extensive raingauge networks. Therefore, it is attractive to design procedures to combine the radar data with satellite data in order to better understand the relationships between variations in visible brightness and infrared or microwave brightness temperatures and rainfall. However, care must be taken as measurements of precipitation made using radar data are subject to various errors which may produce spurious correlations. Hence, most of the work to date has used radar data in a relative sense to locate areas of precipitation. The type of algorithms used range from simple insertion techniques to procedures developed to use radar data to derive visible or infrared rainfall relationships and algorithms using bispectral satellite data.

Perhaps the most widely applicable technique is the bispectral technique. The procedure exploits information on the height of clouds derived from infrared measurements and information on the depth of clouds derived from visible measurements. The visible and infrared data are each divided into a number of classes and a table of 'probabilities' of rain for each class produced by comparison with radar data. Examples of tables for frontal rain are shown in Figure 9. These are then applied to the satellite data beyond the area of radar coverage.

Early work on this type of technique encountered problems arising from registration errors between visible and infrared images, instrument calibration, time difference between images, and illumination geometry. However, these problems can be overcome. Nevertheless, care must be taken to eliminate those areas of poor radar coverage before constructing the look-up table. In addition, average rainfall rates of less than 0.5 mm/h^{-1} that are derived from visible/infrared pairs corresponding to warm temperatures and low visible counts are mainly a result of improper radar-satellite matching caused by the problems mentioned above. The rainfall rates at these pairs are set to zero to avoid the generation of spurious areas of small or trace rainfall amounts.

The rainfall probability maps derived using this technique have been verified with radar and surface data. A useful skill in estimating rain areas has been identified, although experience indicates that the probability of rain table should be changed seasonally. The radar verification shows that the main skill lies in identifying three classes, namely clear, cloudy with low probability of rain, and cloudy with a significant probability of rain. The accuracy of the results diminishes with increasing distance from the training radar.

However, it was pointed out that use of a coarse resolution might allow the introduction of texture matrices. Rainfall over periods from 1/2 to 2 hours may be estimated with an accuracy of around 49% over areas of 10^5 km^2, but more extensive assessments for different rainfall types are required to confirm this performance.

Use of Numerical Models. Given precipitation estimates from satellite data, procedures are being devised to use the estimates as input to numerical models. However, studies to use numerical models to aid the estimation of precipitation from satellite data are limited. Indeed, operational approaches to this problem exist in which the satellite-derived moisture fields are interactively balanced with respect to other model fields, even though relative humidity and vertical velocity should be well correlated. Currently, climatological rainfall fields and orographic rainfall derived from models approach the accuracy of observed fields for periods of a day or longer, but numerical models do not yet reproduce convective rainfall with the spatial accuracy necessary for most hydrometeorological studies.

7. Summary of the Performance of Measurement Techniques and Conclusions

Table 3 gives a summary of the performance of satellite rainfall estimation techniques. It is clear that areal rainfall can be measured with acceptable accuracy from space.

However, this conclusion only applies at present to measurements over areas greater than 10^4 km^2 for integration periods for less than 24 hours. In general, errors over smaller areas are larger, although high accuracy can be achieved by careful 'tuning' on specific datasets.

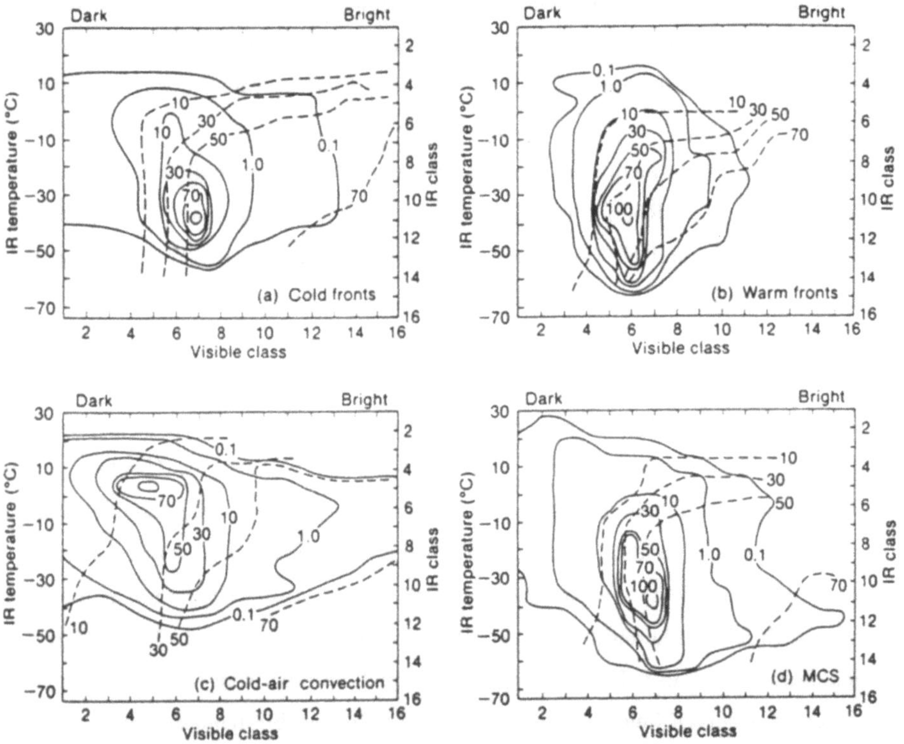

Figure 9. The probability of precipitation as a function of IR temperature and visible class (dashed line). The percentage of precipitation pixels, relative to the total number of precipitation pixels, normalized to a maximum value of 100 (solid lines) (from Cheng et al., 1993).

Ground-based radar offers higher accuracy than satellite techniques over small areas and small time periods. Krajewski et al. (1992) estimated that, when the exponent parameter in the Z and R relationship has small uncertainty (about $\pm 10\%$), radar techniques based upon averaging point observations work better than techniques based upon the area-threshold method. This conclusion is not inconsistent with the expected performance of ground-based radar versus satellite techniques, the best performance of which is achieved with the area-threshold method.

Table 3. Summary of the Performances of Satellite Rainfall Estimation Techniques (partly after Browning and Collier, 1989)

Technique	Area Over Which Estimates are Assessed (km^2)	Period of Integration (hours)	Percent Error
Cloud Indexing	105	24	122
	104	1/2	40
Area-Average	105	Inst to 30 x 24	14
Life History	104	1	85
	105	24	55
	104	1/2	50
	6 x 103	1/2	65
Bispectral visible/infrared	105	1/2-2	50
Passive Microwave	103	24	70
Active Microwave (not yet implemented)	103	12	20 (when combined with bispectral technique)
	103	30 x 24	10

Raingauges remain the best method of estimating point rainfall. Although data from raingauges have been used to calibrate both radar and satellite estimates, there are those who feel it is unsound to force remotely sensed measurements to be compatible with raingauge measurements. Further work on retrieval methods for remotely sensed data is necessary in order to ascertain the correct balance between techniques. Finally, it is most important that techniques using remotely sensed data are tested against numerical model products. There is little point in persisting with a method which does not outperform operational models. Considerable effort is being put into intercomparisons of various satellite techniques with model output as part of the Global Precipitation

Climatology Project (GPCP) Algorithm Intercomparison Projects (WMO, 1990), and efforts to resolve raingauge and satellite measurements of global precipitation are focused on the Global Precipitation Climatology Centre (GPCC) at the Deutscher Wetterdienst in Germany.

8. Acknowledgements

The author has drawn freely on a wide range of literature and acknowledges the contributions of the authors of many papers, the results and analyses from which have been referenced only via specialist reviews.

9. References

Arkin P, P Ardanwy (1989) Estimation of climatological scale precipitation from space: a review, *Journal of Climate*, 2(11):1229-1238.

Atlas D, O Thiele (1981) Precipitation measurements from space, workshop reports, NASA, Goddard Space Flight Center, Greenbelt, Maryland, October.

Atlas D, D Rosenfeld, DA Short (1990) The estimation of convective rainfall by area integrals, part I: the theoretical and empirical basis, *Journal of Geophysical Research*, 95(03):2153-2160.

Auer AH Jr. (1994) Hail recognition through the combined use of radar reflectivity and cloud top temperature, to appear in *Monthly Weather Review*.

Austin PM (1987) Relation between measured radar reflectivity and surface rainfall, *Monthly Weather Review*, 115(5):1053-1070.

Battan LJ (1973) Radar Observation of the Atmosphere, University of Chicago Press, Chicago, Illinois, 324 p.

Browning KA, CG Collier (1989) Nowcasting of Precipitation Systems, *Review of Geophysics*, 27(3):345-370.

Cheng M, R Brown (1993) Estimation of area-average rainfall for frontal rain using the threshold method, *Quart. J.R. Meteorological Society*, 119:825-844.

Cheng M, R Brown, CG Collier (1993) Delineation of rain areas using Meteosat IR-VIS data in the region of the United Kingdom, *Journal of Applied Meteorology*, 32:884-898.

Chiu LS (1988) Rain estimation from satellites: areal rainfall-rain area relations, *Third Conference on Satellite Meteorology and Oceans*, February, Anaheim, California, American Meteorological Society, Boston, Massachusetts, pp. 363-368.

Collier CG (1986a) Accuracy of rainfall estimates by radar, part I: calibration by telemetering raingauges, *Journal of Hydrology*, 83:207-223.

Collier CG (1986b) Accuracy of rainfall estimates by radar, part II: comparison with raingauge network, *Journal of Hydrology*, 83:225-235.

Collier CG (1989) Applications of Weather Radar Systems: A Guide to Uses of Radar Data in Meteorology and Hydrology, Ellis Horwood Ltd, 294 p.

Collier CG, PR Larke, BR May (1983) A weather radar correction procedure for real-time estimation of surface rainfall, *Quart. J.R. Meteorological Society*, 104:589-608.

Crum TD, RL Alberty (1993) The WSR-88D and the WSR-88D Operational Support Facility, *Bulletin of the American Meteorological Society*, 74(9):1669-1687.

Doviak RJ, DS Zrnic (1993) Doppler Radar and Weather Observations, Academic Press.

Duncan MR, B Austin, F Fabry, GL Austin (1993) The effect of gauge sampling density on the accuracy of streamflow prediction for rural catchments, *Journal of Hydrology*, 142:445-476.

Folland CK (1986) A simple numerical model of the loss of rainfall due to wind from a conically shaped collector and a suggested new collector shape, *Proceedings, WMO/IAHS Workshop on Correction of Precipitation Measurements*, Zurich, 1-3 April 1985, pp. 233-238.

Hardaker PJ, AH Auer Jr (1994) The separation of rain and hail using single polarization radar echoes and IR cloud top temperature, to appear in *Met. Apps.*

Hardaker PJ, AR Holt, CG Collier (1994) A melting layer model and its use in correcting for the bright band in single polarisation radar echoes, to appear in *Quart. J.R. Meteorological Society*.

Harrold TW (1974) Ground clutter observed in the Dee Weather Radar Project, *Met. Mag.*, 103:140-141.

Hudlow MD, JA Smith, ML Walton, RC Sheed (1991) NEXRAD-New era in hydrometeorology, in Hydromet, Hydrological App. of Weather Radar, I Cluckie, C Collier (Eds), Ellis Horwood, New York, pp. 151-162.

Hughes C, IC Strangeways, AM Roberts (1993) Field evaluation of two aerodynamic raingauges, *Weather*, 48(3):66-71.

Janowiak JE, PA Arkin (1991) Rainfall variation in the Tropics during 1986-1989 as estimated from observations of cloud-top temperature, *Journal of Geophysical Research*, 16, Supplement, pp. 3359-3373.

Joss J, A Waldvogel (1990) Precipitation measurement and hydrology: a review", in Battan Memorial and Radar Conference, American Meteorological Society, Boston, Massachusetts, D Atlas (ed), Chapter 29a, pp. 577-606.

Kedem B, LS Chui, Z Karni (1900) An analysis of the threshold method for measuring area-average rainfall, *Journal of Applied Meteorology*, 29:3-20.

Kitchen M, R Brown, AG Davies (1994) Real-time correction of weather radar data for the effects of bright band, range, and orographic growth in widespread precipitation, to appear in *Quart. J.R. Meteorological Society*.

Krajewski WF (1987) Cokriging of radar and rain gauge data, *Journal of Geophysical Research*, 92(08):9571-9580.

Krajewski WF, ML Morrisey, JA Smith, DT Rexroth (1992) The accuracy of the area-threshold method: a model-based simulation study, *Journal of Applied Meteorology*, 31:1396-1406.

Krstonovic PF, VP Singh (1992a) Evaluation of rainfall networks using entropy: I. theoretical development, *Water Resources Management*, 6:279-293.

Krstonovic PF, VP Singh (1992b) Evaluation of rainfall network using entropy: II. application, *Water Resources Management*, 6:295-314.

Lovejoy S, GL Austin (1979) The delineation of rain areas from visible and IR satellite data for GATE and mid-latitudes, *Atmos. Ocean*, 17:77-92.

Marshall JS, W McPalmer (1948) The distribution of raindrops with size, *Journal of Meteorology*, 5:165-166.

Mueller EA (1977) Statistics of high radar reflectivity gradients, *Journal of Applied Meteorology*, 16:511-513.

Nystuen JA, DM Farmer (1989) Precipitation in the Canadian Atlantic storms program: measurements of the acoustic signature, *Atmos. Ocean*, 27(1):237-257.

Rosenfeld D, D Atlas, DA Short (1990) The estimation of convective rainfall by area integrals, part I: the Height-Area Threshold (HART) Method, *Journal of Geophysical Research*, 95, D3:2161-2176.

Seed AW, GL Austin (1990a) Sampling errors for raingauge-derived mean areal daily and monthly rainfall, *Journal of Hydrology*, 118:163-173.

Seed AW, GL Austin (1990b) Variability of summer Florida rainfall and its significance for the estimation of rainfall by gauges, radar, and satellite, *Journal of Geophysical Research*, 95, D3:2207-2215.

Sevruk B (1982) Methods of correction for systematic errors in point precipitation measurements for operational use, WMO, No. 589, 91 p., Geneva, Switzerland.

Simpson J, RF Adler, G North (1988) A proposed tropical rainfall measuring mission (TRMM) satellite, *Bulletin of the American Meteorological Society*, 69:278-295.

Strangeways IC (1984) Low cost hydrological data collection, In: IAHS Symp. Colleges in African Hydrology and Water Resources, Harare, 229-233.

Ulaby FT, RK Moore, AK Fung (1986) Microwave Remote Sensing, Vol. 1-3, Addison-Wesley, Reading, Massachusetts, 2162 p.

Weisner CJ (1970) Hydrometeorology, Chapman and Hall, Ltd., 232 p.

Wilson JW (1978) Observation of radome transmission losses at 50 m wavelengths, Preprint Volume, 18th Conference on Radar Meteorology, American Meteorology Society, Boston, Massachusetts, pp. 288-291.

Wilson JW, EA Brandes (1979) Radar measurement of rainfall: a summary, *Bulletin of the American Meteorology Society*, 60:1048-1052.

World Meteorological Organization (1990) The Global Precipitation Climatology Project--Implementation and Data Management Plan, WMO/TD, No. 367, June, Geneva, Switzerland.

Zawadski I (1984) Factors affecting the precision of radar measurements of rain, Preprint Volume, 22nd Conference on Radar Meteorology, Zurich, Switzerland, American Meteorological Society, Boston, Massachusetts, pp. 251-256.

Modeling of Heavy Rainfall

J.A. Smith

Department of Civil Engineering and Operations Research
Princeton University
Princeton, New Jersey 08544

A.A. Bradley

Iowa Institute of Hydraulic Research
University of Iowa
Iowa City, Iowa 52242

1. Introduction

More than any other element of the hydrologic cycle, rainfall exhibits pronounced fluctuations in time and space. Modeling of rainfall entails quantitative representation of the space-time fluctuations of rainfall. Rainfall models are used for a variety of hydroclimatological applications (see Foufoula-Georgiou and Krajewski, 1994 for a detailed review). To narrow the scope of this paper and to focus on an area of particular importance for water resource assessment, attention will center on heavy rainfall events. Modeling issues for heavy rainfall will be illustrated with examples from the central United States. The study region is of special significance for the GEWEX Continental Scale International Project (GCIP; GEWEX is an acronym for Global Energy and Water Cycle Experiment, see WCRP, 1993), which will be located in the central United States.

An objective of this paper is to present conceptual models of heavy rainfall events. The interplay between conceptual models based on observational studies and mathematical modeling has been important for previous work in precipitation modeling. A review of statistical and physical models of heavy rainfall events is presented in Sections 4 and 5. Special emphasis is given to emerging research areas.

NATO ASI Series, Vol. I 46
Land Surface Processes in Hydrology
Trials and Tribulations of Modeling and Measuring
Edited by S. Sorooshian, H. V. Gupta, and S. C. Rodda

2. Climatology of Heavy Rainfall in the Central United States

A fundamental element of rainfall modeling is representation of systematic differences in the spatial and temporal distribution of rainfall. At the extremes, systematic spatial heterogeneities in rainfall define the contrasts between desert and rain forest climates. Similarly, temporal heterogeneities can be as pronounced as the seasonal monsoon regimes of Asia. Although progress has been made in describing heterogeneities in rainfall (for example, Barros and Lettenmaier, 1994), many questions remain concerning mechanisms responsible for spatial and temporal heterogeneities. We will examine prominent heterogeneities in rainfall in the Southern Plains to illustrate common features and draw connections with atmospheric processes.

2.1 Seasonality and Diurnal Effects

There is a pronounced seasonal cycle for occurrence of heavy rainfall in the central United States (see Bradley and Smith, 1994; Maddox et al., 1979). Seasonality of rainfall for the Southern Plains region is illustrated in Figure 1. Mean monthly rainfall is shown in Figure 1a. The monthly occurrence of major rain events during the period 1948-1990 is shown in Figure 1b. The definition of a major rain event is based on a storm magnitude/area criterion introduced by Houze et al. (1990), i.e., 24-hour rainfall accumulation exceeding 25 mm over an area of 12,500 km^2. Results for a subset of storms for which rainfall accumulation at 1 or more gages exceeds 125 mm are displayed in Figure 1c. The most notable feature of seasonality of extreme storm occurrence is the bimodal distribution of occurrence times, with spring and fall maxima.

Precipitable water values (in the prestorm environment) for the 125 mm storms (see Figure 1c) are plotted in Figure 2 relative to the seasonal climatology of precipitable water. Precipitable water estimates are computed using radiosonde data from the Southern Plains. For spring and fall events, prestorm precipitable water values push the upper envelope of climatological precipitable water values. For summer events, when precipitable water values attain their climatological maximum, prestorm precipitable

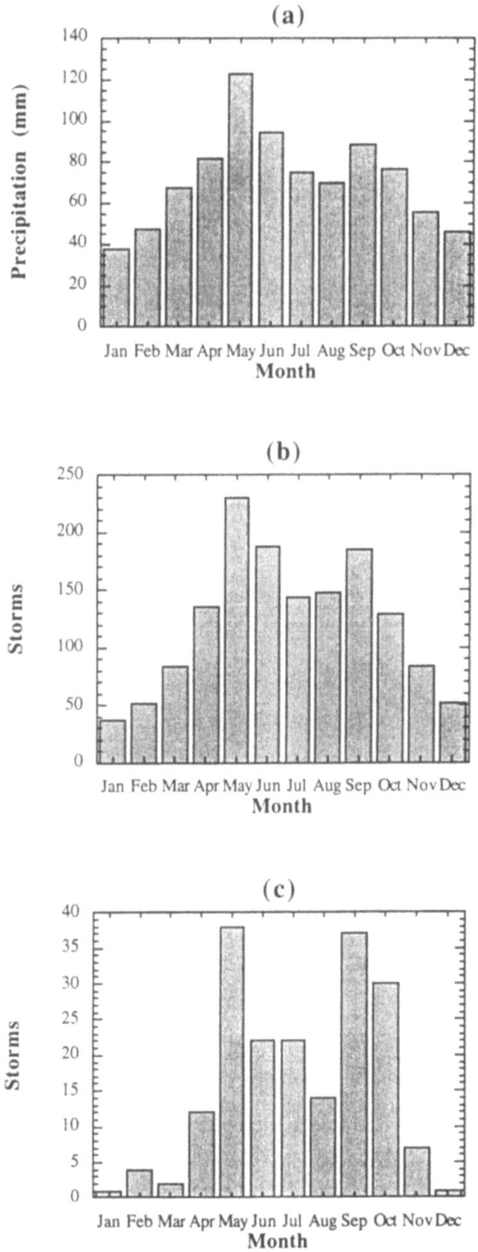

Figure 1. Seasonal cycle of precipitation over the Southern Plains: (a) mean monthly precipitation; (b) occurrence of major rain events (see text) during a 43-year period; (c) occurrence of extreme rain storms (see text) during the 43-year period.

water values are closer to their climatological normal values. The relationship between heavy rainfall events and convective available potential energy (CAPE; see Houze, 1993) for the 125 mm storms is illustrated in Figure 3. Values of CAPE are typically above their climatological normal values for heavy rainfall events. Unlike the situation with precipitable water, however, values of CAPE do not approach their climatological maxima for any season.

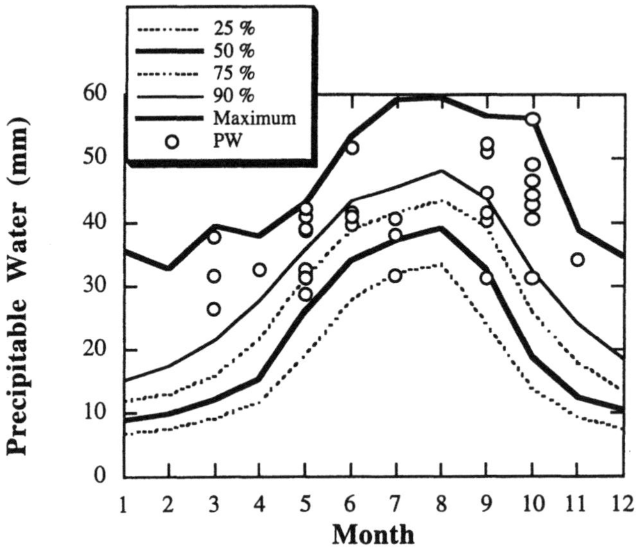

Figure 2. Precipitable water for the 33 extreme rain storms from 1973-1990 and precipitable water climatology. Circles show estimates of precipitable water for individual storms. Lines show quantiles of precipitable water based on the 16-year radiosonde record from Oklahoma City.

Bradley and Smith (1994) described synoptic-scale circulation features associated with heavy rainfall events. They distinguished two principal types of events, which are termed weak dynamic forcing and strong dynamic forcing events. This classification of heavy rainfall bears similarities to Maddox's classification of flash-flood producing storms (Maddox et al., 1979; Hirschboeck, 1987). For strong dynamic forcing events, a strong short-wave trough or cutoff low at 500 mb is a prominent feature of the synoptic scale circulation. Low-level flow is invariably from the south, bringing warm, moist Gulf air into the Southern Plains. Heavy rainfall develops along a near-stationary surface front. For weak dynamic forcing events, heavy rainfall in the Southern Plains is also associated with strong low-level flow from the south. The principal difference with strong dynamic forcing events is the absence of a strong upper-level disturbance. Spring and fall are transition periods between the two types of events, with the summer period represented predominantly by weak dynamic forcing events. The seasonal cycle of precipitable water and convective instability increases to a sharp summer peak. The seasonal cycle of dynamic forcing peaks during the winter with a pronounced trough in the summer. Fall and spring events predominate by virtue of availability of moderate to large values of precipitable water and CAPE and increased frequency of strong dynamic forcing.

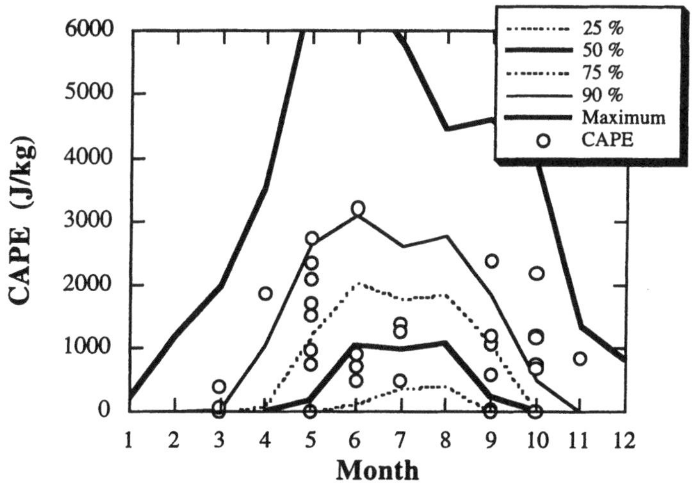

Figure 3. CAPE estimates for the 33 extreme rain storms and climatology of CAPE.

The diurnal variation in extreme rainfall characteristics during the convective season is nearly as large as the seasonal variation (as shown in Figure 4). The nocturnal maximum of heavy rainfall has been related to the nocturnal low-level jet (see Bonner, 1968; Wallace 1975). Often, afternoon thunderstorms grow explosively at sunset as the jet begins to strengthen. In addition, the atmosphere becomes more unstable at night because air above the cloud layer cools more rapidly than near the surface. Baeck and

Smith (1994) illustrated the striking geographic variability in the diurnal cycle of heavy rainfall.

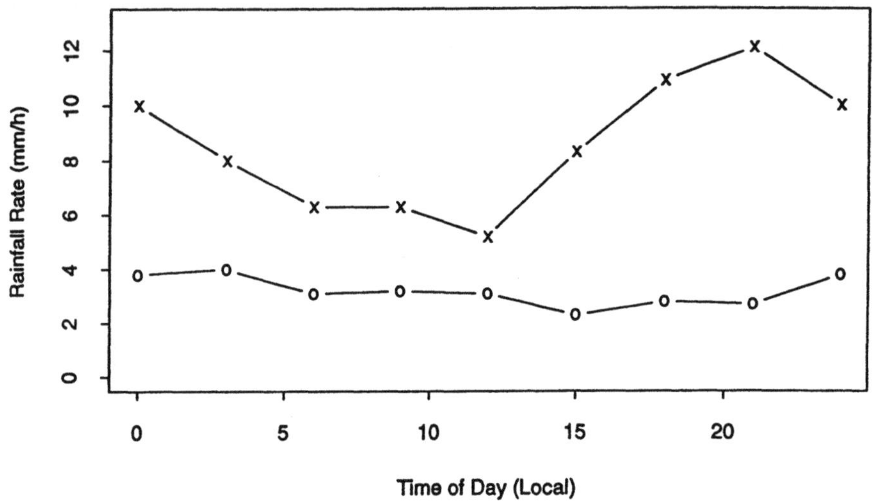

Figure 4. The 0.9 quantile of hourly rainfall accumulation for Oklahoma City, Oklahoma (for hours with positive rainfall) for the summer (June, July, August; denoted by "x") and winter (December, January, February; denoted by "o").

2.2 Spatial Heterogeneities

Figure 5 illustrates the geographic heterogeneity of heavy rainfall over the Southern Plains region (through the maximum annual daily rainfall with a 10-year return interval). There is a pronounced east-west gradient in heavy rainfall across the central United States. The east-west gradients through Texas and Oklahoma, however, also include a bulge of heavy rainfall extending into New Mexico and southwestern Texas.

Availability of moisture and convective instability have been consistently identified as two important elements of heavy rainfall in the central United States (Huff and

Changnon, 1964). Moisture availability and transport clearly play a central role in spatial heterogeneities of rainfall (see Rasmusson, 1971; Bradley and Smith, 1994 for additional details). Figure 6 illustrates geographic variability of precipitable water during the summer season (as computed from the United States radiosonde network for the period 1973-1989). Precipitable water is represented in terms of 0.9 quantile value, i.e., the precipitable water value that is exceeded 10% of the time. Similar features of Convective Available Potential Energy (CAPE) are shown in Figure 7.

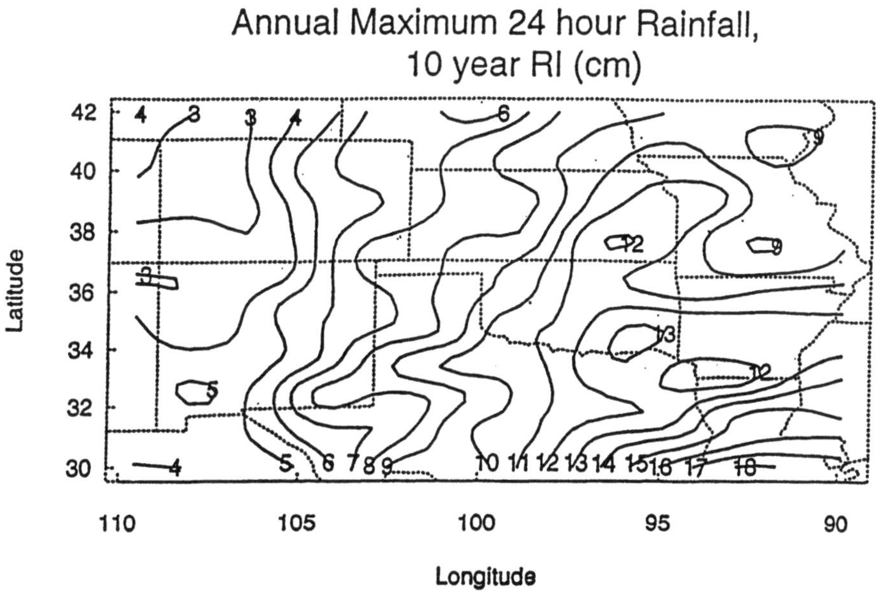

Figure 5. 10-year, 24-hour rainfall accumulation map for the Southern Plains.

Superimposed on the large-scale heterogeneities of heavy rainfall in the Southern Plains are heterogeneities associated with local land-atmosphere interactions. The role of modest topographic relief can be significant. The Caprock Escarpment of the Texas panhandle, for example, plays a prominent role in the initiation of convective systems (Doswell, 1982). Systematic patterns of soil moisture anomalies have also been

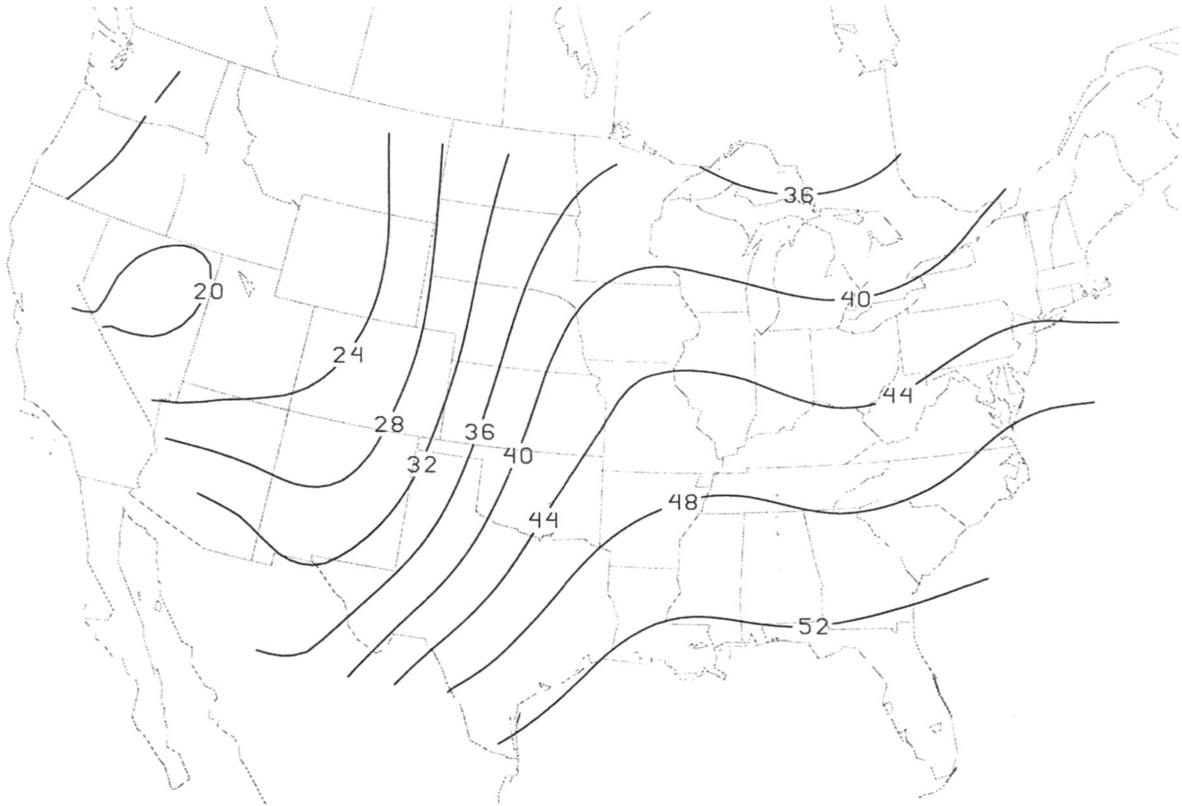

Figure 6. Climatological analysis of precipitable water for summer (June-August). Lines show 0.9 quantile of precipitable water.

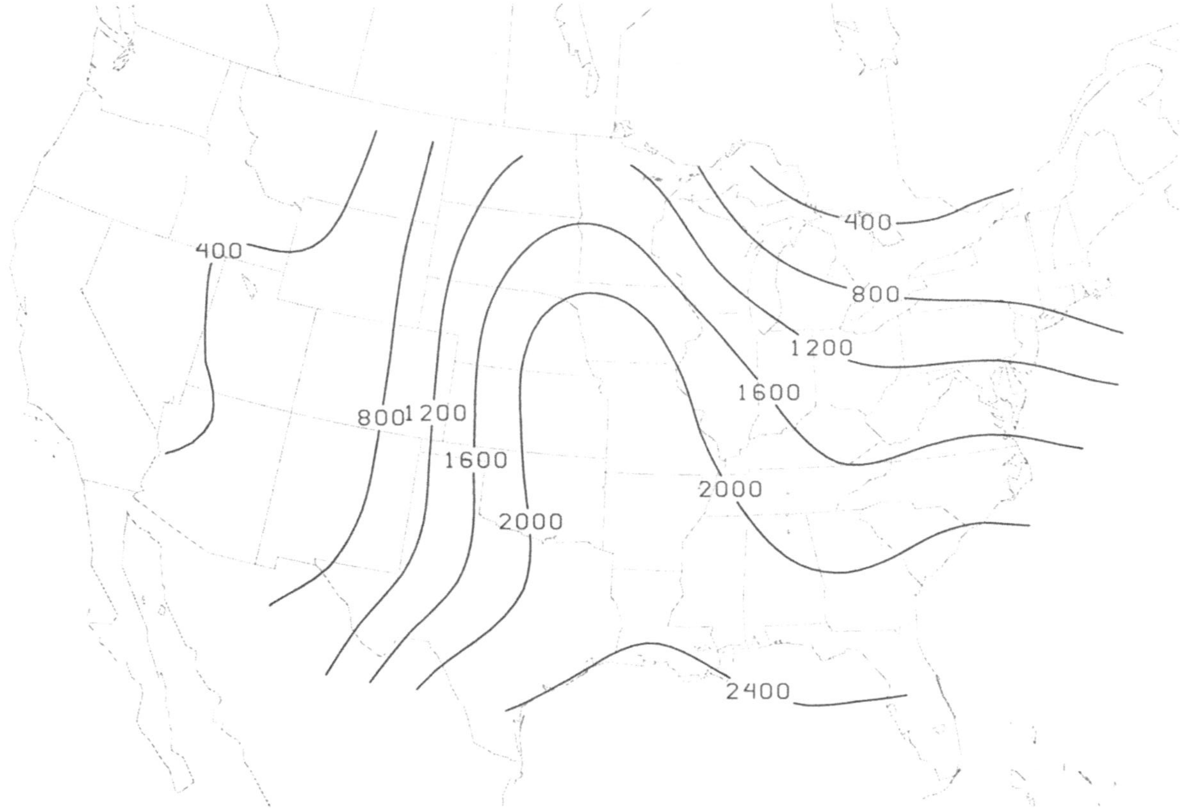

Figure 7. Climatological analysis of CAPE for summer (June-August). Lines show 0.9 quantile of CAPE.

identified as important elements of heavy rainfall in the Southern Plains (Lanicci et al., 1987; Rabin et al., 1990).

2.3 Space-Time Structure of Heavy Rainfall

The space-time structure of rainfall for heavy rain events exhibits characteristic patterns of organization in the Southern Plains (Blanchard, 1990; Houze et al., 1990; Smith et al., 1994). Examples are given in this section.

Considerable attention has been given to mesoscale convective systems which exhibit linear organization (see Houze, 1993, for a comprehensive review). The space-time structure of a storm system that occurred on June 11, 1985 during the PRESTORM experiment (Cunning, 1986) is displayed in Figure 8. Characteristic spatial features include the leading convective line of heavy rainfall and the trailing stratiform region of lighter rain. Due in large part to the rapid movement of the storm, point rainfall accumulations do not exceed 50 mm for the storm. Heavy winds, hail, and lightning were important features of the storm. Because of the special observations available from the prestorm experiment and the occurrence of severe weather, more than 30 research papers have been published on meteorological aspects of this storm (Houze, 1993).

On June 10, 1985, less than 24 hours preceding the squall line shown in Figure 8, a mesoscale convective system produced very heavy rainfall over a small area along the Kansas-Oklahoma border (Figure 9). Blanchard (1990) termed storm systems which exhibit no prominent linear organization "chaotic convective systems". As noted by Smith et al. (1994), these systems often produce locally heavy rainfall (see also Changnon and Vogel, 1981). The June 10 storm shown in Figure 9 had maximum gage accumulations of more than 200 mm (over a period of less than 4 hours). Unlike the June 11 squall line, there has been little study of the June 10 storm. Maddox and Grice (1986) discussed the inclination of meteorologists, both in the forecasting and research community, to place greater emphasis on manifestations of severe thunderstorms, such as damaging winds, lightning, and hail, to the exclusion of heavy rainfall.

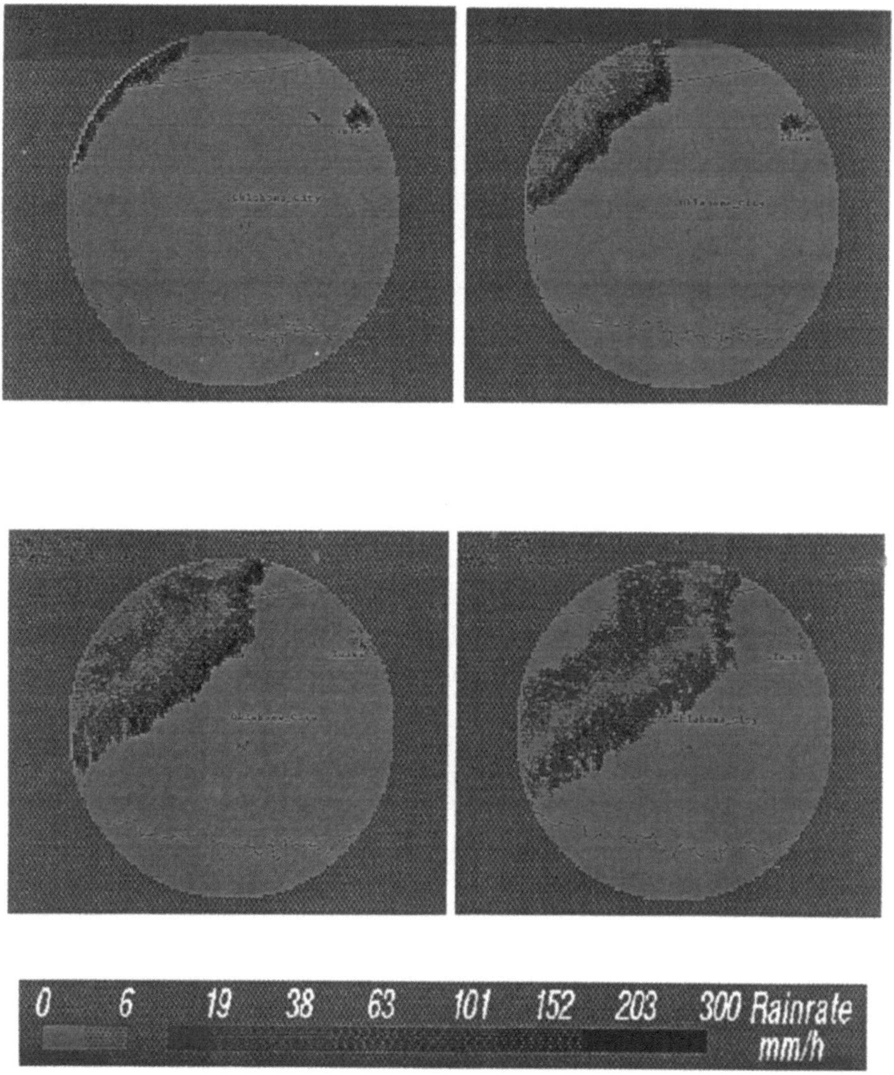

Figure 8. Instantaneous rainfall rates (each figure separated by 1 hour) for a storm on June 11, 1985. Rainfall estimates are from the Oklahoma City RADAP II site.

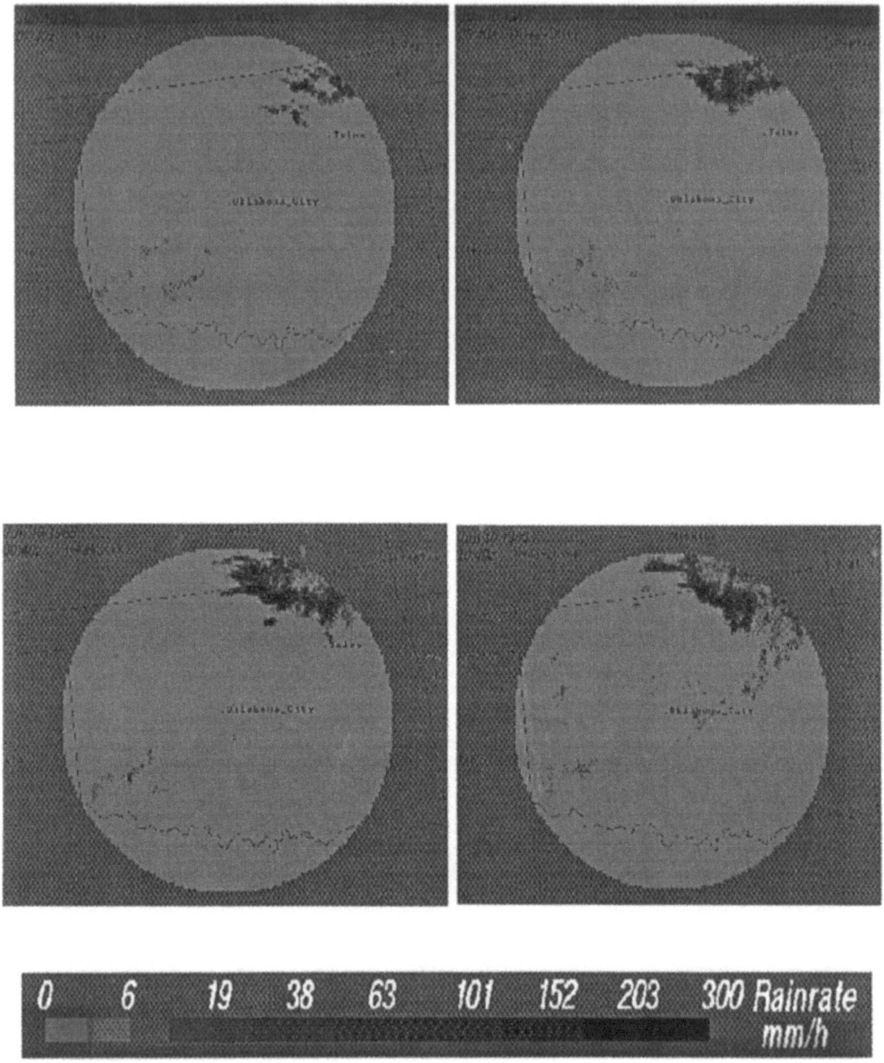

Figure 9. Instantaneous rainfall rates (each figure separated by 1 hour) for a storm on June 10, 1985. Rainfall estimates are from the Oklahoma City RADAP II site.

The spatial organization of rainfall provides important insights into the temporal variability of rainfall at a point. The partitioning of rainfall between infiltration and runoff processes, as represented by the Richards equation, depends in a nonlinear fashion on the fine time-scale variability of rainfall at the plot scale. Recent studies from research watersheds (see, for example, Goodrich and Woolhiser, 1990) have pointed to the need for better characterization of temporal variability of rainfall rate, especially at time scales shorter than 1 hour.

3. Conceptual Models of Heavy Rainfall

Conceptual models of rainfall provide a qualitative link between the space-time distribution of rainfall and atmospheric and land-surface processes. The interplay between conceptual models based on observational studies and mathematical modeling has been important for previous work in precipitation modeling. Austin and Houze (1972) developed a conceptual model for space-time organization of rainfall associated with extratropical cyclones, based largely on radar studies of storm structure. Waymire et al. (1984) combined the conceptual model of Austin and Houze with the mathematical modeling framework of LeCam (1961) to produce sophisticated statistical models of space-time rainfall. Developments along these lines also include the works by Huff and Changnon (1964), Changnon and Vogel (1981), Gupta and Waymire (1979), Amorocho and Wu (1977), Bras and Rodriguez-Iturbe (1985), Gupta and Waymire (1990), Eagleson (1984), Rodriguez-Iturbe et al. (1986), and Smith and Karr (1991). Provided below is brief introduction into some important, emerging areas involving conceptual models of heavy rainfall.

Chappell (1989) described "quasistationary convective systems" as slowly moving systems that are often composed of individual storms that "cross repeatedly over the same region, producing pulsating heavy rains that quickly cause streams and rivers to overflow their banks". The key notion is that motion of the storm system and development of convective elements within the storm system act to maintain heavy rainfall over a limited spatial domain. For example, quasistationary systems can result from propagation of convective cells in the opposite direction to that of the upper-level "steering winds". The interplay of storm motion and cell growth and motion are prominent features of numerous extreme rainfall events (Eisenlohr, 1951; Smith et al., 1994; Maddox et al.. 1978). The conceptual model of Chappell (1989) for quasistationary convective systems succinctly represents the dependence of many heavy rainfall events on storm motion.

One aspect of heavy rainfall events that is poorly understood is their atmospheric water budget. An important element of the water budget is the "precipitation efficiency", i.e., the fraction of water vapor inflow that is converted to precipitation. A common presumption is that storms producing truly prodigious rainfall amounts must have high precipitation efficiencies. This presumption is an implicit element of hydrometeorological procedures used for assessing Probable Maximum Precipitation (PMP; Hansen, 1987). Maddox et al. (1978) and Caracena et al. (1979) presented conceptual models of "cloud-burst" storms in their analyses of the Big Thompson, Colorado storm of July 31, 1976 and the Rapid City, South Dakota storm of June 9, 1972. Their conceptual models of the three-dimensional structure of heavy rainfall events emphasized the distinctive cloud microphysics of heavy rainfall events. Maintenance of high-precipitation efficiencies is an important element of their conceptual models.

Tropical cyclones account for some of the largest storm total rainfall accumulations in regions subject to their occurrence (Hansen, 1987). Case studies of dissipating tropical cyclones over land have provided conceptual models for assessing space-time distribution of heavy rainfall. Davis and Bridges (1972) illustrated the role of storm motion or, more specifically the absence of storm motion, for production of heavy regional scale rainfall. Schwarz (1970) illustrated the distinctive roles of topography and atmospheric moisture flux in producing 6-hour rainfall accumulations greater than 600 mm as Hurricane Camille crossed the Blue Ridge Mountains of Virginia in 1969 (see also Bosart and Carr, 1972; Bosart and Dean, 1991; Funk, 1986; Yi et al., 1980).

Ludlam (1980) noted that most of the world record rainfall accumulations for periods of more than a week are from Cherrapunji, which is located on the south slopes of the Khasi Hills of Assam (see also Dhar et al., 1960). It is on the windward side of the hills and receives the highest amounts of rainfall in the months of the summer monsoon. In this season, conditions may be favorable for almost stationary cumulonimbus over the hills nearly every day (Ludlam, 1980).

4. Statistical Models of Heavy Rainfall

Statistical models serve as the foundation for precipitation frequency analysis and assessment of spatial and temporal trends in rainfall. In this section, recent developments in statistical modeling are reviewed, with particular emphasis on heavy rainfall events.

Considerable attention has been given to development of stochastic models of space-time rainfall (see Foufoula-Georgiou and Krajewski, 1994, for a recent review). As noted above, conceptual models of precipitation have played an important role in guiding the development of these models. A prominent issue in recent research has been the role of scaling (both temporal and spatial) in the representation of rainfall fields (Gupta and Waymire, 1990). This issue is of particular importance for parameterization of precipitation in atmospheric general circulation models (see, for example, Entekhabi and Eagleson, 1989). One of the major practical and theoretical hurdles in this area of research has proven to be representation of the prominent spatial and temporal heterogeneities of rainfall, as illustrated in Section 2.

Relatively little attention has been focused on the development of space-time models for heavy rainfall (Smith and Karr, 1990). This reflects, in part, the lack of maturity of conceptual models of heavy rainfall, in comparison with conceptual models of precipitation associated with extratropical cyclones (Austin and Houze, 1972). Considerable progress has been made in developing nonparametric statistical models for estimating probabilities of extreme rainfall (see Fontaine and Potter, 1989; Foufoula-Georgiou, 1989). These models, which are termed "stochastic storm transposition models", provide an important avenue for enhancing PMP procedures (NRC, 1994).

Significant advances have been made during the past 10 years in development of statistical techniques for at-site and regional precipitation frequency analysis. Advances have been made in: (1) development of regionalization techniques for combining observations from multiple sites, (2) selection of distributional models, and (3) estimation of model parameters (see Guttman et al., 1993, 1994 for discussion). These advances have considerably improved the capability for characterizing the frequency of occurrence of extreme rainfall magnitudes, especially in regions of orographic control.

5. Physical Models of Heavy Rainfall

Numerical weather prediction models have been used successfully for more than 3 decades in predicting large-scale atmospheric circulation patterns, but as Zhang and Fritsch (1986) noted, "quantitative precipitation forecasts have proven to be a particularly stubborn problem to solve, and progress has been slow". In this section, we review research directions in numerical modeling of heavy rainfall.

Numerical models solve the equations of motion, thermodynamic equations, and continuity equations in a three-dimensional coordinate system (Houze, 1993). As Ludlam (1980) noted, heavy rainfall is invariably associated with cumulus convection. Numerical modeling of heavy rainfall events, consequently, revolves around the problem of representing cumulus convection. The traditional approach is to use "cumulus parameterization" for convectively unstable grid elements and explicit condensation computations for convectively stable grid elements. Molinari and Dudek (1992) noted that the fundamental assumptions of cumulus parameterization begin to break down as grid spacing drops below 20 km. These scale limits place important constraints on the range of problems that can be attacked using numerical models.

There have been relatively few numerical model studies of heavy rainfall events. Analyses of the Johnstown, Pennsylvania storm of July 1977 (Bosart and Sanders, 1981; Zhang and Fritsch, 1986, 1988; Stensrud and Fritsch, 1994a,b) using the Penn

State/NCAR community mesoscale model provided an important exception. Important conclusions from these studies include the following:

- representation of rainfall depends crucially on the convective parameterization procedure that is selected;
- boundary layer representations also play a critical role in accurate representation of heavy rainfall; and
- accurate representation of low-level moisture flux is a critical and difficult problem for numerical modeling.

These studies clearly point to the need for more effective integration of atmospheric and land surface models.

6. Summary and Conclusions

The analyses of heavy rainfall in the central United States illustrate and reinforce some well-known facts:

(1) Rainfall exhibits pronounced variability in time and space; assumptions of constant rainfall, in time and/or space, should be closely examined.

(2) Pronounced spatial and temporal heterogeneities are common features of rainfall (rather than the exception); procedures that aggregate rainfall over large areas or lump events from different seasons should be carefully considered.

(3) Synoptic-scale atmospheric processes often exert a dominant control on the spatial and temporal distribution of rainfall.

Some additional conclusions can be drawn concerning rainfall modeling and its role in the broader domain of hydrologic modeling:

(1) Characteristic modes of organization of rainfall at the mesoscale can be identified; these features should play an important role in scaling of land surface hydrologic processes (see Gupta and Waymire, 1990; Smith, 1992);

(2) A major challenge in developing space-time models of rainstorms is simultaneously representing the major structural elements of rainfall fields and the pervasive spatial and temporal heterogeneities of rainfall;

(3) The spatial and temporal distribution of rainfall from distinctive types of heavy rainfall storm systems (for example, dissipating tropical cyclones and quasistationary convective systems) is an important research topic; and

(4) Improved capabilities for precipitation modeling will likely result from an integrated treatment of land surface-atmosphere interactions.

Acknowledgements

This research was supported in part by the National Science Foundation (Grant No. EAR 9204965) and NASA (Grant No. FRS 348420). This support is gratefully acknowledged.

References

Amorocho J, B Wu (1977) Mathematical models for the simulation of cyclonic storm sequences and precipitation fields, *Journal of Hydrology*, 32(3/4):329-345.

Austin P, R Houze (1972) Analysis of the structure of precipitation in New England, *Journal of Applied Meteorology*, 11:926-935.

Baeck ML, JA Smith (1994) Climatological analysis of manually digitized radar data for the United States, Submitted to *Water Resources Research*.

Barros AP, DP Lettenmaier (1994) Dynamic modeling of orographically-induced precipitation, *Reviews of Geophysics*, 32(3):265-284.

Blanchard DO (1990) Mesoscale convective patterns of the southern high plains, *Bulletin of the American Meteorological Society*, 71(7):994-1005.

Bonner WD (1968) Climatology of the low level jet, *Monthly Weather Review*, 96:833-849.

Bosart LF, FH Carr (1972) A case study of excessive rainfall centered around Wellsville, New York, 20-21 June 1972, *Monthly Weather Review*, 106:348-362.

Bosart LF, DB Dean (1991) The Agnes rainstorm of June 1972: surface feature evolution culminating in inland storm development, *Weather and Forecasting*, 6:515-537.

Bosart LF, F Sanders (1981) The Johnstown flood of July 1977: a long-lived convective system, *Journal of Atmospheric Sciences*, 38:1616-1642.

Bradley AA, JA Smith (1994) The hydrometeorological environment of extreme rainstorms in the southern plains of the United States, *Journal of Applied Meteorology*, 33(12):1418-1432.

Bras RL, I Rodriguez-Iturbe (1985) Random Function and Hydrology, 559 p., Addison-Wesley, Reading, Massachusetts.

Caracena F, RA Maddox, LR Hoxit, CF Chappell (1979), Mesoanalysis of the Big Thompson storm, *Monthly Weather Review*, 107(1):1-17.

Changnon SA, JL Vogel (1981) Hydroclimatological characteristics of isolated severe rainstorms, *Water Resources Research*, 17(6):1694-1701.

Chappell C (1989) "Quasistationary Convective Events", in Mesoscale Meteorology, edited by P. Ray, American Meteorological Society, Boston, Massachusetts.

Cunning JB (1986) The Oklahoma-Kansas preliminary regional experiment for STORM-Central, *Bulletin of the American Meteorological Society*, 67:1478-1486.

Davis DR, WC Bridges (1972) Minimal tropical depression produces record rains and unprecedented floods, *Monthly Weather Review*, 100(4) 294-297.

Dhar ON, PR Rakhecha, BN Mandal (1960) Rainstorms which contributed greatest rain depths in India, *Arch. Met. Geoph. Biokl.*, Series A.

Doswell CA (1982) Operational meteorology of convective weather, Vol. I, NOAA Technical Memo, NWS NSSFC-5, Kansas City, Missouri.

Eagleson P (1984) The distribution of catchment coverage by stationary rainstorms, *Water Resources Research*, 20(5):581-590.

Eisenlohr WS (1951) Floods of July 18, 1942, in northcentral Pennsylvania, U.S. Geological Survey Water Supply Paper 1134-B.

Entekhabi D, P Eagleson (1989) Land surface hydrology parameterization for atmospheric general circulation models including subgrid scale spatial variability, *Journal of Climate*, 2:816-831.

Fontaine T, K Potter (1989) Estimating probabilities of extreme rainfalls, *Journal of Hydraulic Engineering*, 1562-1575.

Foufoula-Georgiou E (1989) A probabilistic storm transposition approach for estimating exceedance probabilities of extreme precipitation depths, *Water Resources Research*, 25(5):799-815.

Foufoula-Georgiou E, WF Krajewski (1994) Recent advances in rainfall modeling, IUGG Quadrennial Report, to appear.

Funk TW (1986) The use of water vapor imagery in the analysis of the November 1985 middle Atlantic states record flood event, *National Weather Digest*, 11:12-19.

Goodrich D, D Woolhiser (1990) Rainfall sampling impacts on runoff, *Proceedings of ASCE Symposium on Hydraulics/Hydrology of Arid Lands*, 519-524.

Gupta V, E Waymire (1979) A stochastic kinematic study of subsynoptic space-time rainfall, *Water Resources Research*, 15(3):637-644.

Gupta V, E Waymire (1990) Multiscaling properties of spatial rainfall and river flow distributions, *Journal of Geophysical Research*, 95(D3).

Guttman NB, JRM Hosking, and JR Wallis (1993) Regional precipitation quantile values for the continental United States computed from L-moments, *Journal of Climate*, 6:2326-2340.

Guttman NB, JRM Hosking, and JR Wallis (1994) The 1993 midwest extreme precipitation in historical and probabilistic perspective, *Bulletin of the American Meteorological Society*, 75(10):1785-1792.

Hansen M (1987) Probable maximum precipitation for design floods in the United States, *Journal of Hydrology*, 96:267-278.

Hirschboeck KK (1987) Catastrophic floods and atmospheric circulation anomalies, in Catastrophic Flooding, L Mayer, D Nash (eds.), Allen and Unwin, Boston, Massachusetts, 25-56.

Houze RA (1993) Cloud Dynamics, Academic Press, 540 p.

Houze RA, BF Smull, P Dodge (1990) Mesoscale organization of springtime rainstorms in Oklahoma, *Monthly Weather Review*, 118:613-654.

Huff F, S Changnon (1964) A model 10-inch rainstorm, *Journal of Applied Meteorology*, 3:587-599.

Lanicci JM, TN Carlson, TT Warner (1987) Sensitivity of the Great Plains severe-storm environment to soil-moisture distribution, *Monthly Weather Review*, 115:2660-2674.

LeCam L (1961) A stochastic description of precipitation, IV Berkeley Symposium on Mathematical Statistics and Probability, Vol. 3, pp. 165-186, University of California, Berkeley.

Ludlam, FH (1980) Clouds and Storms, Pennsylvania State University Press, University Park, Pennsylvania, 404 p.

Maddox RA, GK Grice (1986) The Austin, Texas flash flood: an examination from two perspectives--forecasting and research, *Weather and Forecasting*, 1:66-76.

Maddox RA, LR Hoxit, CF Chappell, F Caracena (1978) Comparison of meteorological aspects of the Big Thompson and Rapid City flash floods, *Monthly Weather Review*, 106:375-389.

Maddox R, C Chappell, L Hoxit (1979) Synoptic and meso-alpha scale aspects of flash flood events, *Bulletin of the American Meteorological Society*, 60(2):115-123.

Molinari J, M Dudek (1992) Parameterization of convective precipitation in mesoscale numerical models: a critical review, *Monthly Weather Review*, 120:326-344.

National Research Council (NRC) (1994) Estimating bounds on extreme precipitation events, National Academy Press, Washington, D.C.

Rabin RM, S Stadler, PJ Wetzel, DJ Stensrud, M Gregory (1990) Observed effects of landscape variability on convective clouds, *Bulletin of the American Meteorologial Society*, 71(3):272-280.

Rasmusson E (1971) A study of the hydrology of eastern North America using atmospheric vapor flux data, *Monthly Weather Review*, 99(2):119-135.

Rodriguez I, D Cox, P Eagleson (1986) Spatial modeling of storm rainfall, *Proceedings of Royal Society London*, Series A, 403:27-50.

Schwarz FK (1970) The unprecedented rains associated with the remnants of Hurricane Camille, *Monthly Weather Review*, 98(11):851-859.

Smith JA (1992) Representation of basin scale in flood peak distributions, *Water Resources Research*, 28(11):2993-2999.

Smith JA, AF Karr (1990) A statistical model of extreme storm rainfall, *Journal of Geophysical Research*, 95(D3):2083-2092.

Smith JA, AA Bradley, ML Baeck (1994) The space-time structure of extreme storm rainfall in the Southern Plains, *Journal of Applied Meteorology*, 33(12):1402-1417.

Stensrud DJ, JM Fritsch (1994a) Mesoscale convective systems in weakly forced large-scale environments, II: generation of the mesoscale initial condition, *Monthly Weather Review*, 122:2068-2083.

Stensrud DJ, JM Fritsch (1994b) Mesoscale convective systems in weakly forced large-scale environments, III: numerical simulations and implications for operational forecasting, *Monthly Weather Review*, 122:2084-2104.

Wallace JM (1975) Diurnal variations in precipitation and thunderstorm frequency over the conterminous United States, *Monthly Weather Review*, 103:406-419.

Waymire E, V Gupta, I Rodriguez-Iturbe (1984) A spectral theory of rainfall intensity at the meso-beta scale, *Water Resources Research*, 20(10):1453-1465.

World Climate Research Program (WCRP) (1994) Scientific plan for the GEWEX Continental Scale International Project (GCIP), IGPO Series No. 9.

Yi, HD, HZ Li, JS Li (1980) On the physical conditions of occurrence of heavy rainfall and severe convective weather, *Bulletin of the American Meteorological Society*, 61(1).

Zhang D-L, JM Fritsch (1986) Numerical simulation of the meso-beta scale structure and evolution of the 1977 Johnstown flood, 1: model description and verification, *Journal of Atmospheric Science*, 43:1913-1943.

Zhang D-L, JM Fritsch (1988) Numerical sensitivity experiments of varying model physics on the structure, evolution, and dynamics of two mesoscale convective systems, *Monthly Weather Review*, 45(2):261-293.

Capabilities of Microwave Sensors for Monitoring Areal Extent and Physical Properties of the Snowpack

Helmut Rott
Institut für Meteorologie und Geophysik
Universität Innsbruck
Innrain 52
A-6020 Innsbruck, AUSTRIA

1. Introduction

Active and passive microwave sensors on board of Earth observation satellites show excellent possibilities for monitoring the snow cover from regional to global scales. An important advantage of microwave sensors is the capability to observe the Earth's surface through clouds and thus to provide information on the snow cover in regular time intervals, depending primarily on the satellite orbits. Another advantage is the capability of microwaves to penetrate dry snow, to detect properties of the snow volume, and to estimate the mass of snow lying on the ground.

In spite of these promising capabilities to provide snow cover information for a wide range of applications, the potential of microwave sensors for snow monitoring has been utilized only to a small extent. There are several reasons for this. Among these is the lack of fully operational procedures for using remote sensing information. Research in microwave remote sensing during the last two decades has resulted in a number of different inversion procedures for deriving areal extent and properties of the snow cover from active and passive microwave measurements. Each of these procedures has its particular deficiencies and merits, making the right choice difficult for the nonspecialist. Another problem resulted from the limited access to spaceborne microwave data which, in the past, have not been available in real time or near real time. Improvements in data dissemination have been achieved with recently launched microwave sensors. To foster the development and use of procedures and information systems, which provide areally operational snow cover products derived from spaceborne sensors, improved coordination and collaboration between remote sensing experts and potential users is needed.

NATO ASI Series, Vol. I 46
Land Surface Processes in Hydrology
Trials and Tribulations of Modeling and Measuring
Edited by S. Sorooshian, H. V. Gupta, and S. C. Rodda

In order to illustrate the potential of microwave data for snow applications, basic principles of microwave measurements of snow are explained, and the status of microwave snow retrieval techniques is reviewed.

2. Interaction of Microwaves with the Snow Cover

As the basis for the description of inversion procedures, the main mechanisms of microwave interaction with snow-covered ground are described briefly. The propagation of microwaves through a snowpack is illustrated in Figure 1. Assuming a homogeneous snow layer with depth d_s, the transmittance of the snowpack τ_s can be quantified in the radiative transfer approach by:

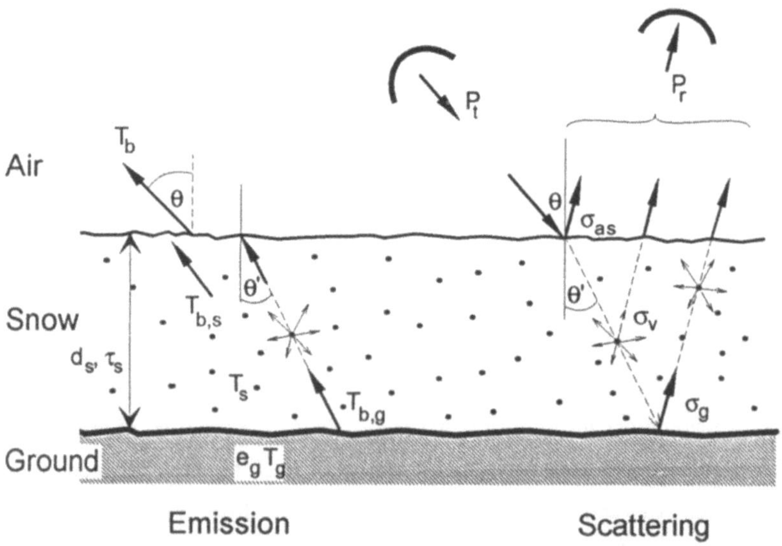

Figure 1. Microwave scattering and emission of snow-covered ground. θ = incidence angle, θ' = refraction angle. Properties of the snowpack: d_s = depth, τ_s = transmissivity, T_s = thermodynamic temperature. Properties of the ground: e_g = emissivity, T_g = thermodynamic temperature. Emission: T_b = brightness temperature of the snowground medium, $T_{b,s}$ = brightness temperature of the snowpack, $T_{b,g}$ = brightness temperature of the ground. Backscattering: P_t = transmitted power, P_r = received power, σ_{as} = scattering coefficient of the air-snow boundary, σ_v = volume scattering coefficient, σ_g = scattering coefficient of the ground.

$$\tau_s = \exp\left(-k_e\, d_s/\cos\theta'\right) \tag{1}$$

where k_e is the extinction coefficient of snow, and θ' is the angle of refraction in the snowpack. Extinction in the snowpack is due to absorption and scattering contributions:

$$k_e = k_a + k_s \tag{2}$$

where k_a is the absorption coefficient, and k_s is the scattering coefficient of the snow volume. k_a can be derived from the complex dielectric constant $\epsilon = \epsilon' - i\epsilon''$ and the wavelength in the vacuum λ_o:

$$k_a = \frac{2\pi\epsilon''}{\lambda_o\sqrt{\epsilon'}} \tag{3}$$

The complex dielectric constant of snow ϵ and, consequently, the absorption losses, are strongly dependent on the liquid water content of the snow volume. For snow with liquid water content V_w of 5% by volume, the imaginary part of the dielectric constant ϵ'' is larger by about 3 orders of magnitude than for dry snow, whereas the real part ϵ' shows less dependence on V_w (Mätzler, 1987). This results in a drastic decrease of the penetration depth δ_p with increasing liquid water content, as shown in Figure 2 for two frequencies {5 GHz (λ = 6 cm) and 19 GHz (λ = 1.56 cm)} used for microwave remote sensing of snow. For wet snow, the absorption losses are clearly dominating over the scattering losses. With the assumption $k_s << k_a$, the penetration depth can be calculated according to $\delta_p \approx 1/k_a$. At 5 GHz, the penetration depth δ_p decreases from values around 20 m for dry snow to δ_p = 3 cm for snow with V_w = 5%. At 19 GHz, the corresponding decrease is from δ_p = 4 m for dry snow to δ_p = 1 cm for wet snow. This means that snow is practically opaque at microwave frequencies $\geq$5 GHz. In order to derive the depth or water equivalent of a snow layer, the measurements need to be made as long as the snowpack is dry.

Microwave radiometers measure the radiance L emitted by a medium in the frequency interval $d\upsilon$. In the microwave region, for $\lambda T >> 1$ [cm K], L can be approximated after Rayleigh-Jeans:

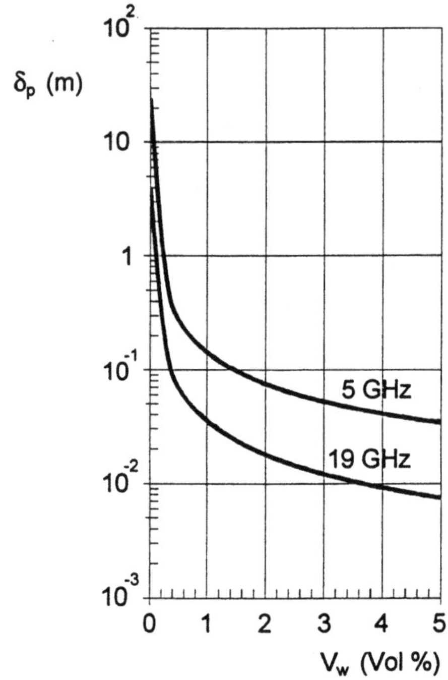

Figure 2. Penetration depth δ_p as function of the liquid water content of snow V_w for snow density $\rho = 0.4$ g cm^{-3}.

$$L(\upsilon)d\upsilon = e(\upsilon,\theta,\phi,p)\,\frac{2kt}{\lambda^2}\,d\upsilon \qquad [Wm^{-2}sr^{-1}] \tag{4}$$

where k is Boltzman's constant, and T is the thermodynamic temperature of the medium in degrees Kelvin. e is the emissivity of the medium ($0 < e \leq 1$) which for natural media varies with frequency υ, incidence angle θ, azimuth angle ϕ, and polarization p. Because the radiance is linearly proportional to temperature, it is convenient to express microwave radiometric measurements in terms of the brightness temperature T_b:

$$T_b(\upsilon,\theta,\phi,p) = e(\upsilon,\theta,\phi,p)T \qquad [K] \tag{5}$$

Assuming the atmospheric contribution to be zero, the brightness temperature T_b above ground covered with dry snow includes contributions from the ground below the snowpack $T_{b,g}$ and from the snow volume $T_{b,s}$ (see Figure 1):

$$T_b = T_{b,s} + T_{b,g}\tau_s(\theta')T_{sa}(\theta') \tag{6}$$

T_{sa} is the Fresnel power transmission coefficient of the snow-air boundary. The transmissivity τ_s of the snow volume is defined in Equation (1). The contribution of the ground $\{T_{b,g} = e_g T_g\}$, where e_g is the emissivity and T_g is the thermodynamic temperature of the ground, is attenuated due to scattering and absorption in the snowpack. Assuming a homogeneous snow layer with smooth surface and neglecting multiple scattering, the emission contribution of the snow volume is:

$$T_{b,s} = T_{sa}(\theta')T_s \frac{k_a}{k_s + k_a}[1 - \tau_s(\theta')] \tag{7}$$

Retrievals of snow depth d_s and snow water equivalent WE from microwave radiometric measurements utilize the spectral region in which the scattering losses are dominating ($k_s > k_a$). For dry winter snow with typical grain diameters $\leq$ 1 mm, this is the case for frequencies υ > 20 GHz. The volume scattering coefficient increases strongly with increasing grain size and with increasing frequency (decreasing wavelength). Calculations of snow depth utilize the fact that the total transmissivity of the snowpack τ_s and, consequently, the brightness temperature T_b of the snow-ground medium decrease with d_s (Equations 1, 6, and 7). For a given snow depth, scattering losses increase also with the number of scattering elements in a given snow volume (with snow density ρ) up to a saturation limit (Fung and Chen, 1989). This enables the retrieval of water equivalent ($WE = \rho d_s$).

Due to high absorption losses, the transmissivity of wet snow is low. For this reason, the emissivity is close to unity, and wet snow radiates almost like a black body.

The right part of Figure 1 illustrates the various contributions due to scattering from snow-covered ground. For clarity of the illustration, the transmitting and receiving antennae were plotted at different locations. On remote sensing platforms, the transmitting and receiving antennae are identical, and the signal reflected or scattered back to the antennae is measured. The radar reflectivity of terrain is usually specified in terms of radar cross-section per unit surface area σ° (also called scattering coefficient) which is a dimensionless number quantifying the radar-scattering cross-section (m^2) per unit area (m^2) of the illuminated surface. σ° is proportional to the ratio of received power P_r to the power transmitted by the sensor P_t:

$$\sigma^\circ = \frac{4\pi^3 R^4}{G^2 \lambda^2 A_{ill}} \frac{P_r}{P_t} \tag{8}$$

where R is the range between the antennae and the object, G is the antennae gain, λ is the radar wavelength, and A_{ill} is the size of the illuminated area. Because σ° in a radar scene typically varies over 2 to 3 orders of magnitude, σ° is usually specified in logarithmic scale (in decibels, dB).

Assuming a homogeneous snowpack and neglecting multiple scattering in the snow volume and multiple reflections between the upper and lower boundary of the snowpack, backscattering from snow-covered ground can be approximated by (Ulaby et al., 1984):

$$\sigma^\circ(\theta) = \sigma^\circ_{as}(\theta) + T^2_{sa}(\theta') \left[\frac{\sigma_v}{2k_e} (1 - \tau^2_s(\theta')) \cos\theta + \sigma^\circ_g(\theta') \tau^2_s(\theta') \right] \tag{9}$$

where σ_v is the volume backscattering coefficient, and $\sigma_g{}^\circ$ is the backscattering coefficient of the underlying soil surface. $\sigma_{as}{}^\circ$, representing the backscattering contribution of the snow surface, is small for dry snow, but is important for wet snow.

For technical reasons, imaging radars on board of satellites operate in the L- to X-band range (frequencies 1 GHz to 10 GHz). At these frequencies, dry snow is highly transparent, and the main part of the radar return signal received at the antennae originates from reflection at the ground surface below the snowpack. For this reason, it is difficult to detect dry snow cover by means of SAR.

Because wet snow is opaque at radar frequencies, the signal scattered back originates from reflection and scattering at the air-snow boundary and from scattering in the top centimeters of the snow volume. With the exception of incidence angles at and close to nadir, σ° of wet snow is low and varies with the dielectric constant of the top snow layer which is related to the liquid water content.

3. Microwave Emission Signatures of Snow

Basic information on the microwave emission response to snowpack parameters has been obtained from ground-based radiometric measurements (Hofer and Schanda, 1978; Mätzler, 1987; Tiuri, 1982; Ulaby and Stiles, 1980). Figure 3 shows the average emission behavior of three typical snow types and of snow-free ground, measured by scientists of the University of Berne at a test site near Davos in the Swiss Alps during several winter seasons (Mätzler, 1987; Mätzler and Hüppi, 1989). The emissivity e at polarization p was calculated from the brightness temperature $T_{b,p}$ measured above the ground:

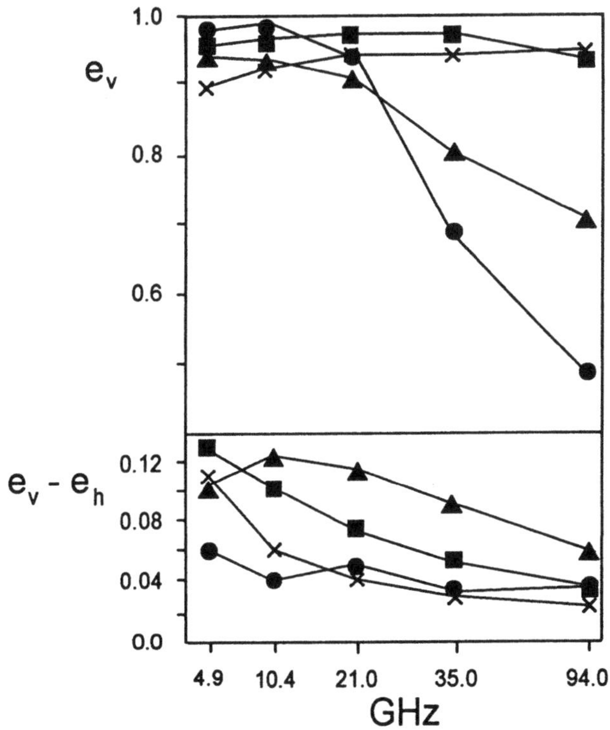

Figure 3. Average emissivities at vertical polarization e_v and polarization difference ($e_v - e_h$) of 3 different snow types and of snow-free ground, measured at an incidence angle of 50° at 5 microwave frequencies. From measurements in the Swiss Alps by Mätzler (1987). x = snow-free ground, ■ = wet snow, ▲ = dry snow, ● = refrozen crust on top of wet snow.

$$e_p = \frac{T_{b,p} - T_b^{\downarrow}}{T_s - T_b^{\downarrow}} \tag{10}$$

where $T_b^{\downarrow}$ is the downwelling atmospheric brightness temperature at the Earth's surface, and T_s is the thermodynamic temperature of the snow-ground medium. The emissivities in Figure 3 were measured at an incidence angle $\theta = 50°$ which is close to θ of satellite-borne imaging microwave radiometers. Radiometric measurements on the ground and by satellite radiometers are frequently made at two linear polarizations: vertical (v), where the incident electrical field vector E is parallel to the plane of incidence, and horizontal (h), where E is perpendicular to the plane of incidence. Polarization differences ($e_v - e_h$) provide useful information on target properties.

Figure 3 illustrates the difficulties in discriminating wet snow from the snow-free site (rocky soil with alpine grass) because both media show high emissivities. Similar problems have been observed for snow mapping with spaceborne radiometers (Künzi et al., 1982). In the 15 to 40 GHz range dry, slightly wet, or frozen soil and vegetated surfaces show emissivities e_v at $\theta = 50°$ between 0.90 and 0.98. The corresponding emissivities of wet snow are between 0.95 and 0.99 (Mätzler, 1987). Polarization differences ($e_v - e_h$) offer the possibility of detecting melting snow in regions where snow-free surfaces show small values of ($e_v - e_h$) (Walker and Goodison, 1993).

The emissivity of dry snow is gradually decreasing with increasing frequency due to increasing scattering efficiency of snow grains. The comparatively high polarization differences of dry snow can be explained by reflections at internal boundaries between layers of different snow densities (Mätzler, 1987). At 35 GHz and 94 GHz, the emissivity of the dry winter snowpack is significantly reduced in comparison to the snow-free site (Figure 3). At the frequencies 10.4 GHz and 4.9 GHz, the measured radiation originates mainly from the ground below the dry snowpack. The lowest emissivities at 35 GHz and 94 GHz were measured for refrozen crusts on top of wet spring snow because the large grains of refrozen snow act as effective scatterers (Reber at al., 1987). This illustrates a basic problem of inverting microwave radiometric measurements in respect to snow-water equivalent WE: increasing the grain size has similar effects on the bulk emissivity of the snow-ground medium as increasing the thickness of the scattering layer. Multi-temporal algorithms monitoring the melt-freeze cycle enable the detection of refrozen snow in satellite measurements (Künzi et al., 1982). Another possibility of separating refrozen snow from dry winter snow is the polarization difference (e_v - e_h), which is low for snow crusts but increases in dry winter snow due to reflections at internal density boundaries related to accumulation layers.

Table 1 summarizes the response of microwave emissivity of snow-covered ground in dependence of snowpack conditions. The dominating response results from snow wetness, causing rapid increase of snow emissivity at all frequencies. Scattering losses are dominant over absorption losses at frequencies larger than 15 to 20 GHz. In the 90 GHz region, the penetration depth of dry snow is only few centimeters (Hallikainen et al., 1987). For this reason, the emissivity responds primarily to the grain size of the top snow layer. The 90 GHz region is useful for mapping thin snow cover which is transparent at lower frequencies (Nagler and Rott, 1992). The best opportunities for sensing water equivalent and depth of snow offers the 20 to 40 GHz region because, in this region, snow is sufficiently transparent to penetrate a layer of 0.5 to 1 m thickness, but not so transparent as to be dominated by the ground signal. Ambiguities for WE retrievals result from the additional dependence of emissivity on grain size. The coefficients for inversion of satellite measurements in respect to WE are usually based on statistical relations between in situ point measurements of WE and satellite data which implicitly account for the average regional variation of snow morphology and grain size.

4. Microwave Backscattering Signatures of Snow

As for microwave emission, the wetness of the snowpack is a dominant factor for backscattering. Figure 4 shows examples of backscattering coefficients σ° measured at 5.2 GHz (C-band) and 10.3 GHz (X-band) in the Alpine Valley Leutasch near Innsbruck, Austria, on several days of the winter season 1990-91 (Rott and Sturm, 1991; Sturm and Rott, 1992). The hh- and vv-polarized measurements and the hv- and vh-polarized measurements, respectively, are averaged in Figure 4, because no significant differences were observed. The differences of σ° between the days when the site was snow-free and the days when it was covered by dry snow were small at both frequencies. On the two days in April 1991, when the measurements of the snow-free conditions shown in Figure 4 were made, the grass at the test site (a cultivated meadow) was low and the soil was humid with water contents of 43 and 53% by volume. The snow depth on the days of the dry snow measurements in December and January was 35 cm (WE 80 mm). Because of the high transparency of dry snow, the main backscattering contribution originated from the soil. In the course of the winter season, σ° of the snow-ground medium gradually changed in time which, in the dry snow cases, was influenced by the freezing conditions of the soil (Sturm and Rott, 1992).

Table 1. Microwave Emission Response of Snow-Covered Soil in Dependence of Snow Conditions. e - emissivity of the snow-ground medium.

Parameter	Microwave Emission Response	Frequencies Mainly Affected
For wet snow:		
Liquid water content V_w	Increase of e with increasing V_w up to saturation at $V_w \approx 2\%$ by volume	All frequencies
For dry snow:		
Snow water equivalent WE	Decrease of e with increasing WE	15 to 40 GHz
Grain size GS	Decrease of e with increasing GS	> 15 GHz
Stratification (layers)	Increase of (e_v-e_h) with layers	5-15 GHz
Soil emission properties	Soil contribution to total emission decreasing with increasing v, WE, and GS	< 40 GHz

At incidence angles larger than 20°, wet snow showed the lowest σ° values of the three surface types. In the angular range from 30 to 70 degrees off nadir, the X-band copolarized σ° values of wet snow were 7 dB to 8 dB lower than the σ° values of the snow-free site. At C-band, the corresponding differences in σ^0 were between 4 dB and 6 dB. This indicates the preference for X-band to discriminate wet snow versus snow-free surfaces which are comparatively smooth (such as meadows) if only copolarized measurements are available. Both the C- and X-band backscatter contrasts of wet snow versus snow-free surfaces are larger if the snow-free surfaces are rough. Copolarized backscatter contrasts of 8 to 10 dB between wet snow areas, and snow-free surfaces were observed in C- and X-band airborne SAR data of the mountainous test site Ötztal in the Austrian Alps (Rott, 1984; Rott and Davis, 1991; 1993).

The measurements in Figure 4 show C-band cross-polarized σ^0 values of wet snow which are 10 dB to 12 dB lower than the corresponding σ^0 values of dry snow and of the snow-free site. Polarimetric SAR data of the test site Ötztal show even larger differences of cross-polarized σ^0 between wet spring snow and snow-free surfaces (Rott and Davis, 1993). In principle, this offers good capabilities for mapping of wet snow. However, imaging radars presently in space or due to be launched in the near future (with the exception of short-term missions on the space shuttle) do not operate cross-polarized channels because of the increased power needs. At X-band, the cross-polarized backscatter of wet snow is higher than at C-band. This is due to increased diffuse scattering of the snow surface and of the top snow layer.

The ability to discriminate wet snow versus snow-free surfaces decreases towards lower microwave frequencies (Rott, 1984; Rott et al., 1988). L-band SAR shows limited capabilities for snow mapping because σ^0 of wet snow is similar to σ^0 of various snow-free surface types.

Table 2 summarizes the response of backscatter at C- and X-band frequencies for snow-covered soil. Because of the high transparency, dry snow only weakly modifies the backscatter signal of the soil surface. Therefore, imaging radars operating at X-band and lower frequencies are not useful for monitoring dry snow. In principle, radars operating at higher frequencies (K-band) should offer better capabilities for dry snow mapping, but for technical reasons such systems are presently not under consideration for satellite missions. On the other hand, signature research indicates good capabilities for monitoring the extent of wet snow at C- and X-band and also for deriving information on the liquid water content of snow.

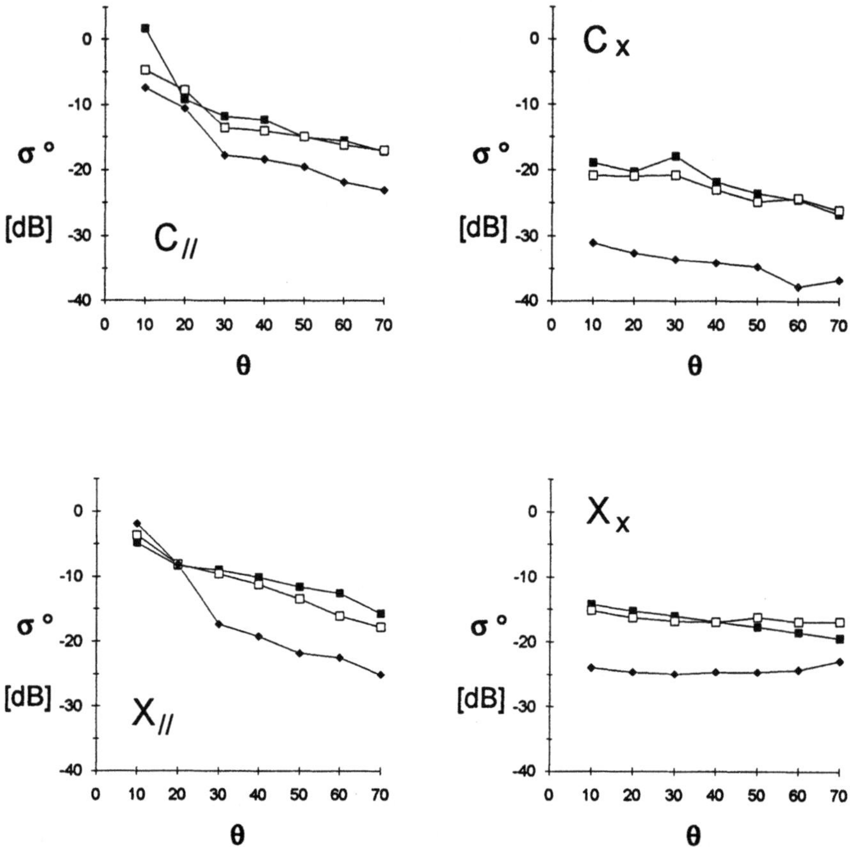

Figure 4. Radar cross-sections σ° in decibels as a function of the incidence angle θ measured at the Alpine Valley Leutasch, Tyrol, in winter and spring 1990-91 for 3 typical situations. Top: C-band (5.2 GHz), $C_{//}$ = mean σ° of hh- and vv-polarized measurements; C_x = mean σ° of hv- and vh-polarized measurements. Bottom: X-band (10.3 GHz). ■ = snow-free, □ = dry snow, ♦ = wet snow.

5. Snow Monitoring by Means of Spaceborne Microwave Radiometry

The brightness temperature T_b measured at satellite altitude H includes contributions of the Earth's surface and of the atmosphere. With the radiative transfer approach, assuming a nonscattering atmosphere in local thermodynamic equilibrium, $T_b(H)$ at a given frequency can be described by:

Table 2. Backscatter Response of Snow-Covered Soil in Dependence of Snow Conditions at C- and X-band Frequencies.

Parameter	Backscatter Response
For wet snow:	
Liquid water content V_w	Decrease of σ° with increasing V_w
Surface roughness S_h	Increase of σ° with increasing S_h
Grain size GS	Increase of σ° with increasing GS (secondary effect)
For dry snow:	
Soil roughness and wetness	Soil dominates backscatter of the snow-soil medium
Stratification (layers)	Increase of copolarized σ° with layers (secondary effect)

$$T_b(H) = \tau(0,H)\left[eT_s + (1-e)T_b^{\downarrow} + (1-e)\tau(\infty,0)T_c\right] + T_b^{\uparrow} \qquad (11)$$

where τ is the atmospheric transmissivity, e is the surface emissivity, T_s is the surface temperature, T_c is the brightness temperature of the cosmic background, and $T_b^{\uparrow}$ and $T_b^{\downarrow}$ are the upward and downward radiating brightness temperatures of the atmosphere. In the frequency range between 18 GHz and 100 GHz, which is used for snow mapping by means of spaceborne microwave radiometry, the propagation losses and emission contributions of the atmosphere show significant variations with frequency. Apart from oxygen absorption in the 50 to 75 GHz region, the main atmospheric attenuators are water vapor, hydrometeors, and clouds. Up to frequencies of 40 GHz, possible errors for snow mapping due to neglecting atmospheric attenuation are small as long as there is no rainfall or dense convective clouds. However, in the 75 to 100 GHz window region, the atmospheric transmissivity reduces to values $\tau \leq 0.75$ in case of stratiform cloud cover and to even lower values for convective clouds and for rainfall (Nagler, 1991; Nagler and Rott, 1992). At these frequencies, corrections for atmospheric effects need to be applied to reduce errors of satellite snow retrievals. Information on atmospheric absorption and scattering properties is required to calculate correction terms. However, spatially distributed information on cloud and precipitation properties is usually not available. In this case, multi-temporal procedures of data analysis which exclude the cases with significant propagation losses help reduce atmospheric effects (Nagler, 1991; Nagler and Rott, 1992).

The first global multi-year data set based on microwave radiometry was compiled from data of the Scanning Multichannel Microwave Radiometer (SMMR) which operated on board of the Nimbus-7 satellite between October 1978 and July 1987 (Chang et al., 1990; Künzi et al., 1982). In June 1987, the first Special Sensor Microwave/Imager (SSM/I) was launched on the US Defense Meteorological Satellite Program (DMSP) Block 5D-2F8 satellite (Hollinger et al., 1990). Up to 1993, two additional SSM/I sensors were launched. It is planned to have operational SSM/Is in space for at least another 15 years. Thus, the sensor is of great interest for a wide range of operational applications and for long-term monitoring of atmosphere, ocean, and land parameters.

Table 3 presents some important characteristics of SSM/I. The channels at 19.3, 37.0, and 85.5 GHz are operating at horizontal and vertical polarization. The 22.2 GHz channel, which is mainly of interest for monitoring atmospheric water vapor, operates only at vertical polarization. The swath width of 1400 km provides complete daily coverage of the Earth's surface at latitudes higher than 42°. Due to overlaps of ascending and descending orbits and of adjoining orbits, high latitudes are covered at least two times daily. For this reason, the sensor is of great interest for snow mapping applications which require information in near real time. The effective field of view on the Earth's surface decreases from 69 km x 43 km at 19.3 GHz to 15 km x 13 km at 85.5 GHz. Gridded data for northern and southern hemispheric areas, with a grid size of 25 km for the 19.3, 22.2, and 37.0 GHz channels, and a grid size of 12.5 km for the 85.5 GHz channel, are provided by the National Snow and Ice Data Center at the University of Colorado, Boulder, Colorado.

Table 3. Characteristics of SSM/I. (EFOV = effective field of view).

Frequency, GHz	19.3	22.2	37.0	85.5
Polarization	h+v	v	h+v	h+v
EFOV, km	69x43	60x40	37x28	15x13
Earth incidence angle			53.1 degrees	
Swath width (km)			1400 km	
Equator crossing time, ascending mode			6:10 mean solar time	

Algorithms have been developed for deriving the following snow parameters from satellite microwave radiometer data: areal extent, water equivalent WE, snow depth d_s, and extent of areas undergoing the melt/freeze cycle (Chang et al., 1990; Goodison et al., 1990; Hallikainen and Jolma, 1986; 1992; Künzi et al., 1982; Nagler, 1991; Nagler and Rott, 1992). WE and d_s can only be determined for dry snow. Permanently melting snow is difficult to detect because of the high emissivities similar to snow-free ground. For the Canadian prairies, an algorithm has been developed for mapping melting snow areas based on 37 GHz polarization differences which are higher than for snow-free areas (Walker and Goodison, 1993). In principle, this algorithm should work also in other regions where the polarization differences of snow-free ground are low.

The algorithms for deriving extent and properties of the snowpack from satellite data were derived empirically, correlating satellite data with ground-based snow measurements, in some cases also with airborne snow surveys. A principal problem for algorithm development is the limited representation of surface point measurements regarding the large areas of satellite radiometer footprints. This is particularly true for development of algorithms to determine the water equivalent of snow WE. Only a few regions of the world provide WE measurements over extended areas with sufficient spatial density to enable accurate algorithm development. Several algorithms have been developed for deriving snow depth d_s because of the better availability of comparative d_s data for algorithm development.

The most frequently applied algorithms for retrieving extent and WE of dry snow by means of microwave radiometry are based on the equations:

$$\text{If} \quad \frac{T_b(\nu_1) - T_b(\nu_2)}{\nu_2 - \nu_1} > -\frac{A_1}{A_2}, \quad \text{Then snow} \tag{12}$$

$$WE = A_1 + A_2 \frac{T_b(\nu_1) - T_b(\nu_2)}{\nu_2 - \nu_1} \quad [\text{mm}], \text{ for } WE > 0 \text{ mm} \tag{13}$$

A_1 and A_2 are empirically derived coefficients. Snow mapping and determination of WE by means of SMMR were usually based on the channels υ_1 = 18 GHz and υ_2 = 37 GHz, vertical and/or horizontal polarization. In the case of SSM/I, the channels υ_1 = 19 GHz and υ_2 = 37 GHz are used. We developed an algorithm which additionally uses the 85 GHz channel and, thus, is also sensitive to thin snow (Nagler, 1991; Nagler and Rott, 1992).

The empirically derived coefficient A_2 in Equations (12) and (13) is indirectly proportional to the sensitivity of the brightness temperature gradient ($\Delta T_b/\Delta \upsilon$) with respect to WE. A_2 shows regional variations depending mainly on morphology of the snowpack and on land-cover type. Forests reduce the response of ($\Delta T_b/\Delta \upsilon$) with respect to WE because the trees are partly masking the radiance emitted from snow. These regional variations can be considered in snow retrievals by including information on the land-cover type such as density of forest cover (Foster et al., 1991).

Interannual variations of the relation between ΔT_b and WE which result from differences of snowpack stratigraphy and grain size as an effect of the meteorological conditions are more difficult to compensate (Hallikainen and Jolma, 1992). Possibilities of overcoming this problem are under investigation. Information on snowpack conditions and metamorphosis can be obtained, at least qualitatively, through studies of the temporal dynamics of the satellite microwave measurements. For instance, temporary melt events which change the metamorphic stage of a snowpack can be clearly recognized. Data of synoptic meteorological stations which are available through the global telecommunication system should also be useful support data for estimating physical conditions of the snow cover in a region.

Maps of snow water equivalent of the Canadian prairies are compiled from SSM/I data on an operational basis (Goodison et al., 1990). The WE maps are generated by the Canadian Atmospheric Environment Service in Downsview, Ontario, on a weekly basis and distributed to the users by fax. The time difference between sensing of the surface by SSM/I and the completion and dissemination of the WE maps can be kept within 4 to 6 hours, if necessary. Figure 5 shows an example of a WE map covering the southern portion of the province of Saskatchewan. The original operational map is larger in size and includes more geographical detail than the map shown in Figure 5. The algorithm for determining WE is based on Equation (13) using the SSM/I channels at 19.3 GHz and 37.0 GHz vertical polarization and the coefficients A_1 = -20.7 [mm], A_2 = 49.3 [(mm GHz/K]. The algorithm was derived experimentally based on relationships of WE determined by means of airborne gamma-ray surveys with brightness temperature gradients measured by means of airborne radiometers, showing a highly significant relationship (R^2 = 0.89) (Goodison, 1989).

The Canadian experience with SSM/I-derived snow cover maps demonstrates the utility of deriving regional maps of snow extent and water equivalent by means of satellite microwave radiometry on an operational basis. Problems arise for determining WE in forested areas, as explained in Section 3. Regional variations of land cover and snow morphology provide difficulties for deriving global maps of snow mass. Chang et al. (1990) used the following equation based on a single coefficient for mapping the extent and depth of the global snow cover with Nimbus-7 SMMR data:

$$d_s = A\left[T_b(\nu_1) - T_b(\nu_2)\right] \quad [\mathrm{cm}] \tag{14}$$

with υ_1 = 18 GHz and υ_2 = 37 GHz, horizontal polarization, and with the coefficient A = 1.59 [cm/K]. Neglecting regional variations of land-cover type and snow morphology results in underestimation of snow depth in forested areas and in overestimation of snow depth in some of the unforested areas (Foster et al., 1991; Rott and Nagler, 1993a). In order to derive accurate global maps of snow mass, it is necessary to develop procedures which account for spatial variations of snow morphology and land cover. Experiences of microwave snow retrievals in various regions of the world provide a good starting point for developing such an algorithm. Further improve-

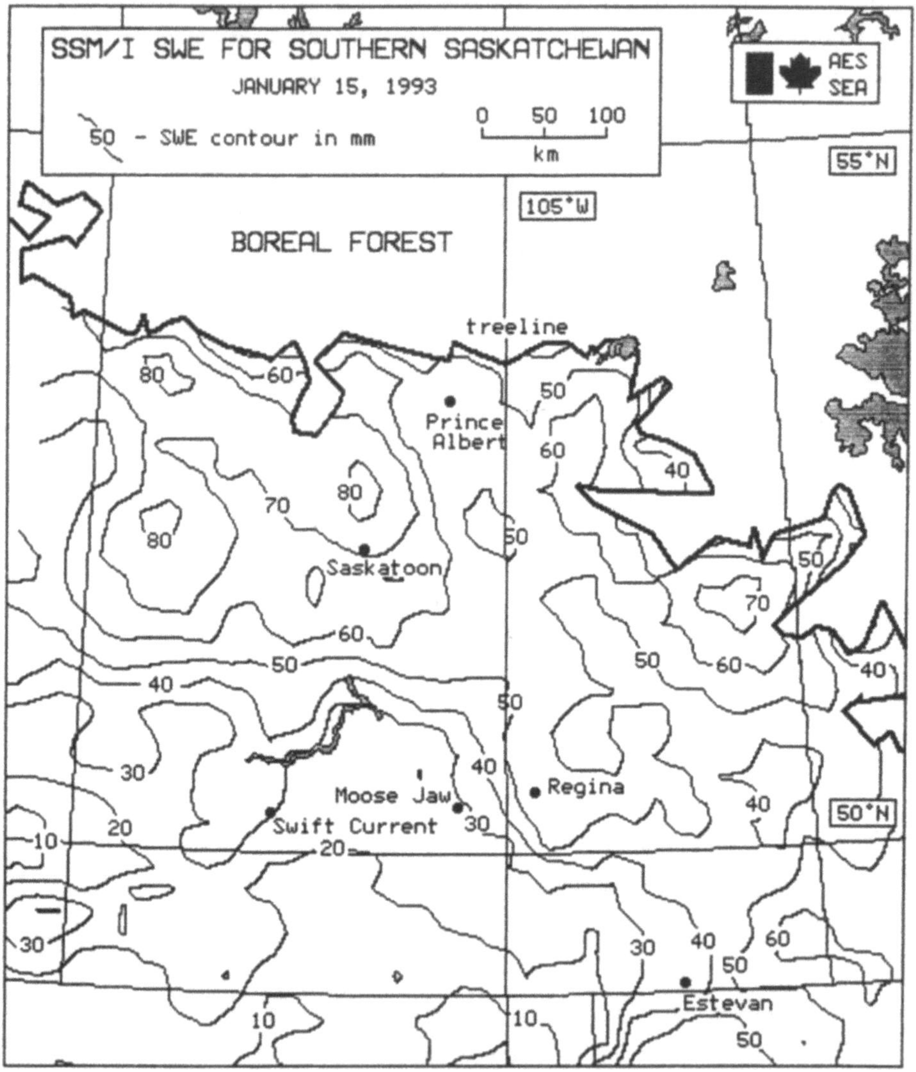

Figure 5. Map of snow water equivalent (in mm) on January 15, 1993, of southern Saskatchewan, Canada, compiled from satellite microwave measurements of SSM/I at the Canadian Atmospheric Environment Service (Goodison et al., 1990).

ments are possible by use of multitemporal procedures for snow classification which reduce effects of atmospheric transmission and temporary melting events (Nagler, 1991; Nagler and Rott, 1992).

6. Snow Mapping by Means of Spaceborne Synthetic Aperture Radars (SAR)

While microwave radiometers are valuable tools for snow mapping at large scales, synthetic aperture radars (SAR) can be used for snow and ice monitoring with comparatively high spatial resolution. A drawback of SAR is the low sensitivity to dry snow, as explained in Section 4. Previous to the launch of the First European Remote Sensing Satellite ERS-1 in July 1991, research on snow mapping algorithms by means of SAR was based on airborne experiments.

The main steps of a classification procedure for generating digital snow cover maps from SAR data are illustrated in Figure 6. Algorithm development was based on airborne SAR experiments in the Ötztal Alps, Austria, with the Canadian SAR-580 operating at C- and X-band and with the 3-frequency polarimetric AIRSAR of NASA/JPL operating at P-, L-, and C-band (Rott, 1984, 1990, 1993; Rott et al., 1988; Rott and Strobl, 1992; Rott and Davis, 1993). The classification procedure for airborne SAR uses data of a single frequency (C-band or X-band), comparing the real SAR image with a simulated image. The first processing steps are generating a simulated SAR image and geocoding the real SAR image. In order to remove SAR image distortions caused by the topographic relief, a digital elevation model (DEM) must be incorporated in the geometric rectification process. Homologue points and linear features such as mountain crests are identified in real and in the simulated SAR images to control the geometric rectification process and to optimize the correspondence of real and simulated image. Low pass filtering is carried out to reduce speckle noise resulting from the SAR imaging process.

The classification rule using SAR data from a single date is based on the comparison of the real image and with a simulated image compiled with angular backscattering

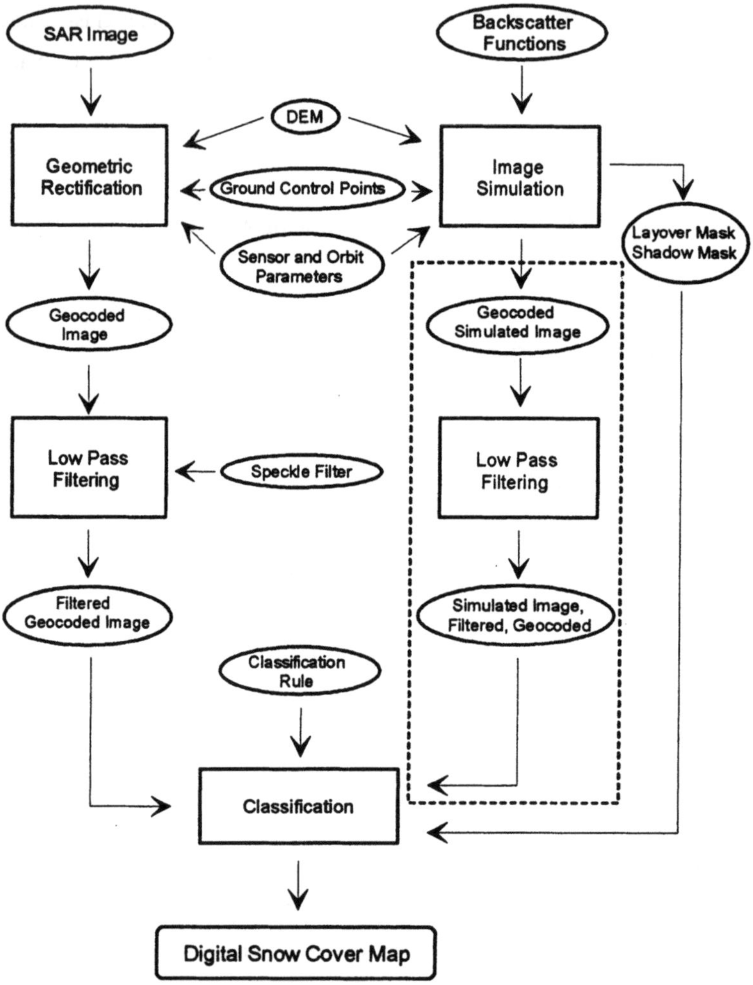

Figure 6. Flow chart of processing steps for snow cover mapping by means of synthetic aperture radar (SAR) imagery.

functions for snow-free terrain (Rott et al., 1988; Rott and Strobl, 1990). In the case of single-frequency, single polarization SAR data, the ratio of backscattering intensity of the real over the simulated image was derived, and a threshold in the ratio image was used to separate snow-covered and snow-free areas. In the case of C-band polarimetric SAR data, thresholds of the hh-polarized backscattering ratio of the real over the simulated image and of the hv/vv-polarized backscattering ratio of the real image were used to separate snow, glacier ice, and ice-free surfaces (Rott, 1994). The classification procedures are applicable for mapping melting snow areas but not for dry snow, because only wet snow shows sufficient contrast to snow-free surfaces.

In order to demonstrate capabilities of spaceborne SAR imagery for snow mapping, an example of SAR data analysis is shown for snow classification in the test site Ötztal using ERS-1 SAR data. Specifications of the ERS-1 SAR are presented in Table 4. Excellent radiometric stability with a standard deviation of 0.36 dB over one year was measured for ERS-1 SAR by means of external calibration (Laur, 1993). This is of particular interest for multi-temporal analysis of SAR images. The absolute radiometric accuracy, determined by means of 3 active calibration devices over one year, was within $\pm$0.7 dB in the worst case.

Precision Images (PRI) were used for our investigations. PRI pictures are produced offline and are radiometrically corrected. Another product type, ERS-1 SAR Fast Delivery Imagery (FD), is processed at ESA ground stations and disseminated by way of high data rate links to user centers in various countries within 3 hours after data acquisition. This makes the use of FD imagery attractive for operational applications. The radiometric accuracy and spatial resolution of FD images come close to those of the PRI pictures.

The high spatial resolution of ERS-1 SAR images is useful for snow mapping in complex terrain. However, the incidence angle θ of the radar beam, which varies for a horizontal surface between 20° and 26° across the observed swath, is far from optimum for applications in mountain areas. The optimum incidence angle range for thematic mapping in Alpine areas is between 40° and 50°, as known from image simulations at various incidence angles and from airborne SAR experiments (Rott, 1994; Rott et al., 1988; Rott and Davis, 1993). Because pixel spacing in range direction is determined by the distance between surface and antennae, slopes facing the radar appear shortened in SAR images (foreshortening) or--if steeper than the radar-look angle--the steep slopes are projected over more gentle slopes (layover) resulting in ambiguous signals. This results in considerable loss of information in ERS-1 SAR images of mountain areas. A good solution to this problem is the combined use of images from opposite passes (ascending and descending).

Table 4. System and Image Characteristics of ERS-1 Active Microwave Instrument (AMI) in SAR Mode (PRI-Imagery) (Francis et al., 1991; Laur, 1993).

Frequency	5.3 GHz (C-band
Polarization	Vertical (vv)
Spatial resolution (azimuth x ground range	22 m x 25 m
Pixel size (azimuth x ground range)	12.5 m x 12.5 x
Number of azimuth looks	3
Radiometric stability (standard deviation)	0.36 dB
Swath width	100 km
Incidence angle	23° at mid-swath
Satellite altitude	785 km

Figure 7 shows a sketch of the central part of the test site Ötztal where field measurements comparative to ERS-1 SAR image acquisitions were carried out between April and October 1992 in 35-day intervals (Rott and Nagler, 1993b). The area shown includes elevations between 1900 m and 3738 m above sea level and is partly covered by glaciers. The firn plateaus of the large glaciers cover comparatively gentle terrain, whereas the glacier tongues are extending into steep, narrow valleys. The ice-free surfaces are made up of bare rock, moraines left from more advanced glacier positions, alpine sedges and grasses, and dwarf shrubs. Altogether, 6 images from ascending orbits (overflight at 22:28 middle European time MET, radar illumination from the west) and

5 images from descending orbits (overflight at 11:08 MET, radar illumination from the east) were acquired in 1992. Between 1 April 1992 and 31 December 1993, ERS-1 has been orbiting with a 35-day repeat cycle. This 35-day cycle has a 16-day subcycle enabling complete SAR coverage of the Earth's surface north of 50° latitude in at least 16-day intervals. Images of descending and ascending passes over Ötztal were acquired with a time difference of 11 hrs 20 min, which is very useful for combined analysis.

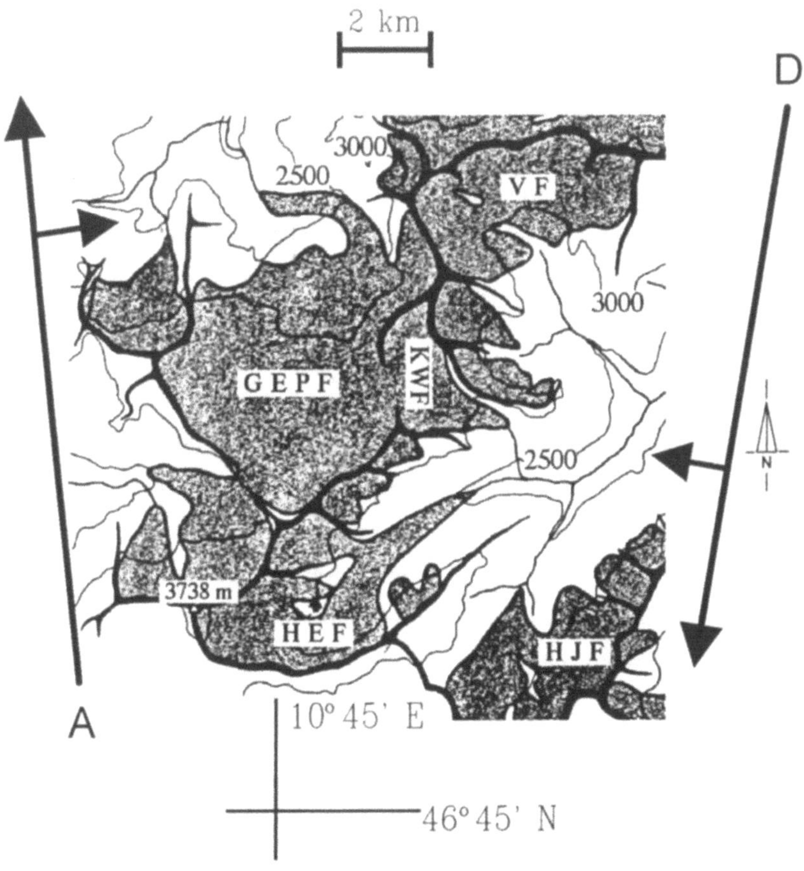

Figure 7. Sketch map of the test site Ötztal, Austria. Shaded areas represent glaciers. Directions of the ascending pass (A) and of the descending pass (D) of ERS-1 and SAR look directions are shown. Glaciers of ground-truth activities: HEF = Hintereisferner, GEPF = Gepatschferner, KWF = Kesselwandferner.

Parts of geometrically uncorrected SAR scenes from ascending and descending orbits acquired over the test site on 6 July 1992 are shown in Figure 8. The comparison with the map (Figure 7) indicates considerable distortions due to SAR imaging geometry. It is evident that terrain-corrected geometric rectification is necessary for appropriate use of the data. The direction of radar illumination of the ascending pass is suitable for imaging the glaciers Hintereisferner and Kesselwandferner because they are sloping away from the radar antennae. In the image of the descending orbit, west-facing slopes are well-detected because they are facing away from the radar (backslopes). The bright spikes in this image are due to specular reflections on slopes perpendicular to the illuminating radar beam. On 6 July 1992, the glaciers were covered by snow, with the exception of the lower part of the terminus of Hintereisferner, where the snow had disappeared. The patchy snowline in the ice-free areas was located at elevations between 2800 m and 3000 m. On the SAR backslopes and on level areas, snow-covered surfaces appear dark because the snowpack was wet at all elevations and consequently showed low reflectivity. In the foreshortening and layover areas, no information on snow extent can be obtained.

Figure 9 shows a geocoded, terrain-corrected SAR image of the test site from the ascending orbit of ERS-1 on 1 June 1992. On this day, the snowline elevation was about 200 m below the snowline on 6 July, and the snow cover near the snowline was also patchy. About 20% of Figure 9 is made up by layover areas; these areas are shown in white. Typically, layover zones in single ERS-1 SAR images cover 20% to 35% of the area in the Central Eastern Alps which show rugged topography. Over areas with more gentle topography, the percentages of layover are smaller.

If SAR images from ascending and descending orbits are combined, the loss of information due to foreshortening is small. This is evident in Figure 10, which is a map of snow extent of the test site Ötztal on 1 June 1992 derived from ERS-1 SAR images of the ascending and the descending pass. After combination of the data from the two passes, the remaining layover zones (shown in black) cover only 2% of the area. Classification of snow extent was based on a multi-temporal procedure. The processing scheme up to the production of low-pass filtered geocoded images and of layover masks

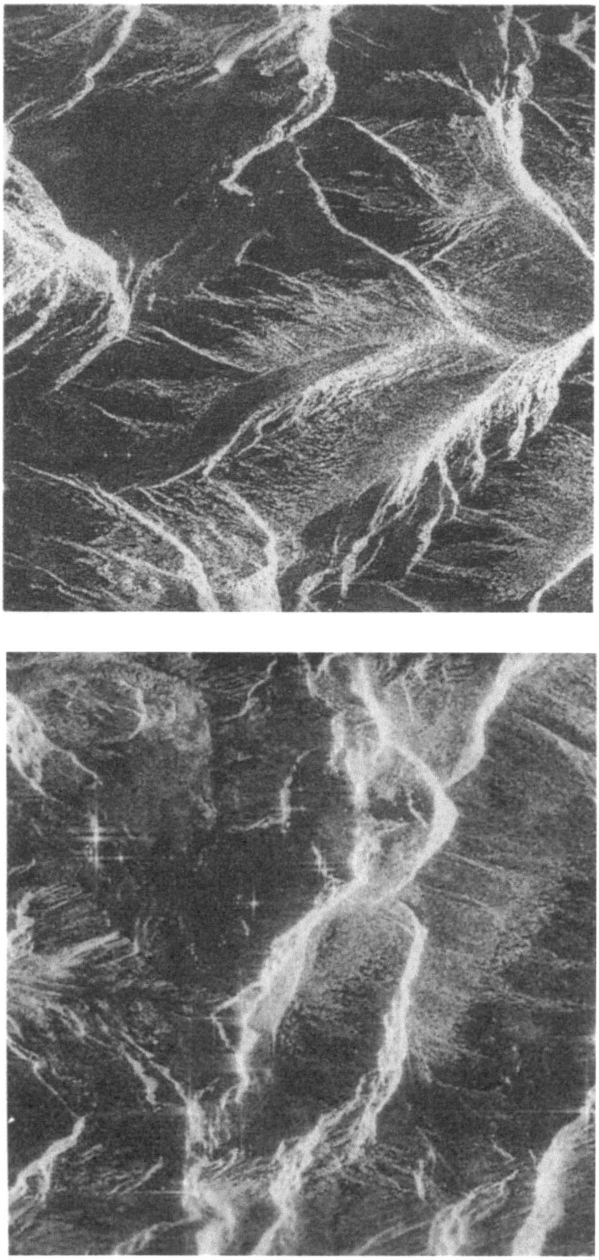

Figure 8. Part of ERS-1 SAR images acquired on 6 July 1992 over the test site Ötztal.
Top: ascending pass (night); bottom: descending pass (day).

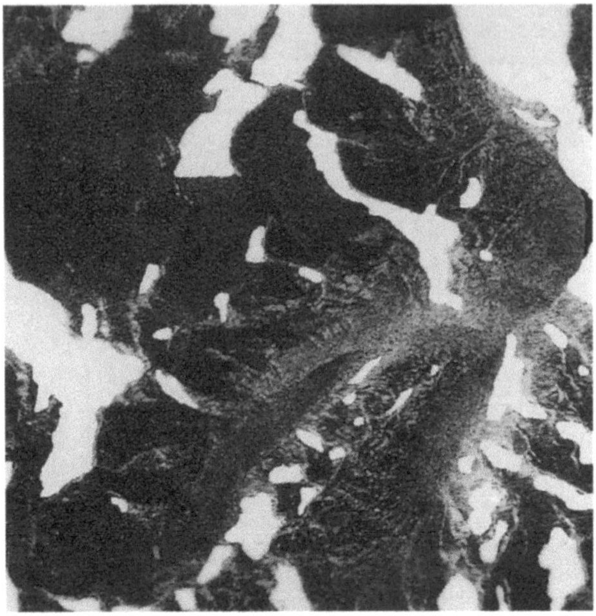

Figure 9. Geocoded, terrain-corrected ERS-1 SAR image of the test site Ötztal on 6 July 1992, ascending pass. The layover mask is shown in white.

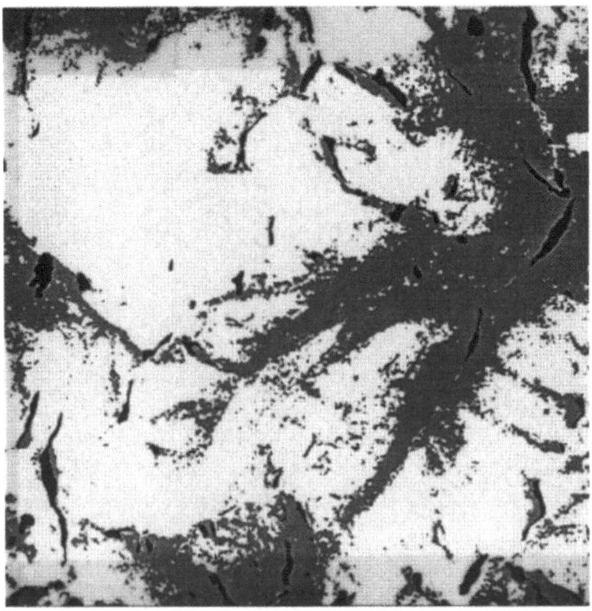

Figure 10. Snow cover map of the test site Ötztal on 1 June 1992, derived from ERS-1 data, ascending and descending passes, by means of multitemporal classification. White = snow, grey = snow-free, black = layover areas in both the ascending and descending passes.

(Figure 6) was applied to the SAR images, ascending and descending passes, from two dates. A 5 x 5 pixel median filter was applied to reduce spatial variability of backscattering resulting from speckle. Image simulation was performed to assist in the geometric rectification process for the real SAR images and to process the layover masks. The processing steps for the simulated image, which are included in the broken box shown in Figure 6, are not needed for multi-temporal classification. After eliminating the layover areas, the following classification rule was applied on pixel-by-pixel basis to separate the two classes: "wet snow cover" and "snow-free areas":

$$\text{If} \quad (A_S/A_R) < TR, \qquad \text{Then snow} \tag{15}$$

A_S is the backscattering amplitude of a pixel in the snow-covered image, and A_R is the backscattering amplitude in the reference image. The amplitude (square root of power) of the return signal was used because the digital numbers of the ERS-1 images are proportional to amplitude. As a reference image, a snow-free image or an image with completely dry snow cover can be taken. For the classification in Figure 10, showing the snow extent on 1 June 1992 when the snow cover was wet at all altitudes, the images of 27 April 1992 were taken as reference. On 27 April, the site was completely covered by dry snow, resulting in high backscattering signals on glaciers as well as on ice-free surfaces (Rott and Nagler, 1993b). The threshold value TR = 0.9 was determined in accordance with ground-based observation of snow extent on 1 June. Satellite or airborne optical imagery was not available for comparison. Qualitative assessment based on field studies and ground-based photography indicate good accuracy of the SAR-derived snow map.

If SAR images from two dates with different extent of wet snow cover are available, the same procedure as described above can be applied to derive the decrease of snow extent between the two dates due to melting. Those areas where the snow cover disappeared show an increase of backscattering above the threshold value.

The example of ERS-1 data analysis shows the usefulness of spaceborne SAR for monitoring the extent of melting snow areas. This is of significant interest for snowmelt runoff modeling and forecasting. The launch of additional SAR systems is planned in the near future. ERS-2, succeeding ERS-1 in December 1994, will be equipped with the same SAR. The Canadian RADARSAT, due to be launched in 1995, will be equipped with a C-band SAR operating in various modes, with the swath width selectable between 45 km and 510 km and with a variety of incidence angles between 20° and 50° (Raney et al., 1991). In the wide swath mode, complete surface coverage will be obtained within 4 days, even at equatorial latitudes.

Multifrequency polarimetric SAR (SIR-C/X-SAR) operated during two space shuttle missions in 1994. In addition to mapping snow extent, polarimetric SAR offers the possibility of obtaining quantitative information on wetness of the surface layer of a snowpack. An inversion model has been developed for calculating the liquid water content of snow based on C-band polarimetric SAR data (Shi et al., 1993). Because backscattering from wet snow is primarily dependent on the roughness of the snow surface and on the liquid water content V_w of the top snow layer, the model separates these two contributions using the following polarimetric parameters: the copolarized backscattering coefficients σ_{hh} and σ_{vv} and the real part of the cross-product of vv and hh complex scattering elements S, Re $[S_{vv} S_{hh}^*]$. The algorithm was tested with AIRSAR data and comparative field measurements of the test site Otztal on two days in 1989 and 1991, showing relative differences between V_w derived from AIRSAR data and V_w derived from the field measurements which were within 25% of V_w. Further verifications are needed to assess the general applicability of the algorithm.

Conclusion

Table 5 summarizes the status of the presently available algorithms for snow cover monitoring by means of microwave sensors. The available sensors offer either low spatial resolution with high temporal repetition rate (microwave radiometers and scatterometers) or high spatial resolution with reduced sampling intervals (SAR). The specified resolutions for microwave radiometers listed in the table are based on SSM/I. The Multispectral Imaging Microwave Radiometer (MIMR) with improved spatial resolution is presently under development in Europe, with its launch currently expected in the late 1990s.

Table 5. Status of Algorithms for Snow-Cover Monitoring by Means of Spaceborne Microwave Sensors. (WE = water equivalent, ★ for ERS-1; ♦ for RADARSAT).

Parameter	Sensor	Spatial Resolution	Delay	Sampling Interval	Status of Inversion Procedures
Extent of dry snow	MW radiometer	$\geq$12.5 km	6h	$\leq$1d	Operational
WE of dry snow	MW radiometer	$\geq$25 km	6h	$\leq$1d	Operational (regionally)
Depth of dry snow	MW radiometer	$\geq$25 km	6h	$\leq$1d	Operational (regionally)
Melting/freezing snow	MW radiometer	$\geq$25 km	6h	$\leq$1d	Method available
Extent of wet snow	MW radiometer	$\geq$25 km	6h	$\leq$1d	Regionally tested, applic. geographically tested
Extent of wet snow	C-band SAR	$\geq$30 m	5h	16d★,4d♦	Basic method available
Liquid water content	Polarimetric SAR	$\geq$30m	--	--	Preliminary method available
Extent of wet snow	Scatterometer	$\geq$25 km	3h	$\leq$3d	Basic principle proven

The repeat coverage for SAR in Table 5 is specified for ERS-1 in the multidisciplinary phase and for RADARSAT in the wide swath mode. The specified time delay indicates average possible time difference between the satellite observation and the dissemination of the snow cover product to potential users. For ERS-1 SAR-derived snow cover information, 3 hours for processing and dissemination of the fast delivery images and 2 hours for image geocoding and classification were assumed. This seems reasonable regarding the time of 3 hours after ERS-1 SAR data acquisition for processing, analysis, and dissemination of sea ice images (Johannessen, 1993). In order to achieve these time intervals, special data links must be established. This presently seems to be a bottleneck for potential users.

In addition to spatial resolution, there are pronounced differences in the type of snow-cover information available from microwave radiometry or from SAR. Microwave radiometers are primarily sensitive to dry snow and also enable mapping of snow depth and water equivalent. Accurate procedures for deriving these two parameters need to account for regional differences in land-cover type and snow morphology. A method for mapping extent of melting snow areas by means of microwave radiometry has been developed over the Canadian prairies. This method, applying the 37 GHz polarization difference, requires low polarization difference of snow-free ground which is not observed in all geographic regions. Improvements for wet snow mapping can be expected from use of microwave scatterometers which, in principle, offer similar capabilities for wet snow mapping as SAR, although with much lower spatial resolution. Our analysis of ERS-1 scatterometer data indicates a strong decrease of backscattering intensities over wet snow areas. The average delay between acquisition of ERS-1 scatterometer data and delivery of the processed data (fast delivery product) through GTS is less than 3 hours (Duchossois, 1993).

Considering the application of SAR, methods are available for mapping the extent of melting snow over open areas. The application over forested areas is questionable and requires further investigation. Presently and in the near future, spaceborne SAR systems operating at C-band, single polarization (ERS-1, ERS-2, RADARSAT) will provide useful data for mapping wet snow areas. Signature research and analysis of SAR data indicate low sensitivity of these systems for identifying dry snow. The potential of polarimetric and multi-frequency SAR systems for monitoring dry snow and for deriving various properties of wet snow is a topic of ongoing research. Advancements in this respect can be expected from the SIR-C/X-SAR experiment, carried out by NASA/JPL in cooperation with the German DLR and the Italian ASI during Space Shuttle Missions in April and August 1994 (Jordan et al., 1991).

Acknowledgements

The author thanks Dr. Barry Goodison and Ms. Anne Walker of AES, Downsview, Ontario, for providing the SSM/I snow cover map of Saskatchewan and related information. The investigations with ERS-1 SAR are supported by the Austrian

Academy of Sciences, National Space Research Program. The European Space Agency made available the ERS-1 SAR data for the AO project A1.

References

Chang ATC, JL Foster, DK Hall (1990) Satellite sensor estimates of northern hemisphere snow. *International Journal of Remote Sensing*, 11(1):167-171.

Duchossois G (1993) ERS-1 one year after: status and perspectives of the ERS programme. *Proceedings of EARSeL Workshop on ERS-1 Data Processing*, Eger, Hungary, September 1992, European Association of Remote Sensing Laboratories: 2-13.

Foster JL, ATC Chang, DK Hall, A Rango (1991) Derivation of snow water equivalent in boreal forests using microwave radiometers. *Arctic*, 44:147-152.

Francis R et al. (1991) The ERS-1 spacecraft and its payload. *ESA Journal*, 65:27-48.

Fung AK, MF Chen (1989) Modeling of microwave emission and scattering from snow and soil. *Advances in Space Research*, 9(1):297-306.

Goodison BE (1989) Determination of areal snow water equivalent on the Canadian prairies using passive microwave satellite data. *Proceedings of IGARSS '89*, IEEE: 1243-1246.

Goodison BE, AE Walker, FW Thirkettle (1990) Determination of snow cover on the Canadian prairies using passive microwave data. *Proceedings of International Symposium on Remote Sensing and Water Resources*, Enschede, The Netherlands, August 1990: 127-136.

Hallikainen MT, PA Jolma (1986) Retrieval of snow water equivalent in Finland by satellite microwave radiometry. *IEEE Transactions on Geosciences and Remote Sensing*, 24:855-862.

Hallikainen MT, FT Ulaby, TE van Deventer (1987) Extinction behavior of dry snow in the 18- to 90-GHz range. *IEEE Transactions on Geosciences and Remote Sensing*, 25:737-745.

Hallikainen MT, PA Jolma (1992) Comparison of algorithms for retrieval of snow water equivalent from Nimbus-7 SMMR data in Finland. *IEEE Transactions on Geosciences and Remote Sensing*, 30:124-131.

Hofer R, E Schanda (1978) Signatures of snow in the 5-95 GHz range. *Radio Science*, 13:365-369.

Hollinger JP, JL Peirce, GA Poe (1990) SSM/I instrument evaluation. *IEEE Transactions on Geosciences and Remote Sensing*, 28(5):781-790.

Johannessen JA, RA Shuchman, K Davidson, O Frette, G Digranes, OM Johannessen (1993) Coastal ocean studies with ERS-1 SAR during NORSEX '91. *Proceedings of First ERS-1 Symposium--Space at the Service of our Environment*, ESA SP-359:113-117.

Jordan RL, BL Huneycutt, M Werner (1991) The SIR-C/X-SAR synthetic aperture radar system. *Proceedings of the IEEE*, 79(6):827-838.

Künzi K, S Patil, H Rott (1982) Snow cover parameters retrieved from Nimbus-7 Scanning Multichannel Microwave Radiometer (SMMR) data. *IEEE Transactions on Geosciences and Remote Sensing*, 20:452-467.

Laur H (1993) ERS-1 SAR products: quality and calibration. *Proceedings of EARSeL Workshop on ERS-1 Data Processing*, Eger, Hungary, September 1992, European Association of Remote Sensing Laboratories: 18-26.

Mätzler C (1987) Applications of the interaction of microwaves with the natural snow cover. *Remote Sensing Reviews*, 2:259-387.

Mätzler C, R Hüppi (1989) Review of signature studies for microwave remote sensing of snowpacks. *Advances in Space Research*, 9(1):253-265.

Nagler T (1991) Verfahren der Analyse der Schneebedeckung aus Messungen des Special Sensor Microwave/Imager (SSM/I). Diploma Thesis, Science Faculty, University of Innsbruck, Austria.

Nagler T, H Rott (1992) Development and intercomparison of snow mapping algorithms based on SSM/I data. *Proceedings of IGARSS '92*, IEEE:812-814.

Raney KR, AP Luscombe, EJ Langham, S Ahmed (1991) RADARSAT. *Proceedings of the IEEE*, 79(6):839-849.

Reber B, C Mätzler, E Schanda (1987) Microwave signatures of snow crusts, modeling, and measurements. *International Journal of Remote Sensing*, 8(11):1649-1665.

Rott H (1984) The analysis of backscattering properties from SAR data of mountain regions. *IEEE Journal of Oceanic Engineering*, 9(5):347-355.

Rott H (1990) Snow and land ice in the climate system: research problems and possibilities of remote sensing. *Remote Sensing and the Earth's Environment*, ESA SP-301: 61-75.

Rott H (1994) Thematic studies in Alpine areas by means of polarimetric SAR and optical imagery. *Advances in Space Research*, 14(3):217-226.

Rott H, C Mätzler, H Miller, KG Lenhart (1988) Study on SAR land applications for snow and glacier monitoring. ESA Contract No. 6618/85/F/FL(SC), Final Report, Institute of Meteorology and Geophysics, University of Innsbruck, Austria.

Rott H, RE Davis (1991) Multi-parameter airborne SAR experiments at an alpine test site. *Proceedings of IGARSS '91*, IEEE:2311-2314.

Rott H, K Sturm (1991) Microwave signature measurements of antarctic and alpine snow. *Proceedings of 11th EARSeL Symposium, Joanneum Research*, Graz, Austria: 140-151.

Rott H, D Strobl (1992) Synergism of SAR and Landsat TM imagery for thematic investigations in complex terrain. *Advances in Space Research*, 12(7):425-431.

Rott H, RE Davis (1993) Multifrequency and polarimetric SAR observations on alpine glaciers. *Annals of Glaciology*, 17:98-104.

Rott H, T Nagler (1993a) Intercomparison of snow retrieval algorithms by means of spaceborne microwave radiometry. *Proceedings of ESA/NASA Workshop on Passive Microwave Remote Sensing Research*, St. Lary, France, January 1993, in press.

Rott H, T Nagler (1993b) Snow and glacier investigations by ERS-1 SAR --first results. *Proceedings of First ERS-1 Symposium--Space at the Service of our Environment*, ESA SP-359:577-582.

Shi JC, J Dozier, H Rott (1993) An improved algorithm for retrieving snow wetness using C-band polarimetric SAR. *Proceedings of Fourth Annual JPL Airborne Geoscience Workshop*, JPL Publication 93-26, 3:61-64.

Sturm K, H Rott (1992) C- and X-band signatures of snow. *Proceedings of URSI Microwave Signature Conference 1992*, Innsbruck-Igls, Austria: 2D27-2D30.

Tiuri ME (1982) Theoretical and experimental studies of microwave emission signatures of snow. *IEEE Transactions on Geosciences and Remote Sensing*, 20(1):51-57.

Ulaby FT, WH Stiles (1980) Microwave radiometric observations of snowpacks. Microwave Remote Sensing of Snowpack Properties, NASA Conference Publication 2153:187-194.

Ulaby FT, WH Stiles, M Abdelrazik (1984) Snowcover influence on backscattering from terrain. *IEEE Transactions on Geosciences and Remote Sensing*, 22(2):126-132.

Walker A, B Goodison (1993) Discrimination of wet snow cover using passive microwave satellite data. *Annals of Glaciology*, 17:307-311.

The Challenges We Face: Panel Discussion on Precipitation

Edited by Hoshin Gupta and Soroosh Sorooshian

1. Introduction

This panel discussion began with introductory remarks by the moderators *David Legates* and *David Goodrich*, which set the tone for a very interesting and enjoyable discussion. The discussion covered a lot of ground and continually reiterated the fact that there needs to be a much better dialogue between the data collectors and the modelers. The following is a summary of the major points of discussion, based primarily on notes prepared by the moderators *David Legates* and *David Goodrich*, and by additional written contributions from participants *Jean Morrill, Xu Liang and Jim Washburne*. Their assistance in documenting the discussions is greatly appreciated.

2. What do we mean by "precipitation"?

Hydrologists have traditionally focused most of their research and data collection efforts on rain and snow, and to some extent on hail, while other hydrometeors have largely been ignored. It was pointed out, however, that other important forms of precipitation are of hydrologic and climatological interest, and agricultural climatologists have recently indicated that they would like to see the measurement and archiving of data on the spatial and temporal variability of hail size and duration, the formation of dew and the occurrence of fog. *Collier* reported that because of its kinetic energy and size, hail can be estimated using conventional radar, with better estimates now provided by multi-parameter radars. However, the importance of measuring dew and fog is dependent on the application. While the moisture contained in dew and fog may, in some ecosystems, be more important than rainfall for vegetation growth, it may be inconsequential for the detection of global change. It was suggested that perhaps an important variable to consider for such purposes is atmospheric humidity. *Dozier* noted that precipitation chemistry is one of the areas where change and effect in ecosystems are most pervasive, and there is a whole set of problems on how one measures concentration in precipitation, how one

NATO ASI Series, Vol. I 46
Land Surface Processes in Hydrology
Trials and Tribulations of Modeling and Measuring
Edited by S. Sorooshian, H. V. Gupta, and S. C. Rodda

translates precipitation concentrations and amounts into loading, and so forth. However, the participants decided to adopt *Dozier's* recommendation that, while issues of atmospheric humidity, precipitation chemistry, and precipitation impact are very important, the panel should simplify the present discussion by restricting attention to just the water molecule, in both solid and liquid form. Also, snow was not discussed, because a separate panel discussion was dedicated to issues surrounding the measurement and modeling of snowfall (see Chapter 10). Therefore, in the following discussions, the word precipitation should be understood as referring to rainfall.

3. What are we really interested in - hydrologic or meteorological precipitation?

Precipitation, in all of its various forms, is the primary driving input to the majority of surface and near-surface hydrologic processes at short time scales and ultimately groundwater processes at longer time scales. However, there appears to be no clear consensus (and none was reached during the discussion) as to whether we are really interested in "hydrologic" or "meteorologic" precipitation. This is related to several issues. For example, most existing precipitation information comes from rainfall gages which provide "point" measurements, and the orifices of such gages are typically oriented in the horizontal position. There is a longstanding debate in the community, however, as to whether gages in mountainous terrain should be set with their orifices oriented parallel to the topography. Also, modern radar based precipitation measurements do not estimate precipitation at the ground surface but instead at some elevation in the atmosphere. These issues were noted but did not receive further discussion.

4. Is there some standard design of rain gage that should be recommended?

A great deal of discussion focused on the problems associated with selection of a raingage. Because they can be directly measured, it is often assumed that precipitation data are reasonably accurate and representative for a wide variety of spatial and temporal resolutions. *Legates* and *Goodrich* reminded the participants that there are several problems inherent with precipitation data that may limit their reliability. These problems include (among others):

(i) inhomogeneities and inconsistencies in the precipitation data due to changes in gage location or siting characteristics, observation practices and/or type of gage used,

(ii) biases in the gage measurement process arising due to wind, wetting losses due to water adhering to the walls of the funnel or collector, evaporation losses, splashing losses, and losses due to the mechanics of automation (such as tipping buckets, recording arm friction, etc.),

(iii) random but systematic effects such as those caused by observational errors and biases, animals, birds, and vandalism (most notably from gunfire),

(iv) uncertainties in estimating areally averaged precipitation from point data, and

(v) issues associated with obtaining an appropriate temporal resolution.

Michaud added also that

(vi) data transmission errors can be quite significant.

Because much of the uncertainty in the data is related to the design of the gage itself, a great deal of discussion focused on the international standardization of raingages. *Rodda* reported that a World Bank initiative called "The World Hydrological Central Observing System" has been proposed that would install and maintain 1000 precipitation gages worldwide with nearly uniform gage and siting characteristics. However, several questions still remain unanswered, and no decisions have as yet been made regarding which gage will be selected if the initiative were to go through. Would standardization and uniformity be a requirement? Would the gages be overseen by the existing hydrometeorology offices in those countries, or by some other organization?

With such developments underway, it would appear to be desirable for the scientific community to recommend the specifications supporting a certain type of gage. *Calder,* noted that the presentations preceding the panel discussion had discussed several different types of rain gages and had implied that the most accurate type that can be installed at present are ground-level gages. He therefore asked whether there was a "standard type" of raingage that should be recommended for general use. The responses to this query were mixed, but the consensus was that there is no single gage design that would eliminate all problems or be amenable to all countries of the world.

Goodison remarked, however, that even were such a "better gage" to be recommended, it is highly unlikely that any national meteorological or hydrological agency would be willing to change their entire network without being given a large sum of money to cover the costs of doing so. Changing an entire network consisting of thousands of gages would be extremely expensive. Moreover, this also would introduce yet another discontinuity into the historical precipitation archive. Clearly, therefore, the practical approach is, instead, to focus on understanding and correcting for the errors (particularly the systematic errors) associated with each national gage and see if we can adjust for the methods of observation and automation as they are changed.

5. Should we focus on obtaining metadata or on expanding the gage network?

Several participants therefore supported the idea that more effort should be focused on improving the accuracy of the data and on identifying and quantifying potential discontinuities in the precipitation record. From the modelers' point of view, station *metadata* should be included in measurement reports that accompany the data. This should include information about the instrumentation and the site at which the precipitation is measured, as well as estimates of uncertainty which can be incorporated into the statistical analysis of precipitation data.

While many participants agreed with this sentiment, *Nicholson* raised the strong objection that the assumptions inherent in a "correction" procedure may itself introduce just as much error into the data as the gages themselves. She pointed out, that for large portions of the world, data is extremely scarce and researchers consider themselves fortunate just to get any sort of rainfall data collected, recorded, and transmitted. In many countries, such as in Africa, South America, and parts of Asia, it is a problem just getting the observations taken. Therefore, to also expect people in such places to provide metadata (of gage locations, measuring procedures, etc.) may be asking too much, and it may optimistically be as much as a century from now before this situation changes significantly. In Chapter 9 of this book, *Nicholson* expands on this argument and suggests that rather than expending time and resources on getting metadata, we should be focusing on expanding the network of gages to cover the vast portions of the globe where rainfall data are scarce or non-existent.

6. What is a hydrologically relevant measure of spatially and temporally aggregated precipitation?

In the application of hydrologic models, researchers must typically convert point estimates to areal averages. In doing this, it is not at all clear what methods are best applicable or even to what spatial resolution interpolations should be done. For most temporal scales of interest, gage networks are nearly always grossly inadequate in resolving the spatial variability of precipitation, and it is likely that aggregate measures seriously underestimate the actual amount. It was pointed out that as the area increases, the accuracy of estimated spatial averages is a function of scale. This is particularly true for precipitation which typically has a rather heavy tailed distribution and for which a mean areal average taken by itself may provide a deceptive idea of hydrologic impact. However, more to the point is the fact that the actual information needed in the simulation of runoff is precipitation excess (the amount that does not infiltrate). The use of mean areal average precipitation in hydrologic models tends to result in serious underestimation of precipitation excess, and hence runoff. So, an "average" may not really be an appropriate aggregate number for use in representing the hydrologic cycle, and some other statistic may be more applicable. *Gupta* pointed out that when one considers the strong nonlinearities involved in hydrologic cycle computations, we may find that we need to replace the use of mean areal averages with some measure that is a function of at least the second order statistics of the spatial and temporal distribution of precipitation.

Goodrich remarked that, once again, the product generated by meteorologists and climatologists may not be consistent with what hydrologists and hydrologic modelers actually need, and that a tremendous need for more dialogue exists. In particular, the hydrological community must make a strong statement about what is required because, without a push to get the information, it will not be collected. Modelers must speak up about their needs so that their models are not limited by inappropriate data. It was noted, for example, that hydrologic models are not presently designed to incorporate additional information related to precipitation, such as second-order statistics.

The problem of obtaining areal aggregate estimates of precipitation is particularly severe for rugged and mountainous terrain where gage networks are typically very sparse. In particular, measurements in mountainous areas are usually made at the lower elevations, even though most measurements in mountainous areas are usually made at the lower elevations, even though most precipitation-elevation relationships indicate that rainfall and snowfall increase with elevation. Thus, these errors usually result in an underestimate of the true area-average precipitation. No suggestions for dealing with this were advanced.

It was, however, proposed that given the poor likelihood that raingage densities will be significantly increased (except in select experimental watersheds), the use of remote sensing techniques (satellites and radars) may, owing to their increased spatial resolution, provide the key to obtaining area-averaged estimates, particularly for remote or rugged regions that are difficult to access. While this does not directly address the problem of obtaining hydrologically relevant aggregate estimates of precipitation, it should (at least in the case of radar) be possible to produce estimates of the standard deviation of remotely sensed areal averages. *Collier* pointed out, however, that this will require changes in operational software and that the community must make a very strong statement about such needs, or nothing will happen. Without a push to get the information, it will not be collected. Conversely, *Sorooshian* raised the question "Is the modeling group really ready to start using information like second-order statistics"?

7. How do we relate radar information to hydrologic models?

While on the topic of remotely sensed precipitation, *Sorooshian* raised questions about the superposition of radar grid based precipitation information over a catchment. In particular he wanted to know whether any thought had yet been given to how the rectangular grid radar estimates can be converted to a hydrologic grid for use within a distributed surface hydrologic model. Radar echoes can be converted into precipitation estimates through the use of a Z-R relationship or the probability matching method. *Smith* reported that the new WSR-88D (NEXRAD) radar uses an 0.9° beam width to produce a resolution of 2 km x 2km which is aggregated to 4 km x 4 km for operational use. These estimates, however, are not without error and bias. *Smith* noted that a rectangular nationally consistent grid was chosen instead of a drainage basin grid system, specifically because there are multiple hydrological and meteorological applications for the data. *Collier* reported that the method used in the United Kingdom to produce catchment precipitation estimates is to integrate over the number of radar pixels in the basin or sub-basin. *Kabat* argued, however, that it is not clear how to weigh the contribution of each radar pixel when some pixels are on the catchment border and some are inside the catchment. While a number of techniques for this purpose have been proposed, it would be useful to have a recommendation for different kinds of situations (e.g., flat versus mountainous terrain). It was concluded that establishing and maintaining consistency within the radar data record is an important concern.

8. Can satellite imagery help?

To study regional or global environmental change we need a precipitation product that covers the entire globe. An active area of current research is to develop methods for estimating areal aggregate precipitation by means of satellite imagery, usually through a relationship between cloud top temperatures and the precipitation intensity. The attractiveness of this approach is the possibility of obtaining data for the entire globe. The cloud top temperature approach is, however, limited to convective situations only. The discussion of satellite-derived precipitation estimates focused on exactly what sort of data are needed. *Collier* commented that the coverage by satellites such as TRMM and AVHRR, for example, are often spatially and temporally limited and the spatial resolution is quite coarse. Other polar orbiters have been proposed, but the projects are still at least 20 years away. However, *Goodison* noted that even polar orbiters will not yield the coverage needed to determine global precipitation, especially at very high latitudes. Current methods cut off between 55 and 60 degrees latitude, and it is in the high latitude areas (where there is the sparsest gauge network) that snow makes up the highest proportion of precipitation. We will not get complete global coverage unless a satellite is placed into a millennia orbit specifically to cover the polar latitudes. As a result, any global product will have to be pieced together, and it will be a challenge to produce a "consistent global product" for use in validation or process modeling. It therefore seems likely that we will have a problem with high latitudes for quite some time into the future. *Rodda* pointed out that this is of particular concern because the high latitudes are one of the areas where researchers believe the biggest changes in precipitation may occur and, as noted by *Becker*, where measurement of snow is difficult.

9. What is an appropriate temporal sampling resolution?

Several questions were raised regarding the appropriate temporal resolution for precipitation sampling, both from the modeling standpoint and from the standpoint of the physics of hydrology. It seems likely that different sampling intervals are important for adequately representing different processes. Further, there are believed to be important, but as of yet poorly understood, interactions between spatial and temporal resolutions. *Goodrich* pointed out that the temporal frequency needed in a catchment modeling scenario is dependent on the scale of the watershed itself and on related factors such as soil type and vegetation. A number of studies have demonstrated that a model's ability to simulate the hydrology of a small catchment degrades as soon as temporal resolution is compromised. For small experimental catchments, it is desirable to have the highest resolution available. However, as catchment sizes increase to very large basins (e.g., the Colorado or Mississippi rivers), temporal resolution has less of an impact.

Goodrich also raised the question of what the temporal sampling needs are for synoptic mesoscale and convective regions. He reported that forecast capability is fairly good for synoptic scale events or when concerned with total storm modeling, and wanted to know whether the scarcity of high temporal (and spatial) resolution data will force us to consider stochastic models in lieu of deterministic approaches, or will we come up with some combination of the two? Also, does our difficulty in modeling single storm events compromise our ability to model long-term climatic change?

Another topic raised was the relationship between precipitation volumes, intensities and time steps. *Goodrich* reported that values of total rainfall depth over a day are insufficient for the application of physically based infiltration models in many environments. For example, in semi-arid climate regimes such as Arizona, even a one-hour time step is insufficient for accurate computation of precipitation intensity. *Calder* reported that a similar comment is applicable to the computation of interception loss, where the average number of rain-days per year, may be more important than precipitation volumes. He also noted that the computation of total number of rain-days will vary with the spatial scale selected for averaging. In many applications, the precipitation frequency distribution, with particular emphasis on the probability of occurrence of extreme events, may be of more concern than the total precipitation.

10. How useful is GCM simulated precipitation?

Of course, a discussion of precipitation modeling would be incomplete without some discussion related to Global Circulation Model (GCMs) simulations, particularly because changes in simulated precipitation may turn out to be important prognosticators of climate change. *Hirschboeck* pointed out, however, that the resolution and accuracy of GCM estimates of precipitation are quite different than those in which hydrologists are typically interested. For example, whereas a hydrologist may be interested in simulating a storm hydrograph, a GCM modeler may be happy just to obtain a reasonable simulation of the spatial mean of precipitation. In particular, GCMs are as yet unable to satisfactorily represent the distribution of precipitation within a computational grid.

Therefore, given that these are two totally different emphases, the question that naturally arose was whether the random and chaotic behavior of precipitation at small spatial scales precludes the possibility of using it to detect global change. Small-scale variability represents local effects, but we are interested in understanding the large-scale circulation and dynamics aspects of such variability. Most spatial and temporal correlation functions exhibit an increase in coherence as space and time scales are increased. By examining the spatial and temporal characteristics and the spatial and temporal correlation functions, we may be able to estimate at what scale we have averaged sufficiently to isolate the large-scale average from the local variability. For example, two stations separated by a distance of 10 km may have the same annual mean but little similarity in precipitation pattern at the daily time scale. As we average up to 10 days, we may begin to see some correlation, and as we average up to a month the means may be essentially the same.

Clearly, the compilation of a "consistent global product" is necessary for validation or process modeling of global environmental change. Because global precipitation climatologies are often based on a variety of data sources and biases, data qualities and availabilities, and estimation and interpolation techniques, a current challenge is for researchers to put all these data together over the globe or for large regions of the globe. *Becker* suggested that we may have approached the point where it is possible to compile sufficiently detailed, more-or-less accurate data that can be used to validate some class of land-surface hydrologic GCM model components. If we can further succeed in adding real-time data to this data base, we can proceed with a rigorous model validation. This is a challenge that should be taken up by the hydrologic community, initially using existing data to validate at least one method. *Becker* also suggested that we should attempt to collect data related to extreme events in certain areas for which climate impact studies have been conducted. By analyzing, for example, a 100-year record, one could identify sub-periods within that longer simulation for which detailed model testing could be performed.

11. What is the usefulness of proxy data?

Sorooshian pointed out that an area that is in desperate need of better data is the study of natural environmental variability at the global scale. While the conventional instrumental data records are available for only about the last 100 years and while they do have various quality related problems, the data are at least in the form of quantifiable numbers available in books and data archives. However, to obtain estimates of precipitation dating back over time periods from centuries to thousands of years in the past, researchers are attempting to develop various forms of proxy data. Such data are being used as inputs to models to investigate what kinds of natural variability have occurred in the past. *Sorooshian* wanted to know what is the state-of-the-art related to acquisition of proxy data and also what degree of confidence should be placed in those data.

Hirschboeck responded that in regard to tree-ring data, the quality of a precipitation estimate or a temperature estimate based on the width or density of tree rings is dependent on the length of the calibration period. For quality control, a split sample method is used where the model is calibrated with a certain amount of data and then checked against the remainder to see if the calibration is correct. While attempts are being made to study climate variability over several thousand years, it should be noted that the reliability of proxy records depends heavily on the longest gage record available for modeling and calibration; hence, it is important that we keep our instrument gages recording for as long as possible. Furthermore, it is important that published proxy data be accompanied by estimates of their accuracy and reliability.

12. Challenges for the hydrological community

The overall message of the discussion was that the products being generated by meteorologists and climatologists are not necessarily what hydrologists and modelers actually need. Therefore, it is extremely important that a better dialogue be established between those who collect and process the data and those who use and model based on the data. *Legates* and *Goodrich* summarized the conclusions and recommendations of the panel as follows:

(a) Gage measurements of precipitation should be continued, with particular attention on their accuracy. These long-term measurements are critical for climate studies. Metadata documenting significant changes in the observation and measurement procedure and the station environment must be made available. This will help to minimize the degradation in data quality due to non-stationarities in the bias caused by environmental changes such as urbanization.

(b) Accuracy and reliability statistics are vital to proper use of the precipitation data. Procedures to interpolate point and area estimates must be improved to include topographic and other available information.

(c) Standard quality checks, such as double mass analysis, are ineffective at identifying discontinuities in the existing precipitation record when national meteorological organizations institute system-wide changes in methodology. Thus, a more comprehensive and objective approach to the estimation and removal of biases and discontinuities in precipitation time series must be developed.

(d) Metadata of radar and satellite estimates should indicate data quality, processing history, and provide a statistical summary. While such techniques represent a significant new source of information, it is critical that ground observation networks continue to collect gage estimates of precipitation.

(e) Multi-source data assimilation and the analysis of long-term climate variability using proxy records must be encouraged.

(f) Precipitation intensity is a critical hydrologic variable at smaller spatial scales, and a greater effort should be made to measure and derive this field using both new and existing data sources.

(g) While the existing precipitation network is very useful for many climate studies, the global distribution of gages is quite inadequate. Additional data collection efforts should be strongly encouraged and supported, particularly in data-sparse regions such as Africa, South America, and parts of Asia.

The Question of Priorities in Rainfall Information and Data: A Minority Viewpoint ?

Sharon Nicholson, Professor
Department of Meteorology B-161
The Florida State University
Tallahassee, Florida 32306-3034

Collection and analysis of station and gage histories and corrections for point rainfall measurements are of low priority for global change studies. Instead, priority must be placed on evaluating data quality, data archiving and dissemination, and on expanding the network of gages in the vast regions of the globe where rainfall data are scarce or non-existent. This includes the high latitudes, high-altitude locations, and arid- and semi-arid regions, all of which are likely to be extremely sensitive to global change. In such regions, if a choice must be made between collecting information on measurement techniques, gages, etc., and expanding the network of rainfall gages, the emphasis must by all means be on the latter.

While factors such as exposure, gage type, etc. certainly affect measurements, it is unlikely that they will systematically affect studies related to large-scale global change. They are of interest primarily in a local context. The exception is a major change of measurement procedures by an agency in charge of measurement for a large region (on scales greater than, for example, 10^3 km). Such a change (e.g., the Canadian Weather Service selecting a new gage for standard use) will introduce a systematic error that is spatially pervasive. This can be easily accounted for in subsequent analyses of the data.

Other changes at individual stations (i.e., height or location of a gage) will be random and cancel out when areal averages are made. Although an individual record will be affected, this is not of great concern, for several reasons. Analyses of individual records should be avoided for rainfall studies because of the tremendous spatial variability of rainfall and because an individual record will reflect many local "microclimatic" effects. Moreover, discontinuities at individual stations become apparent within the course of analysis and from data quality checks.

NATO ASI Series, Vol. I 46
Land Surface Processes in Hydrology
Trials and Tribulations of Modeling and Measuring
Edited by S. Sorooshian, H. V. Gupta, and S. C. Rodda

The quality of long-term records becomes clear enough through careful analysis. If a rainfall fluctuation exhibited by a data set is real, several features of it will be apparent. It will show a certain spatial coherence and will probably exhibit a spatial configuration representing known patterns of teleconnections. It will probably occur at times in which quasi-global anomalies are evident. It will also be reflected in proxy records such as lakes, streamflow and tree rings. Rainfall fluctuations during past centuries have been on the order of ±10 to 30% of the long-term mean. These will show up despite problems associated with measurement gages and station histories.

The Challenges We Face: Panel Discussion on Snow

Edited by Hoshin Gupta and Soroosh Sorooshian

1. Introduction

This panel discussion was moderated in two phases; the first half was led by *Barry Goodison* and the second half by *Roni Avissar*. The following is a summary of the major points of discussion, based primarily on notes prepared by participants *Mark Lynch-Steiglitz, Remigio Galarraga-Sanchez,* and *Jim Washburne*. Their assistance in documenting the discussions is greatly appreciated.

2. How do we specify snow?

Goodison noted that there are two different measures of snow, namely "snowfall" and "snow cover", and that data on both these variables are needed, not just snow cover. With respect to the annual water balance, winter precipitation (snowfall) is required. However, for the validation of General Circulation Models (GCMs), both variables are needed. Furthermore, GCMs are driven by snowfall accumulation. Note that the goals of the Global Energy Water Experiment (GEWEX) cannot be achieved with snow cover alone. The discussion highlighted the fact that the sources and associated measurement difficulties of these data are completely different.

3. Snowfall measurement errors can be very large

Goodison also discussed the errors associated with the myriad kinds of gages that are used around the world by different countries. Errors caused by wind account for most of the total error, with the error being proportional to the wind speed. For some of the gages currently in use, the actual gage catch can be as low as 30-40% of truth; therefore, a correction factor of

NATO ASI Series, Vol. I 46
Land Surface Processes in Hydrology
Trials and Tribulations of Modeling and Measuring
Edited by S. Sorooshian, H. V. Gupta, and S. C. Rodda

approximately 3 may be required. If, as is suspected, it is true that the biggest changes in climate will be experienced in the northern latitudes, then the gages that are used to record this change must be accurately corrected (with an appropriate correction factor) and calibrated to account for the "missing" snowfall. *Rodda* pointed out that there are also difficulties in getting accurate measurements of wind velocity (which is used to estimate the correction factor) due to snow falling on the anometer. *Goodison* noted also that the anometer measures wind velocity at a height that is usually quite a bit above the gage; hence, a further correction must be made to account for this. Another correction that must be made to snow cover data is the effect of vegetation, which causes patchiness and affects the manner of snow accumulation.

4. Should the ground network be degraded in favor of remote sensing?

In addition to snow gages, a major source of information about snow is remotely sensed images. The information that remote sensing may be able to provide includes snow cover extent, snow water equivalent (SWE), depth, density, albedo, and grain size. Current sensors have the following space/time resolutions:

Microwave:

SSM/I resolution is 25-150 km @ 1/day

ERS-1 resolution is 30 m @ every 35 days

SAR resolution is 30 m @ every 1-30 days (depending on location and other factors)

Visible and thermal infrared (VIS/TIR):

AVHRR resolution is 1 km @ 2/day

TM resolution is 30-120 m @ every 16 days

SPOT resolution is 10-20 m @ every 3-14 days

Except for snow cover, significant research is still required to develop and improve algorithms that extract snow parameters from remotely sensed data to an operational level. *Goodison* asked the participants if they thought that modelers have too much faith in remote sensing products. *Collier* did not think so, but emphasized that our faith in the quality and informativeness of remotely sensed snow data should not be at the expense of the data from the ground network. *Goodison* agreed, and felt strongly that the ground network should not be degraded in favor of remote sensing. For example, in the Arctic, the biggest problem is the lack of any ground data with which to validate remotely sensed data; remotely sensed data are being collected and used without any sense of their quality.

Herschy expressed that part of the remote sensing problem is that the satellites have been configured for disciplines other than hydrology and therefore do not necessarily provide the data that we really need. *Collier* stated that one reason for this is the lack of money in hydrology. However, *Goodison* felt that the current satellite configuration, while not perfect, is useful for hydrology, and the information from these satellites has not been exhausted. *Herschy* stated that the satellite products are not at a resolution that is useful, but *Goodison* felt this was not entirely true. Furthermore, *Goodison* felt that modelers and users need to become better educated and more comfortable with these new products. The general consensus was that the remote sensing data cannot (at present) replace the ground system and, therefore, both kinds of data are needed. *Goodison* stated there need to be good, long-term archives for both remote sensing data and ground based data. This, however, requires a good deal of funding and governmental support to be done properly.

5. How do we model snow and what does this effort need?

Avissar raised the perspective of the modeler and brought up two issues for discussion.

(a) What (data) do we need in order to model snow, and

(b) Can we predict long-term hydrologic and climate variability without modeling snow? In particular what is the sensitivity to the snow component of the hydrologic cycle?

With respect to the second question, several participants noted that there have been studies correlating snow cover to atmospheric circulation (e.g., the teleconnection between the extent of snow cover in Eurasia and the behavior of the Asian monsoon) and that when the insulation properties of snow cover are neglected in GCMs, the results erroneously indicate the persistence of frozen ground layers well into late spring. It seems clear that we must include some consideration of snow in these models. *Goodison* felt that in order to conduct a proper hydrological modeling effort, and because snow is dynamic (here today, gone tomorrow), it is necessary to have daily satellite coverage (e.g., passive microwave). This is especially true because, on days where there is melting, it can appear as if the snow cover is gone. However, for the purpose of climate change analyses, *Rott* indicated that a snow product every week to two weeks is sufficient and that today's large data sets were in good shape for this.

6. Challenges for the hydrological community

In summary, the panel discussion highlighted the following facts:

(a) There needs to be a much greater appreciation of the global significance and difficulty in modeling and measuring cold climate hydrology. In general, snowfall models are too simply parameterized, sublimation is ignored, the occurrence and effects of frozen ground on infiltration and soil moisture storage are not modeled adequately, and the complications due to river ice are typically ignored. It is hoped that the critical significance of high alpine areas to water resources and climate change will be better recognized as GCM grid size resolutions become finer.

(b) Snowfall and snowcover are distinct, but complementary, and important components in cold climate hydrologic models at all scales, but their sources of data and associated measurement challenges are completely different.

(c) Remote sensing cannot and should not be viewed as a substitute that will replace conventional data collection. Conventional data sources are critical for basin process studies and to calibrate and validate regional and global remotely sensed products. Other than for snow cover extent, considerable research is still required to develop and improve algorithms for the processing of remotely sensed data to snow parameters to operational levels.

As important as data collection is, the discussions also indicated that the community needs to create and stand behind a well thought-out rationale to justify the maintenance of long-term financial support. A continuing effort is required to maintain and extend global archives of snow (such as NOAA's Snow and Ice Archive in Boulder, Colorado). Realistic time and funding goals need to be set to avoid duplication and to not promise more than can be delivered. It appears, for example, that a reliable passive microwave algorithm for snow cover extent should be quickly achievable (within two years of the workshop).

Finally, as might be expected, this panel discussion raised far more questions than were answered. Of the questions raised by *Avissar* and *Goodison* the following did not receive adequate attention due to lack of time:

(a) Would it would be more productive to direct our efforts towards the creation of large (global) observation data sets at high resolution, or would it be better to compile some limited but high quality data sets and use them to develop and validate models that simulate data for the world at very high resolution?

(b) Should remote sensing and its associated scale of available data play any role in determining the scale of our models?

Part III - Soil Moisture

PROCESS, HETEROGENEITY AND SCALE IN MODELLING SOIL MOISTURE FLUXES

Keith Beven
Centre for Research on Environmental Systems and Statistics
Institute of Environmental and Biological Sciences
Lancaster University, Lancaster LA1 4YQ, UK.

Introduction

We know that hydrological processes are highly complex. We know that the techniques available to assess the complexity of hydrological processes and parameters are inadequate, even for small scale field studies. The observability problem is particularly acute for subsurface flow processes where most measurement techniques can give only a very local indication of the hydrological response. In a series of recent papers I have highlighted some of the problems of subsurface flow modelling in the context of predicting hydrological responses at the hillslope, catchment and larger scales ; the interlinked problems of process, of heterogeneity and of scale (Beven 1987,1989a,b, 1991, 1993). There have also been a number of recent papers that have discussed these problems such as Grayson et al. (1992) but, as yet, very little in the way of progress towards resolving those problems. With some exceptions, hydrological modelling of subsurface processes continues to rely on a Darcian or Richards equation framework coupled to a concept of effective parameters with an a priori assumption that such a concept is valid regardless of the scale of application.

Problems of process arise in such a framework because of certain processes that are neglected or unresolved at the grid scale of a distributed model. In particular attention should be drawn again to the effects of preferential flow, especially in unsaturated soils. A Darcian approach assumes that the pressure at a particular point in the flow domain is equilibrated over some "representative elementary volume" (REV) and that the movement of water takes place in response to a pressure gradient at that scale. At the local scale, within the soil matrix, this assumption may be acceptable but it is applied equally at the grid scale in distributed models. In a heterogeneous unsaturated porous matrix, the application of this concept at the grid scale may be dubious because of the highly non-linear

NATO ASI Series, Vol. I 46
Land Surface Processes in Hydrology
Trials and Tribulations of Modeling and Measuring
Edited by S. Sorooshian, H. V. Gupta, and S. C. Rodda

dependence of the flow velocity on the local moisture content and pressure gradient. In soils in which preferential flow takes place, implying local non-equilibration of pressure and consequently pressure gradients that can be defined only at very small scales, such a concept must surely be rejected. Darcian models assume that the effects of preferential flow can be neglected, or , at best, represented by hysteretic soil characteristics. The evidence of field measurements suggests that this is not so (e.g. Beven and Germann, 1982). In fact it is probable that preferential flow takes place at all scales, even within the matrix and capillary sized pores (e.g. Omoti and Wild, 1979, Ganoulis, 1989). However, no adequate descriptions or parameterisations of non-equilibrium preferential flow is available at any scale despite a number of initial attempts (e.g. Hoogmoud and Bouma, 1980; Beven and Germann, 1981; Germann and Beven, 1985; Gerke and van Genuchten, 1993).

This problem of process is exacerbated by the effects of vegetation, not only in the way in which root channels may promote preferential flow, but also in the way in which the pattern of rainfall reaching the canopy in space and time may be modified by the vegetation into a much more heterogeneous distribution at the soil surface. The processes involved are interception, throughfall, drip and stemflow. Both drip and stemflow can lead to locally enhanced effective input intensities to the ground surface. These may, in themselves, help promote the occurrence of preferential flows.

Problems of heterogeneity are caused by the variability in soil characteristics within the flow domain. A distributed hydrological model can, in principle, take some account of soil heterogeneity by allowing the specified soil characteristics to vary in each grid element of the model. Thus it is only necessary, in principle, to resort to the concept of effective parameter values at the grid scale (if the information were available to be able to define effective values for each and every grid element in the flow domain). The primary difficulty is then one of defining the grid element effective values given sub-grid heterogeneity of soil characteristics. Starting with the pioneering work of Nielsen et al. (1973) a great deal has been learned about the nature of soil heterogeneity in the last 20 years for both flow and transport processes. Numerical experiments suggest that grid scale effective values may, at least under certain conditions, provide reasonably acceptable results for the prediction of Darcian flows, both saturated and unsaturated (e.g. Binley et al., 1989), but that the relationship between an apparent effective value and the underlying stochastic distribution may vary with the flow conditions. Fully stochastic models are also available now which can take account of the effects of uncertainty in grid scale values on the predicted variables (e.g. Jensen and Mantoglou, 1992). I am not aware, however of any explicit methods of estimating grid scale effective values given a model of the underlying heterogeneity, except for the essentially linear methods of block integration, such as block kriging (see Beven, 1989). Such linear methods should be expected to be more acceptable for saturated flow domains than for unsaturated flows or hillslopes in which there is an interaction between surface runoff and subsurface flow. The results of Binley et al. (1989) show quite unequivocally that for such coupled flow processes the concept of effective parameters will not be acceptable, despite the fact that it continues to be used in all distributed Darcian catchment models to date.

Problems of scale are clearly closely related to these problems of heterogeneity, since the larger the grid scale used in a model the greater the effects of sub-grid-scale heterogeneity and coupled processes are likely to be. Indeed, it has been quite common for grid scales that are larger than the scale of individual hillslopes or small catchments to be used (e.g. Refsgaard et al.,1992), still with the assumption that effective parameters are valid at the grid scale. An extreme case is that of the hydrological components of General Circulation Models, in which effective parameters are used at the GCM grid scale, which may be of the order of 300km. Early models were conceptual bucket models (e.g. Manabe and Wetherall, 1980) but more recent formulations of Soil-Vegetation-Atmosphere transfer (SVAT) models have made use of quasi-Darcian vertical flow models in the soil component (e.g. Sellars et al., 1986). Such models cannot, of course, be justified by any physical argument or theory for the definition of effective parameter values, it is purely a convenient assumption. It is also clear that, at the catchment scale and above, factors other than pure soil characteristic heterogeneity assume an importance in governing hydrological responses. In particular spatial variability of input rates and the effects of geology and topographic forcing on downslope transfers of subsurface flow and the control of surface and subsurface "variable contributing areas" will become important.

An interesting concept in this respect is that of the "representative elementary area" (REA) for runoff production introduced by Wood et al. (1988, 1990) in analogy with the REV of soil physics. The REA is meant to represent the scale at which the variance in runoff production, given heterogeneity in soil characteristics, topography and rainfall distributions (and perhaps other variables), between areas of a similar scale is minimised. In numerical experiments using a simplified distributed model and assumed heterogeneity of rainfall, soil and hillslope form Wood et al. (1988) suggested that the scale of the REA may be of the order of 1 km^2. It is important to note here, however, the difference between the REV and REA concepts. The REV is similarly meant to represent a scale at which microscopic variability can be integrated into representative quantities, but carries with it an assumption that at that scale macroscopic variables and parameter values can be defined. The REA is a more nonlinear concept in that it represents the scale at which the pattern of variability may no longer be important, but it may still be necessary to take the statistical nature of the underlying variability at some (as yet unspecified) scale of sampling into account. Wood et al. (1988) suggest that at scales greater than the REA it may be possible to represent the macroscopic response with sufficient accuracy using distribution functions to represent the underlying variability. They do not suggest that the response can necessarily be represented by effective parameter values at that scale.

In practical terms the difficulty of defining appropriate large scale descriptions of hydrological processes lies not only in the nonlinearity of the unsaturated flow and coupled surface/subsurface flow processes but also in the mismatch between the scale of measurement techniques and the scale at which hydrological predictions are required. Discharges can be measured relatively easily at the "catchment" scale and provide a useful integrated measure of hydrological response at that scale. Nearly all other measurement techniques, from observation wells and tensiometers to micrometeorological measurements of evapotranspiration provide measurements at the "point" scale, thereby highlighting the underlying variability of hydrological processes, without the detail that would be required to develop a theory of spatial integration. The lack of measurement techniques appropriate to a given scale of prediction has undoubtedly hindered the development of larger scale hydrological theory. If it were possible, for example to measure patterns of water storage or dynamic contributing areas at the catchment scale, in the same way that radar has allowed the assessment of large scale rainfall patterns, then "physically-based" hydrological models might look very different. Lacking such techniques, however, has resulted in the simplistic and totally unjustified use of small scale physical theory at the grid and catchment scale. It is possible that in the future remote sensing techniques might help in the assessment of internal state variables at larger scales, either for surface soil moisture (see Schmugge, this volume) or for saturated contributing areas (see Brun et al., 1990), while tracer techniques could provide an assessment of patterns of lateral inflows to stream channels (see for example Huff et al., 1982; Genereux et al., 1993). Experimental techniques for assessing the water table configuration or storage in the saturated zone seem remote but would be very useful indeed in formulating a catchment or larger scale model.

Integration of process, heterogeneity and scale in a distribution function model of hillslope response.

The problems involved can be illustrated by a thought experiment. The starting point for this thought experiment is an extension of the linear distribution function model of hillslope processes presented in Beven (1989b), augmented here to take account of some other sources of variability. In essence this experiment is a particular representation of one person's perceptual model of hillslope response (see Beven, 1989b). In its formulation it is also an extension of the catchment scale theory of the geomorphological instantaneous unit hydrograph (GIUH, see for example Rodriguez-Iturbe, 1993) in its form derived from the residence time distribution of an instantaneous input of raindrops falling on a catchment. The GIUH included an assumed temporal distribution function for the hillslope response, followed by distribution functions for the residence time in channel reaches of different orders. The emphasis was on the role of the channel network on the form of the hydrograph. However, as noted in Beven (1986), this is a relatively simple hydrological problem, a far greater problem is being able to predict how much of the incoming rainfall to route to form the discharge hydrograph. This problem is particularly difficult in relatively dry climates, or in a wetting up sequence of storms following a prolonged dry spell. The experiment here concentrates on the distribution functions controlling the hillslope response and how they might be characterised.

Figure 1 shows a schematic representation of the distribution functions controlling the response of a hillslope or collection of hillslopes in a small catchment to an isolated unit input of rainfall in a single time step of unit duration. The distribution functions are labelled as types E, P, D or T. Distributions of type E are external forcing functions that must be specified. Distributions of type P are parametric functions that must also be specified. Distributions of type D and T are derived from distributions specified earlier. Distributions of type E, P and D may be considered, to a first approximation to be distributions representing spatial heterogeneity; distributions of type T are distributions in time or transfer functions.

The only external distribution considered here, E1, is the spatial distribution of the input rainfall rates. This would of course vary with the nature of a rainstorm and the scale of the collection of hillslopes being considered. Models for the prediction of spatial rainfall distributions are available while the nature of such distributions can be determined from radar measurements at least at the scale of resolution of current radar systems. It will be assumed for the moment that the effects of other external forcing variables (temperature, humidity, solar radiation, wind, antecedent rainfall) can be subsumed into parameterisations of the initial state of the catchment prior to the time step in which this input occurs.

The first parametric distribution, P1, is that associated with the effect, I, on interception, throughfall and stemflow of the vegetation. This affects both the amount of the rainfall input reaching the soil surface and its effective intensity. This results in the first derived distribution, D1, the spatial distribution of input rates at the soil surface, g. To a first

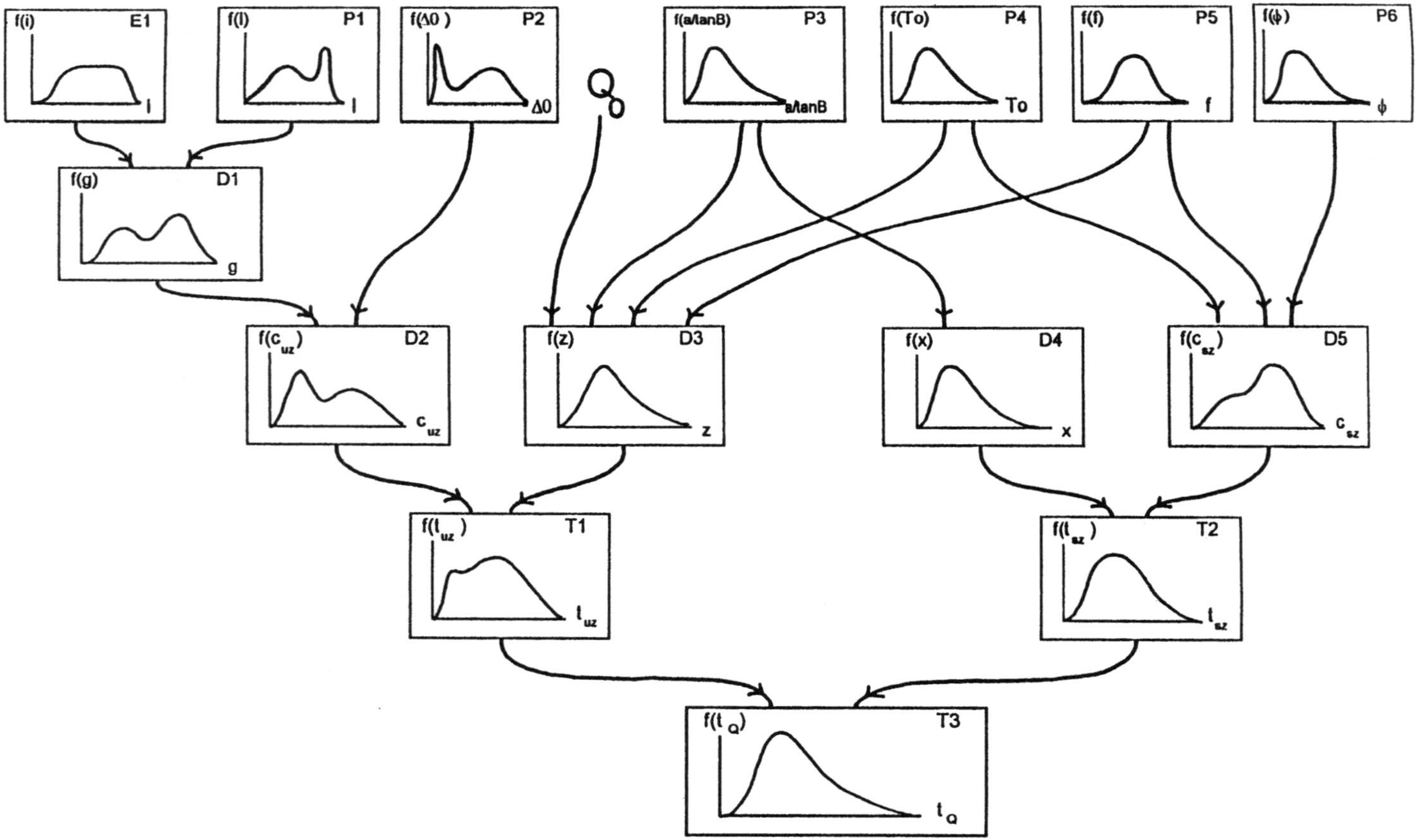

Figure 1: A Distribution Function Model of Hillslope Responses to a Unit Input (see text for details of variables)

approximation we assume that D1 is produced from E1 by a process of convolution with the distribution P1, noting that because of interception storage P1 will have a gain (or integral value) less than unity (the nature of this assumption will be reconsidered below). D1 may already have a complex multimodal form. For certain vegetation canopies, stemflow may result in 10 to 15 % of the incoming rainfall being concentrated over a relatively small proportion of the soil surface resulting in high effective intensities. However, the major part of the rainfall, as a consequence of both interception and diversion to stemflow, may be effectively reduced in intensity in comparison with the incident rainfall rates.

The distribution D1 then provides the driving force for infiltration at the soil surface and the propagation of wetting into the soil. As already noted above, the wetting of an unsaturated soil is a complex process involving matrix infiltration in soils of variable conductivity characteristics, preferential flow in pathways of complex geometry and connectivity, and perhaps some areas of saturated soil, surface runoff production and run-on. In predicting the storm response of the collection of hillslopes we will assume here that flow in the unsaturated zone is predominantly vertical and that we will be most interested in that part of the input that does not infiltrate and becomes surface runoff, and that part of the infiltration that reaches the water table causing a downslope response. In the latter case, we are consequently primarily interested, not in the flow velocity through the unsaturated zone but in the wave velocity with which disturbances to the soil moisture profile can propagate downwards to the water table. This will then allow for the fact that the input may initiate a displacement of already stored water further down the profile. The wave speeds will depend both on the magnitude of the inputs and on the antecedent moisture status of the soil. In the simplest kinematic wave analogy for wetting front propagation, the wetting front is treated as essentially a shock front and propagates as $i/\Delta\theta$ where i is the input intensity and $\Delta\theta$ is the moisture content change across the wetting front. Taking D1 as the distribution of input intensities and specifying (for the moment) P2 as a distribution of moisture content changes, a second derived distribution D2 representing the distribution of wave speeds through the unsaturated zone, c_{uz} can be derived. There will also be some loss associated with each of these multiple wetting front flow pathways, representing storage retained as the unsaturated zone wets up. This part of the infiltrated water will consequently have very low or zero vertical flow velocities. Other parts of the input associated with either high effective input intensities (stemflow) or small $\Delta\theta$ values (macropores or near saturated soil and the capillary fringe) may have very fast wetting front velocities. Again the occurrence of flow in preferential pathways relative to the slower flows through the soil matrix may lead to a bimodal or multimodal distribution.

The distribution D2 may also be interpreted in terms of an effective time of travel distribution for unsaturated flow per unit of depth, although it is not necessary to assume that all the moving water is event water, some displacement may also be occurring. To derive the time distribution of inputs to the saturated zone it will then also be necessary to have an idea of the distribution of depths to the water table. This could either be treated as an externally specified distribution or, under an assumption of quasi-steady state flow, it may be possible to derive this distribution, D3, from a knowledge of an antecedent flow, soil transmissivities at different depths and the topographic form of the hillslopes. One formulation for such a prediction is provided by the TOPMODEL theory outlined in Beven (1986) and adapted for water table depths rather than storage deficits in Sivapalan et al. (1987, 1990); another, based on Dupuit-Forchheimer theory, is given by Troch et al. (1993).

Following the TOPMODEL framework, for a catchment average steady input rate to the saturated zone of r, the drainage through any point of the catchment is given by $q = a\, r$ where a is the upslope area draining through that point per unit contour length. If it can be assumed that water table lies within the soil and the soil is relatively thin compared with the hillslope length then the effective hydraulic gradient at that point can be approximated by the slope of the soil surface, tanB. Thus the flow through that point can also be described by $q = T(z) \tan B$ where T is the downslope lateral soil transmissivity which will be a function of the depth to the water table, z. Let the effects of soil heterogeneity be simplified so that the soil transmissivity can be specified as a simple function of the transmissivity at the soil surface T_0 and depth to the water table, allowing that the T_0 values may vary in space so that $q = f(T_0,z)\tan B$. At the scale of the REA and above it may be possible to treat the soil and topographic variables as parameterised distribution functions as used, for example, by Beven (1986) and in the macroscale model of Sivapalan et al. (1987) and Wood et al. (1990) Thus:

$$q = ar = f(T_0,z) \tan B$$

Taking one particular function $f(T_0,z)$ as $T = T_0 \exp\{-fz\}$ (other forms are equally possible) then

$$z = -\ln(ar/T_0 \tan B)/f \qquad (1)$$

where the a, T_0 and tanB values are all local values in the collection of hillslopes being considered.

If the equivalent antecedent catchment discharge Q_0 is assumed to derive from a steady inflow to the saturated zone over the area of the hillslopes then $Q_0 = r\,A$ or $q = Q_0\,a / A$ everywhere. Thus substituting into (1)

$$z = -\ln(aQ_0/AT_0\tan B)/f$$

$$= -\{\ln(a/\tan B) + \ln(Q_0/A) - \ln(T_0)\}/f$$

from which it can be seen that a distribution function of local values of depth to the water table can be inferred from a measure of antecedent wetness (Q_0/A), a distribution of a topographic index (a/tanB) and distributions of soil characteristics, in this case T_0 and f. Assuming independence of the soil and topographic distributions for simplicity, the derived distribution of water table depths, D3, can be calculated from the parameterised distributions of a/tanB, (P3), T_0 (P4), and f (P5). Of these distributions that of (a/tanB) is easily calculated from a digital terrain map (e.g. Quinn et al., 1991, see figure 2A) and has sometimes been parameterised as a gamma distribution (e.g. Sivapalan et al., 1987). The distribution of the soil parameters may be more difficult to characterise since downslope transmissivity values are not very often measured. This is, however, a very important parameter in the hillslope response. The expectation that there may be lateral, as well as vertical preferential flow pathways in natural pipes or zones of higher hydraulic conductivity (see discussions in Beven and Germann, 1982, McDonnell, 1990) suggest that these parameters may exhibit considerable spatial heterogeneity.

Given the distributions D2 and D3, a distribution can be derived for the time of arrival of flow through the unsaturated zone at the water table, t_{uz}. Because D2 is a distribution of velocities the transformation to a time variable results in a transfer function, T1, representing the response at the water table of the remaining part of the original input variable, i. This input to the saturated zone will cause a response at the base of the hillslopes in a way that is controlled by the wave speeds through the saturated zone. Again using a simple kinematic approach, and assuming that the effective hydraulic gradient can be approximated by the slope of the soil surface, this wave speed will be given by $K_s\tan B/\phi$ where K_s is an effective lateral hydraulic conductivity at the water table given by dT/dz and ϕ is an effective storage coefficient. A combination of the distribution of transmissivity and a parameterised distribution of effective storage coefficients, P6, will yield a derived distribution of wave speed travel times in the saturated zone, c_{sz}, distribution D5.

Since ϕ is always less than one, the wave speed will always be faster than the equivalent Darcian flow velocity. Where ϕ is very small, for example where the water table is rising

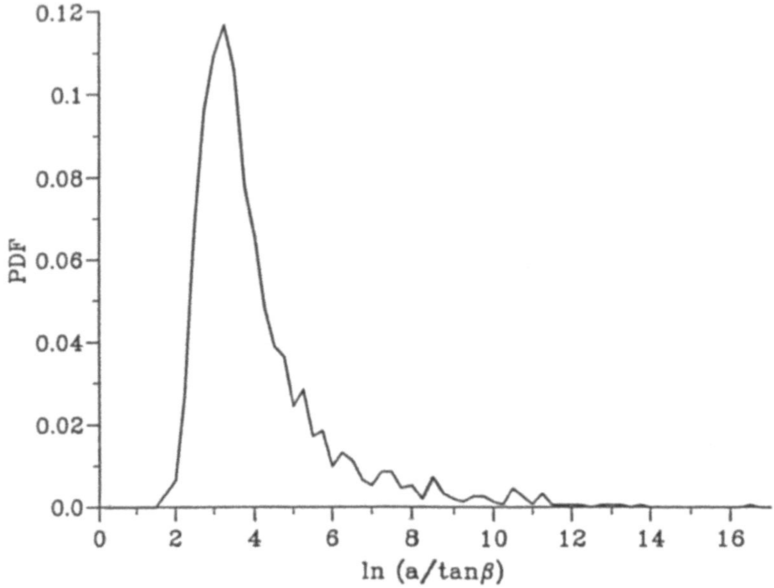

Fig 2A. Distribution Function of ln(a/tanβ) Index Derived from Topography of Maimai Catchment

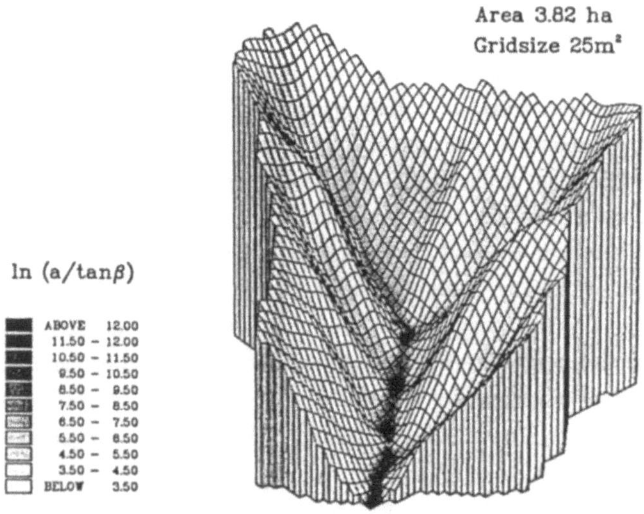

Fig 2B. ln(a/tanβ) Index Overlain onto Elevations of Maimai Catchment

into a capillary fringe, the wave speed will be much faster than the effective downslope flow velocity. The timing of outflows from the base of the hillslopes will depend on the pattern of inputs to the saturated zone and the way in which their effects are propagated downslope by the wave speeds. This is one explanation for the displacement mechanism inferred from the commonly observed high proportion of old water in hydrographs derived from hydrograph separations using environmental isotopes (see for example Sklash, 1992). We are interested in the way in which inputs to the saturated zone are "processed" to become outputs at the base of the hillslopes. This will depend on the complete configuration of the water table, the profile of moisture storage above the water table (which will determine ϕ) and the temporal and spatial pattern of inputs to the water table. Assuming for the moment independence of the soil and topographic distribution functions, the derived distribution of water table depths, D3, also defines, on average, a distribution of travel distances, x (distribution D4), in the saturated zone given the plan of the associated a/tanB values on the hillslopes (e.g. figure 2B). A convolution of the distributions of saturated zone travel distances, D4 and effective wave speeds, D5, will lead to a derived distribution for the effective time of travel in the saturated zone. This is the transfer function, T2, for inputs to the saturated zone to reach the channel.

A final convolution of the transfer function distribution in the unsaturated zone, T1, and transfer function the saturated zone, T2, will create the time distribution of outflows from the hillslopes, T3, resulting from the initial pulse of rainfall input. D8 represents an effective unit hydrograph, in that it is a transfer function from a unit of rainfall input in a unit time step, although we note that, being derived from the total rainfall input, it will normally be associated with a gain considerably less than one. Such a model could then feed into a routing algorithm of similar complexity such as the transfer functions derived from GIUH theory (Rodriguez-Iturbe, 1993) or, preferably, from the width function of the actual channel network in the catchment (see for example Kirkby, 1976a; Beven, 1979; Beven and Wood, 1993).

"True" parameters, internal state variables, and antecedent conditions.

The transfer function model described above has most of the elements of my perceptual model of hillslope response to a single isolated input pulse, albeit in a simplified linear form. It is clear that the parameterised distributions P1 to P6 can be divided into three classes: those that might be called "true" parameters and those where it has been convenient to parameterise distributions of internal state variables. In the first category are included the distributions of vegetation parameters, topographic parameters (a/tanB), and soil parameters (T_0, f); in the second the soil moisture change parameters ($\Delta\theta,\phi$). There is also the specified flow rate Q_0 which provides a measure of the antecedent state of the catchment. These distributions of internal state variables will undoubtedly change in a nonlinear manner with time, with the antecedent state of the catchment and also with the magnitude of the input pulse. They have been used here as distributions of effective values. With an appropriate choice of the distributions, this description is, perhaps, better than many lumped conceptual storage representations, such as the simple exponential transfer functions used in the GIUH (Rodriguez-Iturbe, 1993). However, a general model would need to be able to describe the way in which such effective values, or their equivalent in terms of soil moisture profiles, vary dynamically as the catchment wets and dries. The description above therefore fails to provide one of the primary functions of the so-called physically-based distributed hydrological models in that it cannot predict the dynamic and continuous changes in storage that affect the actual "transfer function" of any particular sequence of inputs. The point of presenting this thought experiment has been to point out that the distribution function description contains elements that are equally lacking, but might also be important, in current physically-based models.

In fact, this distinction between constant "true" parameters and dynamic state variable distributions is a distinction more of degree than of type since all the parameters should be considered as dynamic variables. The vegetation distribution, for example, would vary with the state of the interception store and the intensity of the input as well as the physical characteristics of the vegetation canopy itself. The topographic index at a point might also be dynamic in that the effective upslope drainage area, a, might not extend to the divide as assumed by DTM analysis but may rather vary with antecedent conditions. There is certainly evidence that even in humid temperate areas subsurface responses depend upon the connectivity of saturated zones on the hillslopes (e.g. Burt and Butcher, 1985a,b). In climatic regimes subject to strong seasonal wetting and drying some parts of the hillslopes may only rarely contribute recharge to a saturated zone providing lateral flow towards a stream channel, resulting in markedly reduced subsurface effective contributing areas. Finally the soil transmissivity parameters may vary seasonally and over longer periods of time.

Hydrologists have for too long thought in terms of trying to describe catchments by means of constant parameter values, even though in many catchments the effects of natural and anthropogenic processes result in continual change. The same process continues today at the scale of the GCM grid square. It might be ultimately much more productive to take the more radical view that there are no constants in hydrology (save gravitational acceleration) and to develop techniques that cope with change. There may be another marginal benefit of such an approach: it might encourage the idea that man can have a beneficial effect through managed change, both at the catchment and global level.

Indeterminacy and Uncertainty

Consider now the implications of the conceptual distribution function model outlined above. Despite its assumptions of linearity, crude approximations to antecedent conditions, and limitation to an isolated input event, it has some important characteristics. In particular it is not a reductionist model but aims to provide a description appropriate to the scale at which predictions are required. It also does so by explicitly recognising the heterogeneity in both process representations and effective catchment parameters that are characteristic of the real system, using a "minimal" parameterisation. However, even so , this very simplified "physically-based" model of the hillslope response under one specific set of antecedent conditions will require specification of one external distribution function, the antecedent discharge, Q_0, and 6, perhaps complex, parametric distribution functions. Since the minimum number of parameters to describe a distribution function is one (for example the exponential or uniform functions) and unrealistically ignoring any covariation between the distributions, a minimum of 7 parameters must be specified in this "model". It is clear, however, that many of these distributions could not be represented adequately by an exponential or uniform distributions and at least two parameters would be required to characterise the modal behaviour of each distribution, and in some cases their covariation, a total of perhaps 15 or more parameters. This is a large number in terms of parameter identifiability but not so different from many conceptual catchment models and much less than any physically-based distributed model.

It is clear that a continuous simulation model would require a more complex parameterisation to determine the distributions of $\Delta\theta$, and ϕ which depend on the spatial and temporal evolution of soil moisture profiles. It is not, at the moment clear how this would be done in a non-reductionist way but certainly it is likely that such a description would involve more rather than less parameters than the isolated event descriptions used here. Yet, normally, the only measurements that would be available from which to estimate the distribution parameter values involved in this highly parsimonious process description would be the discharge hydrograph. It has been pointed out many times (e.g. Kirkby, 1976b; Beven, 1989) that rainfall-runoff data alone may have information to specify perhaps 4 or so parameters: a gain function, two timing parameters and some antecedent condition parameter for continuous simulations. It may also be possible to approximate the topographic driving forces, at least for shallow systems, from an analysis of a topographic map or digital terrain model (here used to characterise the distribution of a/tanB, P3). This is still well short of the information content that would be required to characterise all the distributions. One implication of this is that any reasonably realistic description of hydrological processes must be essentially indeterminate unless some additional sources of parametric information can be provided. Any predictions based upon such a poorly identified model would consequently be inherently uncertain, even if the model structure were adequate.

There are a number of possible responses to this situation. One is to simplify the model to be used, either by finding alternative model structures or by carrying out a sensitivity analysis to determine which parameters may be fixed a priori and which should demand most attention in the calibration process (see for example Hornberger et al., 1985). Such simplification will, however, also tend to reduce the accuracy with which the variables of interest can be simulated and also result in uncertain predictions. It has been argued elsewhere (see particularly Beven 1989a,b) that hydrological models should properly be posed in a framework in which a rigorous and realistic estimate of predictive uncertainty arises naturally. Such a framework should allow for the fact that due to parameter interactions there may be no "optimal" parameter set (see discussion in Beven, 1993), but in which different model structures can be compared in a way similar to that already well developed in linear systems theory. One such methodology, the Monte Carlo simulation based Generalised Likelihood Uncertainty Estimation (GLUE) technique, has been described by Beven and Binley (1982). Some example results, derived from the application of one particular model, TOPMODEL (Beven and Kirkby, 1979) to the Maimai catchment in New Zealand, are shown in figure 3.

A second strategy to cope with indeterminacy is clearly to introduce more information into the modelling processes by obtaining better estimates of the distributions required. It is not, however, at all clear what sort of measurements would be of most use to the model calibration process. Since, for a distributed model, there are no techniques available to estimate grid-scale effective parameter values directly, and little or no theory for deriving such values from knowledge of parameters at smaller scales it will be necessary to rely on indirect means of deriving parameter distributions from measurements. For a spatially

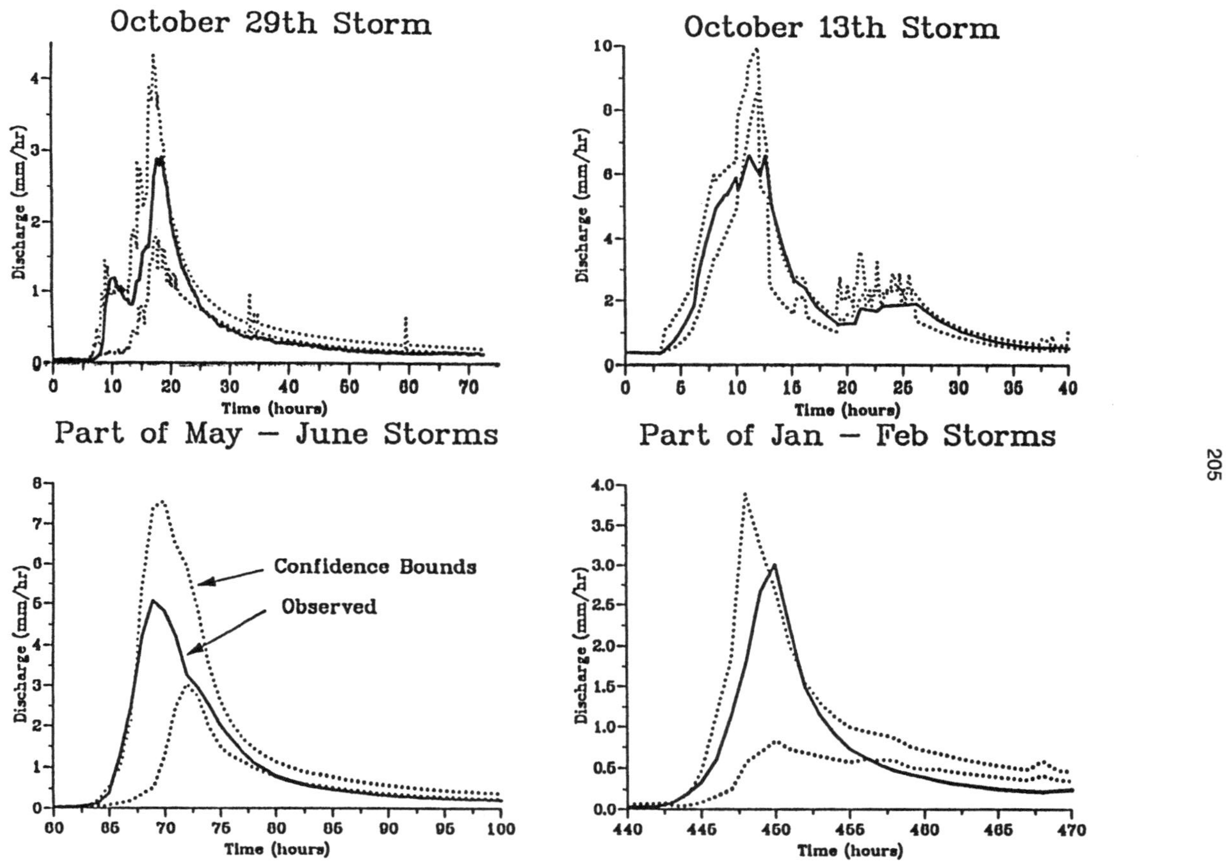

Fig 3. Discharge Predictions for 4 Storms in the Maimai Catchment with 90 % Confidence Limits

distributed model, even one that represents heterogeneity in terms of distribution functions rather than some elemental grid, some spatial information would undoubtedly be useful in parameter identification. There are, however, very few sets of measurements, or even measurement techniques, that have attempted to assess the spatial patterns or distributions of hydrological parameters or variables at the catchment scale. This will be discussed further in the following section in which the role of data in constraining predictive uncertainty will be considered.

Constraining uncertainty

The GLUE methodology starts from the premise that we are really so unsure of hydrological theory at the catchment scale that all model structures and parameter sets within a given model structure are, a priori, equally likely simulators of the catchment of interest. If we then take random choices of parameter values across some reasonable range within an available model structure the results can be compared with the available measurements and some measure of possibility or likelihood given to that particular realisation. Since it is more than likely that there will many simulations that achieve some acceptable level of likelihood ("behavioural" simulations in the sense of Hornberger and Spear, 1981), this methodology rejects the concept of an "optimum" simulation or set of parameter values in favour of retaining many simulations with their associated likelihoods (see Beven, 1993). These likelihoods can then be used in different ways to provide a measure of predictive uncertainty.

Intrinsic to the methodology is the Bayesian idea that the prior (usually non-informative) likelihoods are conditioned on the available evidence. Such evidence may not always be quantitative measurements but may also be the result of qualitative assessments by the modeller (for example if Hortonian overland flow is predicted in a catchment where it would not be expected). It is also possible to combine different likelihood weights associated with different bodies of evidence (such as the predictions of both discharges and water table levels) if the measurements are available. Refinement of the likelihood measures and uncertainty estimates as new data become available is easily implemented in this Bayesian framework.

One result of approaching model structure assessment and calibration in this way is that attention is focused on the availability and <u>value</u> of data. Beven and Binley (1992) have shown how uncertainty measures can be used to assess the value of adding new rainfall and discharge data to the model calibration process. The same procedure could be used to assess the value of adding different types of data. For example, Binley and Beven (1991) examined the effect of having water table information available at a small number of locations in the calibration of a physically-based rainfall-runoff model (the Institute of Hydrology Distributed Model, IHDM, of Beven et al., 1987) to data from a hypothetical 3-dimensional Darcian simulation of a small headwater catchment. They concluded that the water table information was of very little use, because the internal state simulations were dominated by the effects of the reduced dimensionality of the IHDM. In effect, although a Darcian physically-based quasi-3-dimensional model was attempting to reproduce the behaviour of a Darcian fully-3-dimensional model, the simpler model structure was an inadequate representation if internal state predictions were required. It might not be unreasonable to extend this conclusion to applications of such models to real catchments.

But, that is just what is required in many situations to which "physically-based" models are being applied. So it is worth considering what types of internal measurements might have value to the modelling process. Referring back to the conceptual distribution function model, the data required are the distributions (or even better patterns) of input intensities at the ground surface, lateral transmissivities of the soil (and even better the way they change with depth to the water table), soil water content profiles (including the possibilities for any preferential flows), the pattern of water table depths and extent of surface and subsurface contributing areas. This list can be compared with the types of hydrologically relevant data that might be available: a classification of land use; a digital elevation map (and indices derived from it); point measures of infiltration capacities; point measurements of hydraulic conductivity (biased towards surface layers); point profiles of soil moisture; point measurements of water table depth; remotely sensed measures of near surface soil moisture (at least in ideal conditions); resistivity and geophysical transects; and possible some maps of surface saturation (e.g. Dunne et al., 1975; Beven, 1978; Ambroise, 1988). Different types of measurements may have different utility for different purposes of carrying out a simulation study, but it should be noted that this list does not include many of the requirements of the model outlined here. This suggests, therefore, that there may be a considerable degree of residual uncertainty associated with both catchment scale and internal state predictions.

In terms of predicting the broad spatial heterogeneity in catchment responses, what type of data would be of greatest value? There is some evidence that even intensive spatial measurements of point-scale parameter values or moisture content values may be of limited value in catchment scale predictions using distributed models (Loague, 1990, 1992). In some circumstances, where vegetation influences are small and surface runoff processes dominate, remotely sensed surface soil moisture measurements may be useful in refining model prediction. For areas dominated by subsurface flow it would appear that current measurement techniques may have little to offer. What is needed, perhaps, is some way of sensing the pattern of depths to water table on the hillslopes. This would be both directly useful (such as to define distribution D3 in the model presented here) but also will reflect the broad pattern of water balance components on the hillslope in terms of both water reaching the saturated zone and patterns of lateral flow on the hillslopes. Because of the great effort required to do this by point measurements even on very small areas such data sets are very rare. Some techniques, such as ground probing radar transects, are being tried to measure patterns of moisture storage and may be useful in the future. It may also be that water table measurements, or surface contributing area measurements at certain critical locations may provide the sort of discrimination between models and/or parameter sets necessary to constrain predictive uncertainty. The proper use of different types of data must be the subject of continuing research in hydrology; the continuing and unthinking assumption that effective values can be used in point-scale equations at grid scales an order of magnitude larger is just not good enough.

Conclusions

This exploration of a particular conceptual framework based on distribution functions to describe the heterogeneity of catchment responses has been intended to highlight some of the issues that remain in representing hydrological processes in a realistic way; in problems of model definition at different scales; in parameter identification and estimation of predictive uncertainty; and in the role of field observations in constraining predictive uncertainty. This framework has the advantage of being able to represent some of the complexities of the real response, in a very generalised way. It has the disadvantage that, even in the restricted form for the quasi-linear response of an isolated rainfall input presented here, it is virtually unusable in that the distribution functions involved could not easily be quantified. In addition issues such as covariation between distribution functions and the scale of sampling that they are intended to represent have been neglected here. It is hoped, however, that the reader will draw an analogy with the "physically-based" representations that are being used for quantitative predictions and recognise equal problems of identifiability and process representation.

The distribution function model has also been used here to illustrate the relationship between flow velocities and wave speeds in the effective transfer function between inputs and outputs. The resulting transfer function should not be interpreted as a residence time distribution of water in the system, since it is possible that much of the output will be made up of "pre-event" water, that may have very different geochemical characteristics. In fact, transfer functions have proven very useful in hydrology as a way of masking lack of knowledge about the processes of hydrological response, from the original unit hydrograph concept of Sherman (1932), through the geomorphological unit hydrograph (Rodriguez-Iturbe, 1993) to the recent studies using the DPFT methodology to derive patterns of "effective rainfall" (Sempere-Torres et al., 1992) and the parallel transfer function models of Littlewood et al. (1989) and Young and Beven (1991). Transfer functions allow predictions of the modal response at an appropriate scale providing input and output data are available. Without such data they are not easily identifiable.

It is worth noting in this respect that a transfer function model similar to that proposed here but in the form of transition probability matrices between different internal states, was being developed at Stanford University for monthly water balance calculations by Jay Bagley (1964) at the very same time that Crawford and Linsley were developing the Stanford Watershed Model. The use of transition probability matrices is equivalent to using discrete increment transfer functions with a consequent increase in the effective number of parameters in comparison with parameterised transfer functions. Thus, the Bagley model had many more "parameters" than the Stanford Watershed model, and could not easily be identified without data on the internal states themselves. The Stanford Watershed Model, and many other conceptual catchment scale models since, have parameterised their transfer functions in the form of simple storage elements as a way of making the parameter estimation problem tractable,

It has been argued here that these problems of identifiability of model structures and parameter values are best posed in the context of predictive uncertainty in a way that focuses attention on the availability and value of data. One particular framework for doing this, the GLUE technique, has been briefly outlined and is discussed in detail elsewhere (Beven and Binley, 1992). Hydrological science appears to reached something of a plateau of achievement within the current dominant reductionist paradigm. It is really very difficult to see how significant further advances are going to be made within this paradigm by theorising alone. Advances in hydrology in the near future are going to be driven by the availability of new measurement techniques and new data sets on the internal operation of catchments. The value of data in constraining predictive uncertainty will provide one essential justification for new data collection programmes. The need to compare different types of data in terms of their value in this respect might allow the collection of comprehensive data sets and the development of ideas for the new, larger scale, measurement techniques that are needed desperately now.

There may, however, be a problem with this proposed methodology for parameter calibration and uncertainty estimation. It depends on being able to find parameter sets and model structures that have some likelihood of being acceptable simulators of the system of interest. It is possible that the more that internal state data is examined in comparison with model predictions, the more it will appear that current models cannot be considered as acceptable simulators. It may be necessary to reject all the models tried (or give very low weightings to the simulations of internal state data). If this is the case then hopefully such data will again lead to new and more thoughtful approaches to hydrological simulation.

Acknowledgements

The calculations of figure 3 were carried out by Jim Freer as part of work supported by the EEC ENCORE project. The work has benefited from support under Grants GST/02/491 and GR3/8950 of the U.K. Natural Environment Research Council. Andy Binley and Renata Romanowicz have also contributed to the development of the GLUE concepts which are the subject of continuing work within CRES at Lancaster.

References

Ambroise, B (1988) Interactions eaux souterraines-eaux de surface dans le bassin de Ringelbach a Soultzeren (Hautes-Vosges), France: role hydrologique des surfaces saturees. In P Dahlblom and G Lindh (Eds.), Interaction between Groundwater and Surface Water, Proc. Int. Symp. IAHR, Ystad, 231-238, Univ. Lund, Sweden.

Anderson, M G and T P Burt (1978) The role of topography in controlling throughflow generation, Earth Surface Processes, 3, 331-344.

Bagley, J M (1964) An application of stochastic process theory top the rainfall runoff process, Technical Report No 35, Dept. Civ. Eng. Stanford University, 118 pp.

Beven, K J (1978) The hydrological response of headwater and sideslope areas, Hydrol. Sci. Bull., 23, 419-437.

Beven, K J (1979) On the generalised kinematic routing method, Water Resources Research, 15, 1238-1242.

Beven, K J (1986) Runoff production and flood frequency in catchments of order n: an alternative approach. In V K Gupta, I Rodriguez-Iturbe and E F Wood (eds.), Scale Problems in Hydrology, Dordrecht, Reidel, 107-131.

Beven, K J (1987) Towards a new paradigm in hydrology, Int. Assoc. Sci. Hydrol., Pubn. No. 164, 393-403.

Beven, K J (1989a) Changing ideas in hydrology: the case of physically-based models, J. Hydrol., 105, 157-172.

Beven, K J (1989b) Interflow, in H J Morel-Seytoux (Ed.), Proc NATO ARW on Unsaturated Flow in Hydrologic Modelling, Dordrecht, Kluwer, 191-219.

Beven, K J (1991) Spatially distributed modelling: conceptual approach to runoff prediction, in D S Bowles and P E O'Connell (eds.), Recent Advances in the Modelling of Hydrologic Systems, 373-387, Dordrecht, Kluwer.

Beven, K J (1993) Prophesy, reality and uncertainty in hydrological modelling, Advances in Water Resources, 16, 41-51.

Beven, K J, and A M Binley (1992) The future of distributed models: model calibration and predictive uncertainty, Hydrol. Process., 6, 279-298.

Beven, K J, A Calver and E M Morris (1987) The Institute of Hydrology Distributed Model, Institute of Hydrology Report No. 98, Wallingford.

Beven, K J and P F Germann (1981) Water flow in soil macropores. II. A combined flow model. J. Soil Sci., 32, 15-29.

Beven, K J and P F Germann (1982) Macropores and water flow in soils, Water Resources Research, 18, 1311-1325.

Beven, K J and P F Germann (1985) A distribution function model of channelling flow in soils based on kinematic wave theory, ILRI Publ. 37, Wageningen, 89-100.

Beven, K J, D E Henderson and A D Reeves (1993) Dispersion parameters for undisturbed partially saturated soil, J. Hydrology, 143, 19-44.

Beven, K J and M J Kirkby (1979) A physically-based, variable contributing area model of basin hydrology, Hydrol. Sci. Bull., 24, 43-69.

Beven, K J and E F Wood (1993) Flow Routing and the hydrological response of channel networks, in K J Beven and M J Kirkby (eds.), Channel Network Hydrology, Chichester, Wiley , 99-128.

Binley, A M, K J Beven and J Elgy (1989) A physically-based model of heterogeneous hillslopes. II Effective hydraulic conductivities, Water Resour. Res., 25, 1227-1233.

Binley, A M and K J Beven (1991) Physically-based modelling of catchment hydrology: a likelihood based approach to reducing predictive uncertainty, in D G Farmer and M J Rycroft (Eds.), Computer Modelling in the Environmental Sciences, Clarendon Press, Oxford, 75-88.

Binley, A M, K J Beven, A Calver and L G Watts (1991) Changing responses in hydrology: assessing the uncertainty in physically-based predictions, Water Resour. Res., 27, 1253-1262.

Brun, C, R Bernard, D Vidal-Madjar, C Gascuel-Odoux, P Merot, J Duchesne and H Nicholas (1990) Mapping saturated areas with a helicopter-borne C Band scatterometer, Water Resources Research, 26, 945-955.

Burt, T P and D P Butcher (1985a) Topographic controls of soil moisture distribution, J. Soil Sci., 36, 469-486.

Burt, T P and D P Butcher (1985b) On the generation of delayed peaks in stream discharge, J. Hydrology, 78, 361-378.

Dunne, T P, T R Moore and C H Taylor (1975) Recognition and prediction of runoff producing zones in humid regions, Hydrol. Sci. Bull., 20, 305-327.

Ganoulis, J G (1989) Multiphase flow in porous media: description at the pore and macroscopic scale. in H J Morel-Seytoux (Ed.), Unsaturated Flow in Hydrologic Modeling, 27-52, Dordrecht, Kluwer.

Genereux, D P, H F Hemond, and P J Mulholland (1993) Spatial and temporal variability in streamflow generation on the West Fork of Walker Branch watershed, J. Hydrology, 142, 137-166.

Gerke, H H and M T van Genuchten (1993) A dual-porosity model for simulating the preferential movement of water and solutes in structured porous media, Water Resources Research, 29, 305-320.

Grayson, R B, I D Moore and T A McMahon (1992) Physically based hydrologic modeling. 2. Is the concept realistic?, Water Resources Research, 26, 2659-2666.

Hoogmoed, W D and J Bouma (1980) A simulation model for predicting infiltration into cracked soil. Soil Sci. Soc. Amer. J. 44, 458-461.

Hornberger, G M, K J Beven, B J Cosby and D E Sappington (1985) Shenandoah Watershed Study: calibration of a topography-based, variable contributing area model to a small forested catchment, Water Resour. Res., 21,1841-1850.

Hornberger, G M and R C Spear (1981) An approach to the preliminary analysis of environmental systems, J. Environ. Manag., 12, 7-18.

Huff, D D, R V O'Neill, W R Emanuel, J W Elwood and J D Newbold (1982) Flow variability and hillslope hydrology, Earth Surf. Process. Landf., 7, 91-94.

Jakeman, A J, I G Littlewood and P G Whitehead (1990) Computation of the instantaneous unit hydrograph and identifiable component flows with application to two small upland catchments, J. Hydrology, 117, 275-300.

Jensen, K H and A Mantoglou (1992) Application of stochastic unsaturated flow theory: numerical simulations and comparisons to field observations, Water Resources Research, 28, 269-284.

Kirkby, M J (1976a) Tests of the random network model and its application to basin hydrology, earth Surface Processes, 1, 197-212.

Kirkby, M J (1976b) Hydrograph Modelling Strategies, in R Peel, M Chisholm and P Haggett (Eds.), Processes in Physical and Human Geography, Heinemann.

Loague, K M (1990) R-5 revisited. 2. Reevaluation of a quasi-physically-based rainfall-runoff model with supplemental information, Water Resour. Res., 26, 973-987.

Loague, K M (1992) Soil water content at R-5. Part 2. Impact of antecedent conditions on rainfall-runoff simulations, J. Hydrology, 139, 253-261.

Loague, K M and R A Freeze (1985) A comparison of rainfall-runoff modelling techniques on small upland catchments, Water Resour. Res., 21, 229-240.

Manabe, S and R T Wetherall (1980) On the distribution of climate change resulting from an increase in CO_2 content of the atmosphere, J. Atmospheric Science, 37, 99-118.

McDonnell, J J (1990) A rationale for old water discharge through macropores in a steep, humid catchment, Water Resources Research, 26, 2821-2832.

Nielsen, D R, J W Biggar and K T Erh (1973) Spatial variability of field measured soil properties, Hilgardia, 42, 215-260.

Omoti, U and A Wild (1979) Use of fluorescent dyes to mark the pathways of solute movement through soils under leaching conditions. 2. Field Experiments, Soil Science, 128, 98-104.

Quinn, P F, K J Beven, P Chevallier and O Planchon (1991) The prediction of hillslope flow paths for distributed hydrological modelling using digital terrain models, Hydrological Processes, 5, 59-80; also in K J Beven and I D Moore (Eds.), Terrain analysis and distributed modelling in hydrology, 63-84, Chichester, Wiley.

Refsgaard, J C, S M Seth, J C Bathurst, M Erlich, B Storm, G H Jorgensen and S Chandra (1992) Application of the SHE to catchments in India. Part 1. General Results; Part2. Field experiments and simulation studies with the SHE on the Kolar subcatchment of the Narmada River, J. Hydrology, 140, 1-47.

Rodriguez-Iturbe, I (1993) The Geomorphological Unit Hydrograph, in K J Beven and M J Kirkby (eds.), Channel Network Hydrology, Chichester, Wiley , 99-128.

Schmugge, T (1993) Passive microwave sensing of soil moisture, this volume.

Sellars, P J, Y Mintz, Y C Sud and A Dalcher (1986) A simple biosphere model (SiB) for use within General Circulation Models, J. Atmospheric Science, 43, 505-531.

Sempere-Torres, D, J Y Rodriguez and Ch Oiled (1992) Using the DPFT Approach to improve flood forecasting models, Natural Hazards, 5, 17-41.

Sherman, L K (1932) Streamflow from runoff by the unit-hydrograph method, Engin. News. Rec., 108, 801-805.

Sivapalan, M, K J Beven and E F Wood (1987) On hydrological similarity II. A scaled model of storm runoff production, Water Resources Research, 23, 2266-2278.

Sivapalan, M, E F Wood, and K J Beven (1990) On hydrological similarity III. A dimensionless flood frequency model using a generalised geomorphological unit hydrograph and partial area runoff generation, Water Resources Research, 26, 43-58.

Sklash, M G (1990) Environmental isotope studies of storm and snowmelt runoff generation, in M G Anderson and T P Burt (Eds.), Process studies in Hillslope Hydrology, 401-436, Chichester, Wiley.

Troch, P A, F P De Troch and W Brutsaert (1993) Effective water table depth to describe initial conditions prior to storm rainfall in humid regions, Water Resources Research, 29, 427-434.

Wood, E F, M Sivapalan, K J Beven and L Band (1988) Effects of spatial variability and scale with implications to hydrological modelling, J. Hydrology, 102, 29-47.

Wood, E F, M Sivapalan, and K J Beven (1990) Similarity and scale in catchment storm response, Reviews of Geophysics, 28, 1-18.

Wilson, G V, P M Jardine, J D O'Dell and M Collineua (1993) Field-scale transport from a buried line source in variably saturated soil, J. Hydrology, 145, 83-109.

Young, P C and K J Beven (1991) Computation of the instantaneous unit hydrograph and identifiable component flows with application to two small upland catchments - Comment. J. Hydrology, 129, 389-396.

Approaches for Measuring and Modeling Soil Moisture

Ian R. Calder
Institute of Hydrology
Wallingford
UNITED KINGDOM

1. Introduction

Soil moisture is determined by the various factors which regulate the entry of water through the soil surface and the removal of water from the soil through evaporation from the soil surface, root abstraction, redistribution, and drainage.

These factors are themselves influenced by virtually all the factors and processes operating in the hydrological cycle. Rainfall amount, spatial variability, intensity, and drop size affect interception, surface crusting, infiltration rates and surface runoff. Atmospheric demand affects interception, transpiration and surface evaporation. Vegetative properties affect interception, transpiration, and soil surface evaporation rates and can also influence soil properties through macropore development associated with roots and the development of surface crusts through drop size modification of net rainfall from canopies. Soil properties influence surface crusting characteristics, infiltration rates, water retention and redistribution, and drainage and root abstraction. Topography influences the generation of surface ponding and runoff, soil moisture drainage and redistribution, and vegetation type.

Although to a greater or lesser extent information on all these processes and interactions exists, knowledge of many of the parameters, particularly those relating to vegetation and soils, is not generally available.

This paper is concerned with identifying the key processes which determine soil moisture in different regions of the world. It is concerned primarily with vegetation and soil property determinants of soil moisture. Other papers in this volume deal specifically with atmospheric, topographic, and rainfall spatial variability issues.

NATO ASI Series, Vol. I 46
Land Surface Processes in Hydrology
Trials and Tribulations of Modeling and Measuring
Edited by S. Sorooshian, H. V. Gupta, and S. C. Rodda

2. Principal Limits on Evaporation and Soil Moisture

Soil moisture plays a crucial role in partitioning available energy into sensible and latent heat. Within GCMs, soil moisture accounting procedures are used, together with some type of evaporation/infiltration/drainage model, to estimate the changing soil moisture status.

Although the complexity of the model may range from simple "Budyko bucket" models (Manabe et al., 1965) to those using up to 50 model parameters such as BATS and SiB (Dickinson, 1983; Dickinson et al., 1986; Sellers, 1986), it is perhaps worth bearing in mind, particularly with respect to the more complex models, that the relationships used to estimate both surface and aerodynamic resistance terms in relation to soil and vegetation properties are, to a greater or lesser extent, empirical in nature.

It may also be worth injecting a cautionary note that there is likely to be a point where increasing model complexity results in diminishing returns in terms of improved accuracy of model prediction.

From experimental studies carried out by the Institute of Hydrology, UK, using neutron probe estimates of soil moisture at a wide range of sites covering both tall and short crops in both wet and dry climates and in temperate and tropical conditions, it appears that the point of diminishing return (in terms of the number of model parameters required for soil moisture prediction) is quickly reached. These studies also illustrate that different factors are important in determining soil moisture deficits (SMDs) at the different sites. It is suggested that, from a knowledge of these controlling or limiting factors, it may be possible to devise simpler, few-parameter models calibrated for the vegetation and climate characteristics of the site. Even if it may not be possible in the design of GCMs to ensure that models are "matched to sites", the studies may, nevertheless, be useful in identifying the particular parameters which are likely to be most sensitive for SMD prediction at different sites and point the direction where further refinements in model description and parameter identification are required.

Results from five studies carried out by the Institute of Hydrology have been chosen as examples. These were: (1) studies into the accuracy of soil moisture predictions for grassland in the UK (Calder et al., 1983); (2) the Plynlimon and Scottish Consortium studies of the effects of upland afforestation in the UK (Institute of Hydrology, 1991; Calder, 1991); (3) studies of the evaporation and soil moisture deficits under broadleaf forests in the UK (Harding et al., 1992); (4) studies of the hydrological impact of tropical afforestation in India in both the wet and dry zones (Calder, 1992); and (5) studies of the water relations of tropical rain forest in Indonesia (Calder et al., 1986). To assist in identifying the principal controls on soil moisture in relation to site variables, the sites have been classified according to vegetation type as short or tall crop and under climate as wet or dry and tropical or temperate.

Soil Moisture Studies Under Grassland, UK: Short Crop, Wet, and Dry Temperate

The ability of different soil moisture models to predict SMDs at six different grassland sites in the UK was assessed by comparison with neutron probe measurements of SMD.

The models were configured with different levels of complexity with respect to the calculation of evaporative demand and with respect to the soil moisture regulating functions.

The structure of the models is shown in Figure 1; an example of model predictions at one of the sites is shown in Figure 2. A full description of the procedures and the results obtained is given by Calder et al. (1983). The main conclusions from this study were that:

1. For all sites, the best SMD predictions were obtained with either the Penman potential transpiration estimate for short grass calculated from daily meteorological observations made close to the site or from a climatologically derived Penman potential transpiration estimate derived by fitting a sinusoidal curve through monthly estimates derived from a central location in the UK (Northamptonshire) (Figure 3).

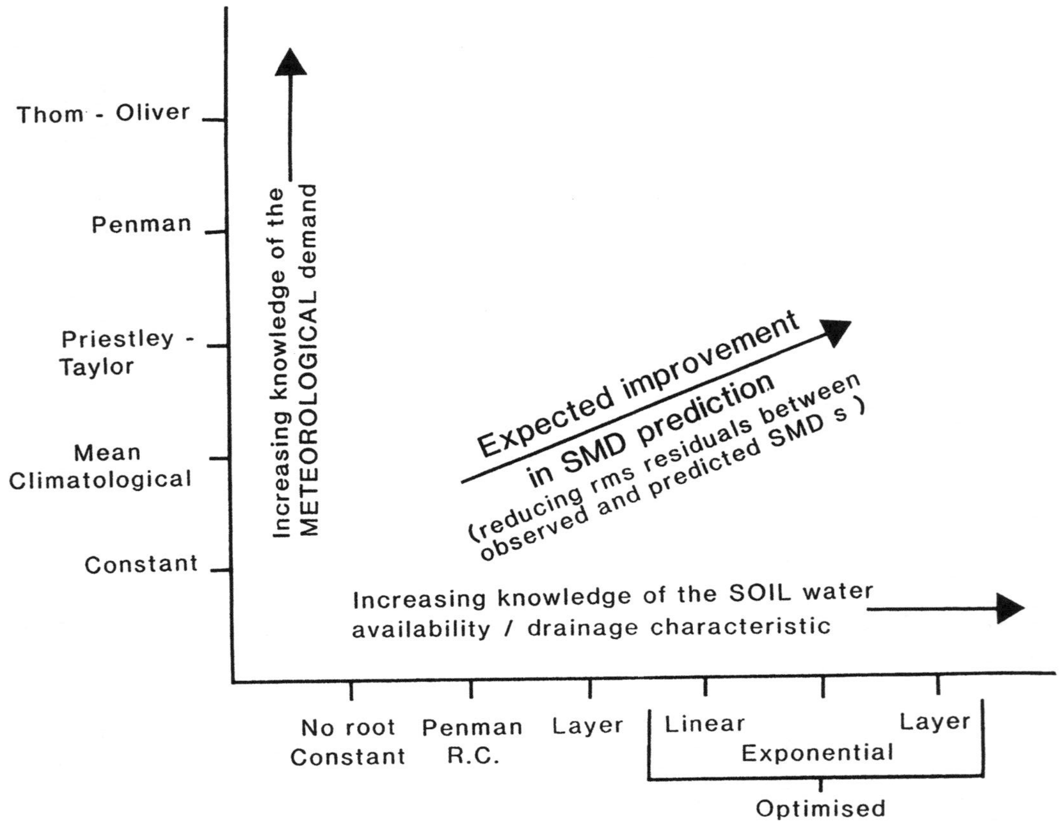

Figure 1. The expected improvements in model fit as a function of increasing knowledge of the root constant functions and potential evaporation estimates (from Calder et al. 1983).

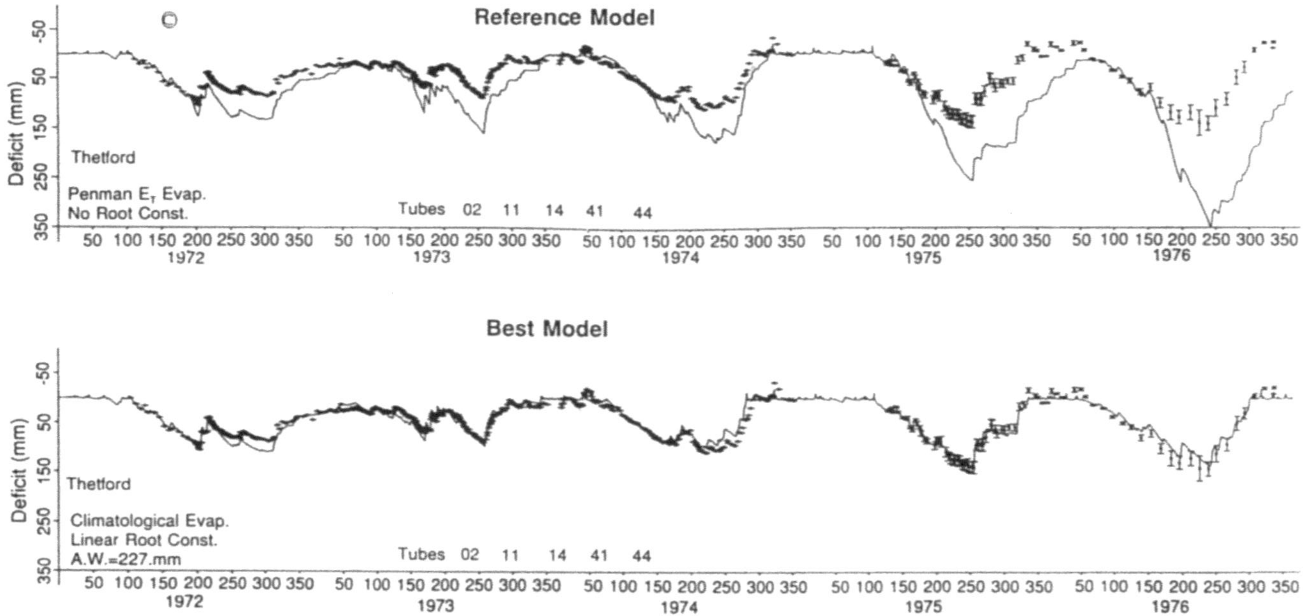

Figure 2. Example of the mean observed and the predicted SMD-values obtained using the "best" model and the standard reference model at one of the six sites investigated. The error bars denote the standard error of the mean observed SMD (from Calder et al., 1983).

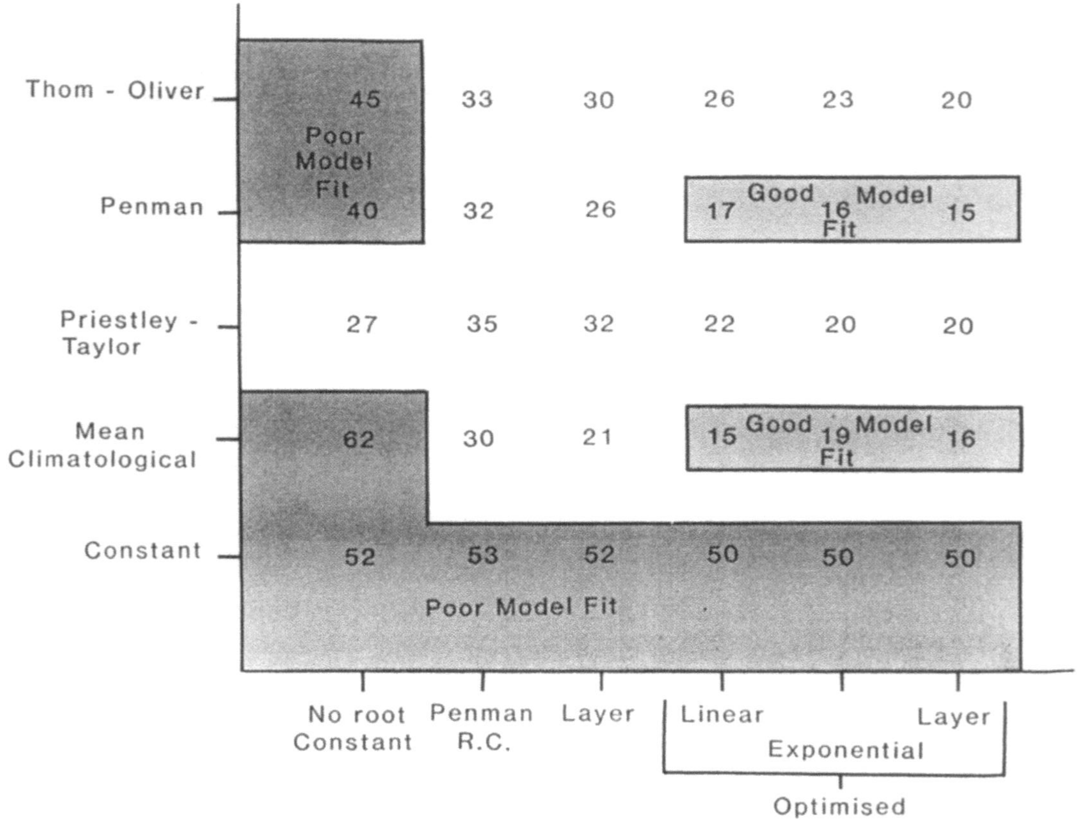

Figure 3. The mean of all sites model error (mm) for all combinations of root constant functions and evaporation estimates (from Calder et al., 1983).

2. Soil moisture estimation could be greatly improved if the available water parameter, a_w, was optimized for each site. Generally, much larger values for a_w were found for chalk sites, a result that was also found for trees growing on chalk soils (see below).

3. Soil moisture predictions were not especially sensitive to the form of the regulating function chosen and, although certain functions performed better at specific sites, no function performed significantly better than the others over all sites.

It is interesting to note that, for the purposes of SMD prediction increasing sophistication in the description of evaporative demand, using, for example, the Thom and Oliver formulation (Thom and Oliver, 1977) did not result in improved model fits. Neither did the use of local daily meteorologic data in the Penman calculation, as opposed to using a climatologically derived estimate, result in improved estimates. Although atmospheric demand is clearly the driving mechanism for evaporation, the relative insensitivity of SMD prediction to the precise estimation of atmospheric demand is a common feature that has been observed in other studies described below. In general, this suggests that, for operational purposes of predicting SMD, better knowledge of the available water/soil moisture regulating function is required for improved SMD prediction rather than more detailed meteorological information.

Soil Moisture Studies Under Upland Forest and Moorland, UK: Short and Tall Crops, Wet Temperate

From studies carried out at the Plynlimon experimental catchments in mid Wales, the Balquhidder catchments in central Scotland and at the Crinan Canal catchments in west Scotland, some general conclusions can be drawn regarding the important factors determining soil moisture deficits under different upland land uses.

Figure 4 shows soil moisture deficits recorded under coniferous forest, heather, *Calluna vulgaris*, and grass at the Balquhidder and Crinan sites. Modeling studies have established (Calder, 1990) that, at these wet upland sites (annual rainfall >1500 mm), with vegetation growing on generally deep peaty soils, soil moisture availability is not

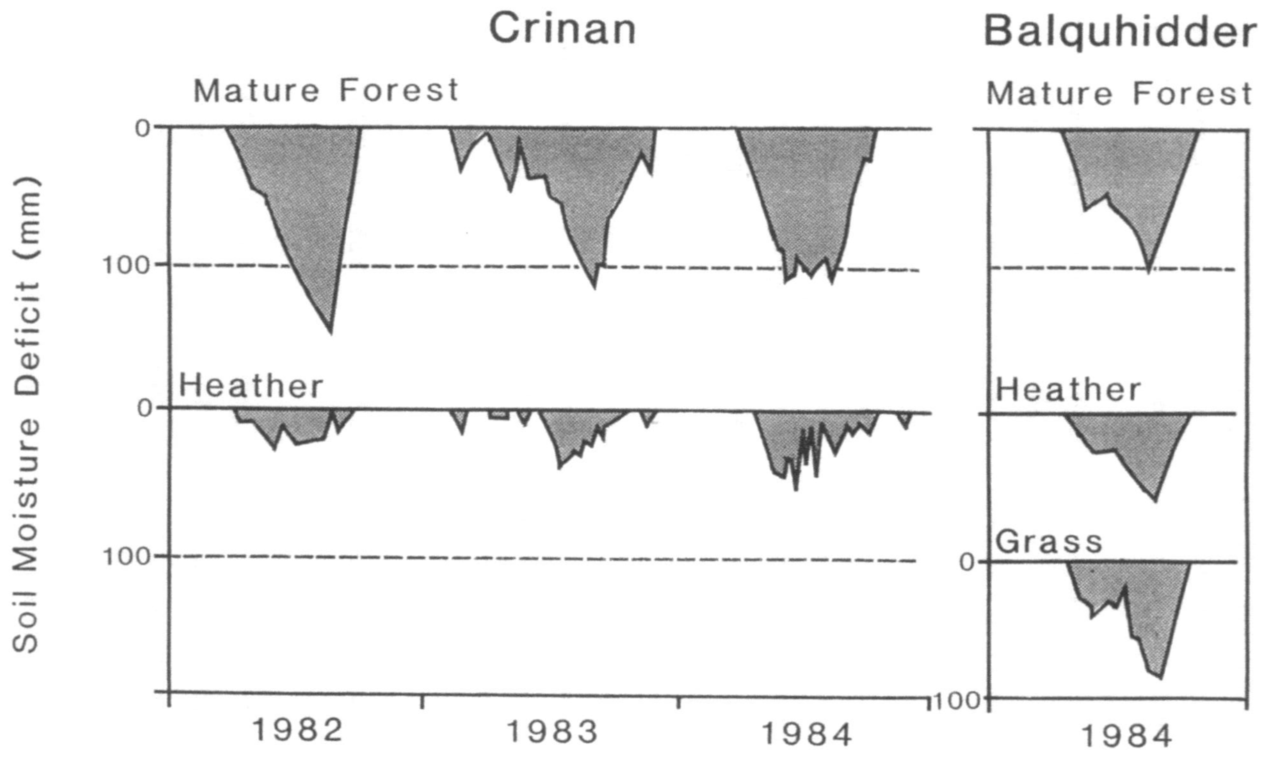

Figure 4. Soil moisture deficits measured beneath mature spruce forest, heather and grass moorland at Crinan and Balquhidder (from Calder, 1990).

usually important in determining SMDs. Of more importance at these sites are the physical and physiological controls on interception and transpiration from the different vegetation types. The higher interception losses from forest, because of its increased aerodynamic roughness, is the reason for increased SMDs from forest as compared with shorter vegetation, grass, and heather, in these conditions. Higher SMDs under grass than heather are a reflection of the increased physiological controls on transpiration imposed by heather as compared with grass; heather generally has a greater surface resistance than grass.

The evaporation and SMDs under these different land uses can be predicted using an evaporation equation that assumes that, in dry conditions, transpiration from the different land uses is a constant fraction of the Penman estimate, and interception losses can be determined as a function of daily rainfall.

Soil Moisture Studies Under Lowland Broadleaf Forest, UK: Tall Crop, Dry Temperate

Evaporation and soil moisture studies were carried out for ash and beech as part of an investigation (Harding et al., 1992) into the hydrological impact of increased broadleaf plantations, which were expected as a consequence of the EC's set-aside policy in southern England.

At these sites, soil moisture deficits were strongly influenced by physiological controls on transpiration but were also influenced by soil water availability and interception (Figure 5). Differences in available water to the forests was strongly related to soil type. For ash and beech growing on soil overlying chalk, the available water was essentially infinite, whereas at the clay soil site the value of the available water parameter was on the order of 280 mm.

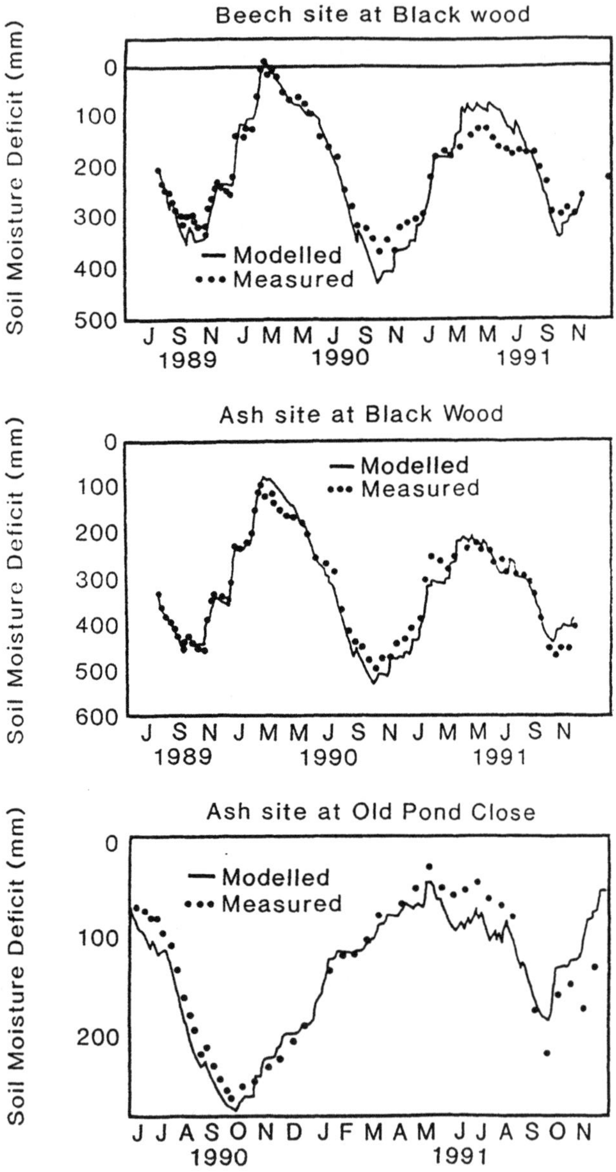

Figure 5. Soil moisture deficits, measured (solid squares), and modeled (solid line): (A) for the beech site on chalk at Black Wood; (B) for the ash site on chalk at Black Wood; (C) for the ash site on clay at Old Pond Close (from Harding et al., 1992).

Soil Moisture Studies Under Forest and Agricultural Crops, India, Indonesia: Short and Tall Crops, Dry and Wet Tropical

As part of a study to investigate the hydrological impact of eucalypt plantations in the dry zone of southern India, comparative studies were carried out on the evaporative characteristics and soil moisture deficits under eucalypt plantations, indigenous forests, and agricultural crops. For all of the four sites where investigations were carried out, soil moisture availability was found to be a major constraint on SMD for both agricultural crops and trees. However, the shallower rooting depth of the agricultural crop compared with that of the forest resulted in much reduced soil water availability. For the agricultural crop studied, finger millet, *Eleusine coracana*, the available water was found to be 160 mm as compared with 390 mm for most of the forest sites (at one of the *Eucalyptus* sites, roots extend to at least 6 m and probably 10 m, and the concept of available water cannot be simply applied at this site). This reduced soil water availability was the principal reason why the annual evaporation from the agricultural crop was about half that from the forest.

A further major determinant of evaporation and SMD at forest sites was found to be tree size. The eucalypt plantations are usually grown with a rotation period of between three and seven years and at no time during this period does the canopy approach closure. From deuterium tracing studies of single tree transpiration, it has been established that there is a very "tight", linear, relationship between transpiration rate and basal cross-sectional area. This relationship, which can be interpreted as the Darcy sap flow velocity, implies that Darcy sap velocities are independent of tree size but are modified by soil moisture deficits (Figures 6 and 7). This relationship has been used as the basis for operational models for estimating both evaporation and growth from these plantations (Figure 8) (Calder, 1992). It would appear that, although evaporative demand is clearly the driving mechanism for evaporation, for most of the year, it is not limiting; the primary controls are soil moisture availability, and, for the tree crops, some index of tree size. (It is interesting to note that, although there is clearly a positive correlation between tree basal cross-sectional area and tree leaf area, a higher

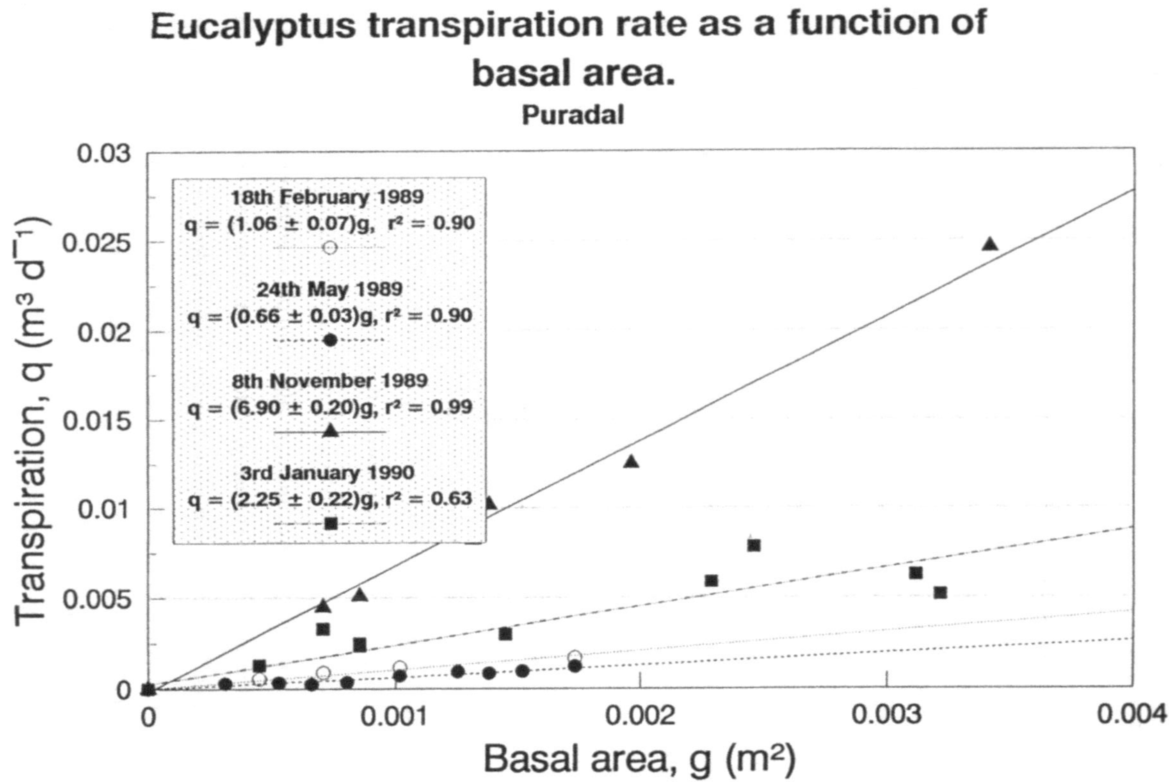

Figure 6. Measured transpiration rates as a function of tree basal area (g) at the Puradal site prior to and after the monsoon (July-September) 1989 (from Calder, 1992).

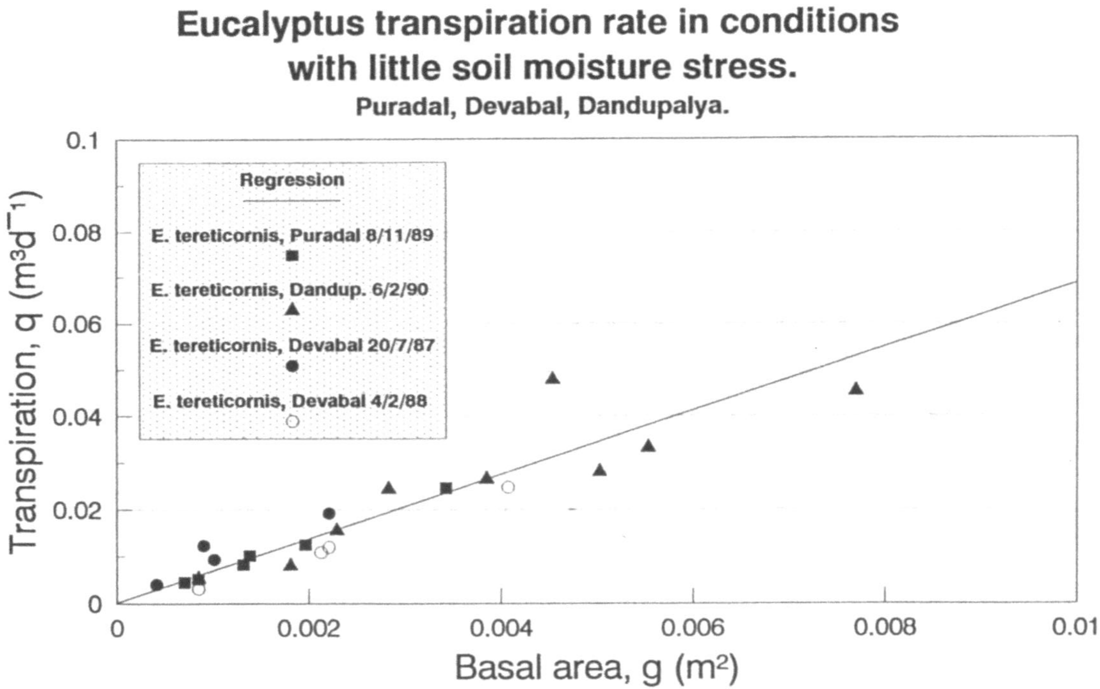

$q = (6.63 \pm 0.33)g \quad (m^3\ d^{-1})$

$q = (0.663 \pm 0.033)G\ (mm\ d^{-1})$ [G units: $m^2\ ha^{-1}$]

$r^2 = 0.85$

Figure 7. Measured transpiration rates as a function of tree basal area (g) for all sites for measurements made under (relatively) non soil-moisture stressed conditions (from Calder, 1992).

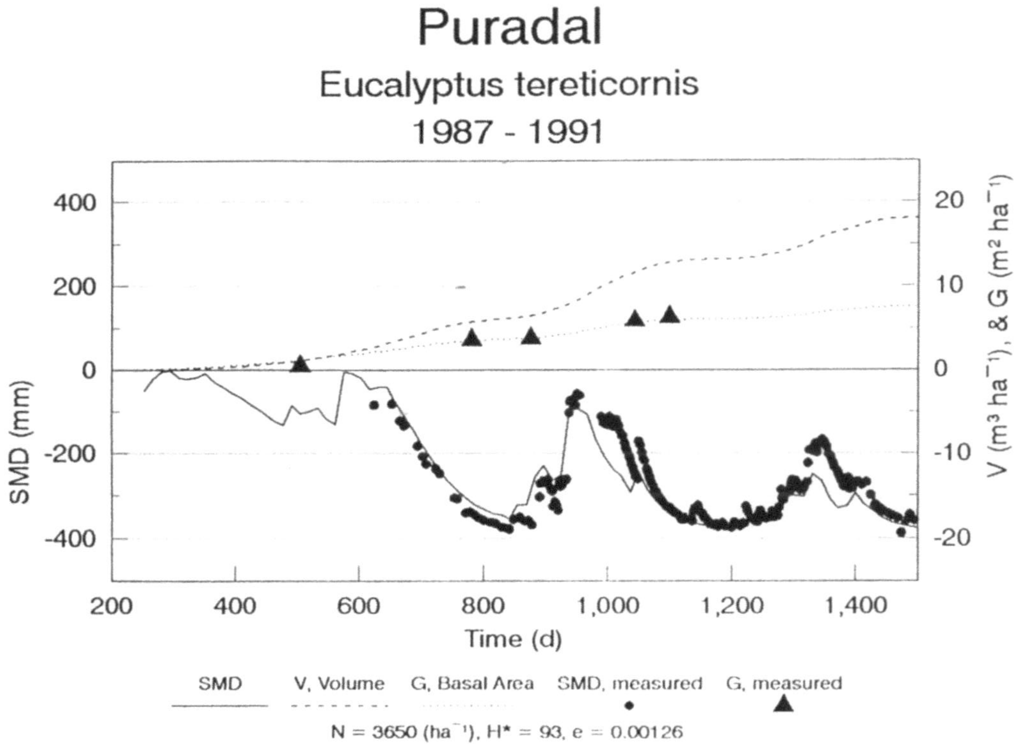

Figure 8. Measured and predicted stand basal area, *G*, predicted stand volume, *V*, and the measured and predicted soil moisture deficit, for *Eucalyptus tereticornis*, 1987 planting, at the Puradal experimental plantation. All day numbers are with respect to January 1, 1987 (from Calder, 1992).

correlation is obtained between transpiration rate and basal cross-sectional area than with leaf area.)

At these dry zone sites in India, which experience an extended dry season, interception losses, which amount to less than 13% of the annual 800 mm rainfall, are not important in determining soil moisture deficits.

At wetter sites, forest interception is a more significant component of the annual forest evaporation and a greater determinant of SMD. From studies carried out in Indonesian rain forest (Calder et al., 1986) and in India, the importance of rain drop size in determining interception losses from forest was realized. Application of a "stochastic" interception model, which was required to describe the interception process in these conditions, has shown that up to ten times as much rain is required to achieve the same degree of canopy wetting for tropical convective storms, with large drop sizes, than would be necessary for the range of drop sizes usually encountered for frontal rain in the UK. There are also indications, from preliminary results carried out with rainfall simulators, that the final degree of canopy saturation also varies with drop size being greater for drops of smaller size.

Vegetation canopies also influence drop size of the net rainfall; deeper layers in the canopy will be more influenced by the modified drop size spectrum falling from canopy layers above than by that of the incident rain. Recent studies have shown that different vegetation canopies have very distinct canopy spectra (Figure 9) (Hall and Calder, 1993).

For canopies with a low leaf area index, the interception characteristics would therefore be expected to be related more to the drop size of the incident rain, whereas for canopies with higher leaf area index the characteristics would be less dependent on the drop size of the rainfall.

Although the mathematical description of the wetting process in the stochastic model is more complex than in the conventional "bucket" description of the interception store, few extra parameters are required for potentially large improvements in model efficiency.

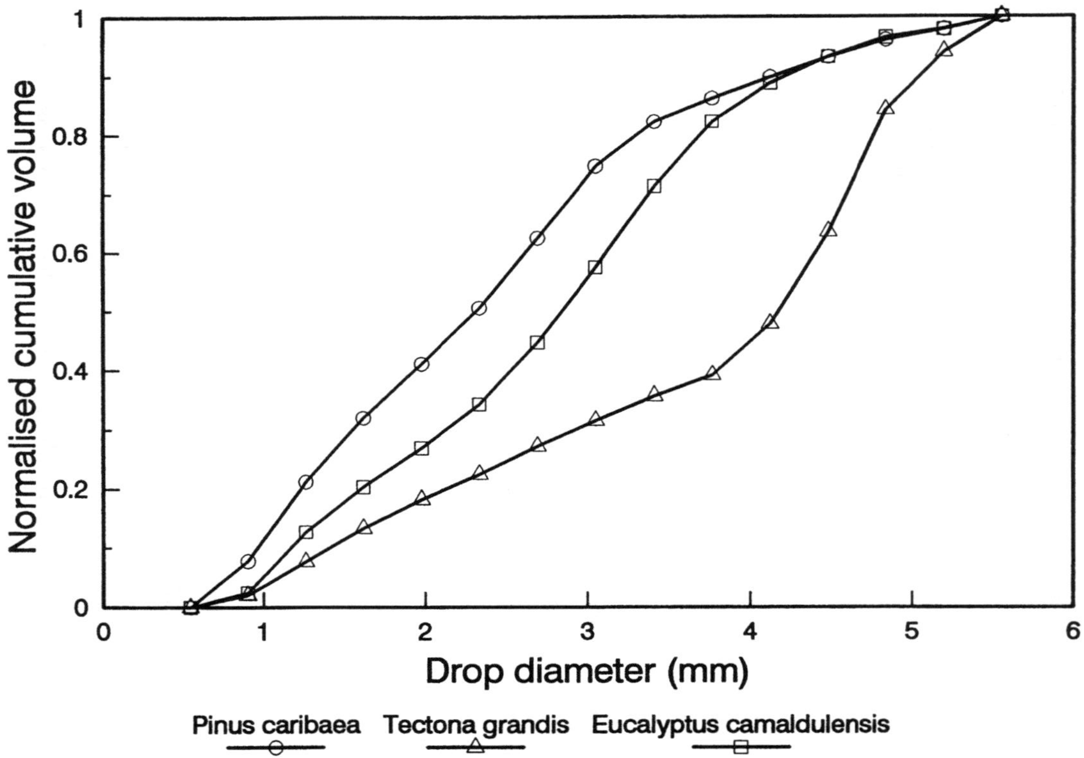

Figure 9. The cumulative frequency distribution (by volume) of the subcanopy drip spectra for three tree species (from Hall and Calder, 1993).

If this model description was to be incorporated within GCMs, information on the drop size spectra of the incident rain could be obtained from model-predicted rainfall intensities via the Marshall Palmer distribution.

Principal Controls on Soil Moisture

A summary of the principal controls on SMD expected for both short and tall crops in wet and dry climates of the tropical and the temperate world is shown in Table 1.

3. Soil Moisture Models for Application in GCMs

From the above discussion, it is suggested that certain simplifications may be possible in the formulation of SMD models based on the primary limits on SMD in particular configurations of land use (vegetation type) and climate.

In certain areas, topography may also be an important determinant of SMD, but this question is addressed by other papers in this volume.

The soil moisture observations referred to above are, generally, the mean of observations from a number of neutron probe soil moisture access tubes (typically ten to 20). Therefore, the measured SMDs are not point measurements but more "plot" measurements, representative of an area of the order of an hectare. The suitability of such models for estimating SMDs over larger areas is linked with the variability of the factors which are important in determining the SMDs. A major factor is always is the spatial variability of rainfall.

The usefulness of such an approach would be related to the ease with which model parameters could be estimated. The principal parameters and the suggested methods for determination are shown in Table 2.

Table 1. Primary Controls on SMD for Different Land Uses in Different Climates

Dominant Controls on SMD

TEMPERATE CLIMATE

Land Use	DRY	WET
Short crop	available water	meteorological demand
Tall crop	physiological controls available water interception	interception physiological controls

Dominant Controls on SMD

TROPICAL CLIMATE

Land Use	DRY	WET
Short crop	available water	meteorological demand
Tall crop	basal area/LAI available water	interception, drop size physiological controls

In addition to ground-based, plot scale methods of the type described in the field studies outlined above for parameter estimation, inverse and remote sensing methods have a large role to play in upscaling parameter values over large areas. Remote sensing methods can indicate land use; they can also be used to determine indices of vegetation density (Choudhury, 1993) and can sense soil moisture, although the depth over which the moisture content is detected is, using current methods, limited to the surface layer. Soil moisture measurements, even though limited to the surface layer, may be of value in validating soil moisture accounting models when applied over large areas and could, in principle, be used as a check on operational, real-time, soil moisture accounting models and to "reset" them to field capacity conditions where indicated by remote sensing.

Table 2. Principal Soil Moisture Model Parameters and Suggested Methods for Their Determination

Principal Parameters	Method of Estimation
a_w, soil moisture availability in the root zone	Neutron probe studies under representative soil/vegetation types; inverse methods
Vegetation cover	Remote sensing of LAI and stand basal area; ground-based measurements of the same
Interception parameters for tall crops	Interception studies under different vegetation types, relating interception characteristics to GCM predicted rainfall intensity and, hence, drop size.
Physiological controls	Physiological studies of the major biomes

Comparisons of neutron probe measurements of moisture contents in the top soil layer (which could be determined by satellite radar) with measurements of total soil profile water contents indicate that, for many vegetation/soil types, for most of the time, very good correlations exist (Ragab, 1992). Although the ratio between top layer moisture content and profile water content is largely a reflection of total available water in the rooting zone, a parameter which is itself not easy to determine, the method may be of value in checking the operation of GCM soil moisture predictions for extensive areas.

Another approach for estimating parameters such as soil water availability for large areas is through the use of inverse techniques. For example, on a catchment scale, soil moisture model parameter values can be optimized to give the best fit to predicted and measured flow data. This approach may be particularly relevant in calibrating models which consider subgrid variability of vegetation types and soil types.

The usefulness of this approach for investigating the effects of land use change on the water balance of Lake Malawi is currently being investigated. Assuming that model parameter values for forest and agricultural crops are the same as those determined in southern India, and assuming appropriate values for lake evaporation, good fits can be obtained to runoff from the catchment and lake levels, simply by optimizing the proportion of the catchment under the forest (Figure 10). The optimized value of 76% is probably higher than reality, which may indicate either inadequacies of the approach and the inappropriateness of using the Indian parameter values or uncertainties in the estimation of other components of the water balance terms. The rainfall falling over the lake is a component which is particularly poorly known. At present, it is assumed that the rainfall over the lake is the same as that recorded at gauges near the lake shore. If this component was, say 15%, less than the lake shore value, the optimized forest cover parameter would become a more realistic value of 59%. Studies are currently in hand to improve rainfall and land use estimation. Even though parameter interdependence may limit the usefulness of the inverse approach, the model that has been developed does have a practical value as it is now being used to assist in the management of the barrage which regulates Lake Malawi's level.

The combination of all approaches, process studies for determining parameter values at the plot scale, remote sensing for determining land use and vegetation cover, remote sensing for estimating surface soil moisture, and inverse methods using large catchment flow, may prove to be most effective in developing and calibrating soil moisture models to operate at the GCM grid scale.

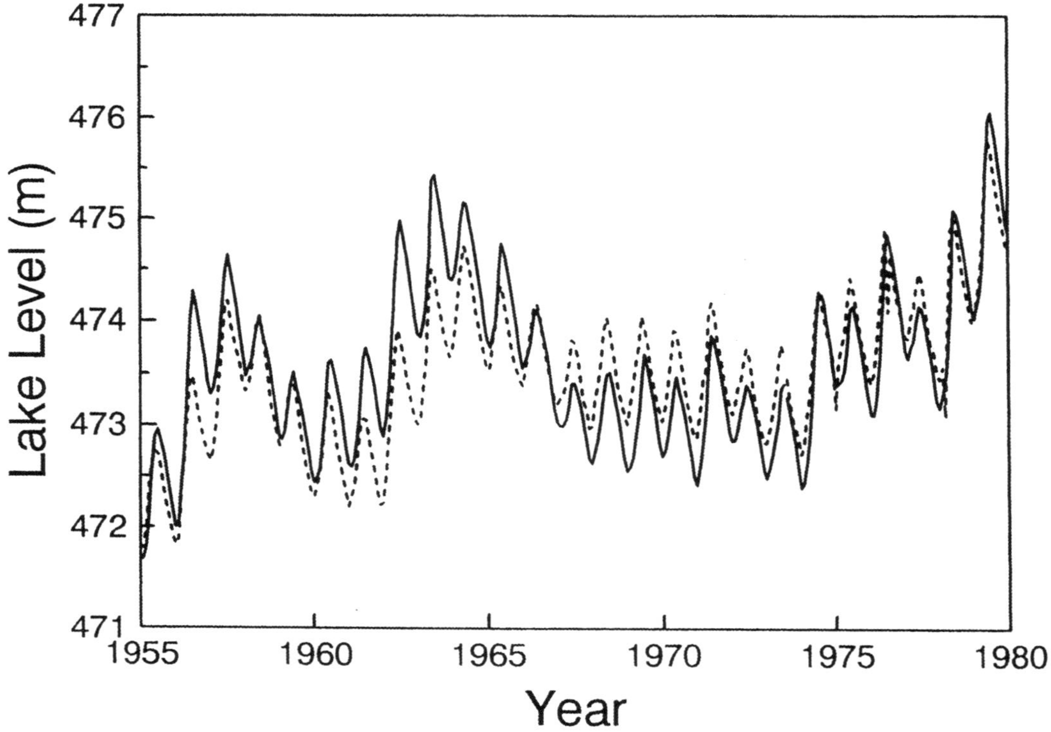

Figure 10. Levels of Lake Malawi, observed (dotted), and predicted (solid) with an evaporation /soil moisture model assuming 76% forest cover of the land catchment, 24% agricultural crop.

References

Calder IR (1986) A stochastic model of rainfall interception, *Journal of Hydrology*, 89:65-71.

Calder IR (1990) Evaporation in the Uplands, John Wiley Publishers, ISBN 0-471-92487-3, p. 148.

Calder IR (1992) A model of transpiration and growth of *Eucalyptus* plantation in water-limited conditions, *Journal of Hydrology*, 130:1-15.

Calder IR, RJ Harding RJ, PTW Rosier (1983) An objective assessment of soil-moisture deficit models, *Journal of Hydrology*, 60:329-355.

Calder IR, IR Wright, D Murdiyarso (1986) A study of evaporation from tropical rainforest--West Java, *Journal of Hydrology*, 89:13-33.

Choudhury BJ (1993) Synergism of multispectral observations in the context of estimating regional evaporation, This volume.

Dickinson RE (1983) Land surface processes and climate-surface albedos and energy balance, In: B. Saltzmann (ed.), Theory of Climate, *Advances in Geophysics*, 25:305-353.

Dickinson RE, A Henderson-Sellers, PJ Kennedy, MF Wilson MF (1986) Biosphere Atmosphere Transfer Scheme (BATS) for the NCAR Community Climate Model, NCAR Technical Note, NCAR, TN275 + STR, 69 p.

Institute of Hydrology Report No. 109 (1991) Plynlimon research: the first two decades, Edited by C Kirby, MD Newson, K Gilman, Institute of Hydrology, Wallingford, United Kingdom, 188 p.

Hall RL, IR Calder (1993) Drop-size modification by forest canopies--measurements using a disdrometer, Submitted to *Journal of Geophysical Research*.

Harding RJ, RL Hall, C Neal, JM Roberts, PTW Rosier (1992) Hydrological impacts of broadleaf woodlands: implications for water use and water quality, Institute of Hydrology, British Geological Survey Project Report 115/03/ST for the National Rivers Authority, Institute of Hydrology, Wallingford, United Kingdom.

Manabe S, J Smagorinsky, RJ Strickler (1965) Simulated climatology of a general circulation model with a hydrological cycle, *Monthly Weather Review*, 97:739-774.

Ragab R (1992) Assessment of the relationship between remotely sensed topsoil moisture content and profile moisture content, In: FJ Eley, R Granger, L Martin (Editors), Soil Moisture Modelling and Monitoring for Regional Planning, *Proceedings of the NHRC Workshop*, March 9-10, 1992, National Hydrology Research Centre, Canada, pp. 141-153.

Sellers PJ (1986) The simple biosphere model (SiB), In: C Rosenzweig, R Dickinson (Editors), Proceedings of Workshop held at NASA/Goddard, Greenbelt Maryland, January 1986, UCAR/OIES Report OIES-2, pp. 87-90.

Thom AS, HR Oliver (1977) On Penman's equation for estimating regional evaporation, *Quart. J.R. Meteorological Society*, 103:345-357.

Passive Microwave Remote Sensing of Soil Moisture

Thomas Schmugge and Thomas Jackson
USDA/Hydrology Laboratory
Beltsville, Maryland 20705
USA

1. Introduction

Satellites provide stable platforms necessary for the long-term monitoring of climate change. For example, this has been demonstrated with the long-term observation of the vegetation patterns in the Sahelian region of Africa using the visible and near-infrared data from the AVHRR sensor on board the NOAA series of polar orbiting satellites (Tucker et al., 1991). Short wavelength ($\lambda \sim 1$ cm) microwave sensors in space on the Nimbus-7 satellite have also been used to monitor vegetation conditions in sparsely vegetated areas susceptible to desertification (Choudhury, 1989). Another example is the atmospheric temperature observations made with a microwave sounding unit, also on the NOAA series of polar orbiters (Spencer and Christy, 1990). We propose that a long wavelength ($\lambda \approx 20$ cm) microwave radiometer could provide a similar data set for surface soil moisture observations. While case studies with existing microwave sensors in space have shown sensitivity to soil moisture conditions (Owe et al., 1992), there is no satellite sensor in space or planned that will adequately monitor surface moisture conditions. In this paper, we will present the case for such a sensor and present a brief description of a possible satellite implementation.

Water stored in the soil serves as a reservoir for the evapotranspiration (ET) process on land surfaces; therefore, knowledge of the soil moisture content is important for partitioning the incoming solar radiation into latent and sensible heat components. While there is no remote sensing technique which directly observes the amount of water in this reservoir, microwave remote sensing at long wavelengths ($\lambda > 10$ cm) can give estimates of the moisture stored in the surface 5-cm layer of the soil (Jackson and Schmugge, 1986; Schmugge, 1989). At microwave frequencies, the most striking feature of the emission from the Earth's surface is the large contrast between water and land. This emissivity contrast is due to the large difference between the dielectric constant (k) of water (~80) and that of dry soils (~3.5). The former arises from the ability of the electric dipole of the water molecule to align itself in response to the electric field at microwave frequencies. The dielectric constant for ice is as large at low frequencies but, because of the binding of the water molecule in the solid, the dipole rotation is inhibited at low frequencies, about 10^4 Hertz. For liquid water, this reduction in the

NATO ASI Series, Vol. I 46
Land Surface Processes in Hydrology
Trials and Tribulations of Modeling and Measuring
Edited by S. Sorooshian, H. V. Gupta, and S. C. Rodda

molecule's ability to rotate does not occur until about 10^{10} Hertz, i.e., in the microwave range. The dielectric constant is of importance here because it describes the propagation characteristics of an electromagnetic wave in the medium. These characteristics include the velocity of propagation, the wavelength in the medium, and the absorption of energy in the medium. The square root of the dielectric constant is the index of refraction (n) for the material. It is the contrast in n at the boundary between two media which determines the reflection and transmission coefficients of electromagnetic wave at such a boundary.

When water is added to dry soil, the real part of the dielectric constant changes from about 3.5 to more than 20 for the wet soil. The resulting emissivity (ϵ) changes from 0.96 for a dry smooth soil to less than 0.6. This dependence of ϵ on moisture content has been observed with radiometers operating from tower, aircraft, and spacecraft platforms. The sensitivity of ϵ to soil moisture depends on surface parameters such as roughness and vegetation cover and soil properties such as density and texture. The most important of these is vegetation, because it can totally mask the soil surface by absorbing the radiation emitted from the soil surface and re-emitting at its temperature. At the 21-cm wavelength, the amount of vegetation required for this screening is about 6 kg/m^2 in terms of its plant water content (W) (Jackson and Schmugge, 1991).

2. Microwave Radiometry

A microwave radiometer measures the thermal emission from the surface and, at these wavelengths, the intensity of the observed radiation is proportional to the product of the thermodynamic temperature of the soil and the surface emissivity (Rayleigh-Jeans approximation to the Planck radiation law). This product is commonly called the brightness temperature (T_B). The factors affecting radiometric observations over land are represented schematically in Figure 1. The left side of this figure presents the equations representing these factors for measurements made by a microwave radiometer on a remote platform, e.g., an aircraft or a satellite. For simplicity, the equations in Figure 1 are given for nadir observations only. At the bottom, the emissivity for the soil is determined by its dielectric properties and the surface roughness. The radiation from the soil passes through the vegetation layer in which some is absorbed and re-emitted at the temperature of the vegetation. There is a similar, but much weaker, effect by the overlying atmosphere. The magnitude of several of these is given in the right portion of the figure. These factors will be discussed in some detail in the succeeding sections.

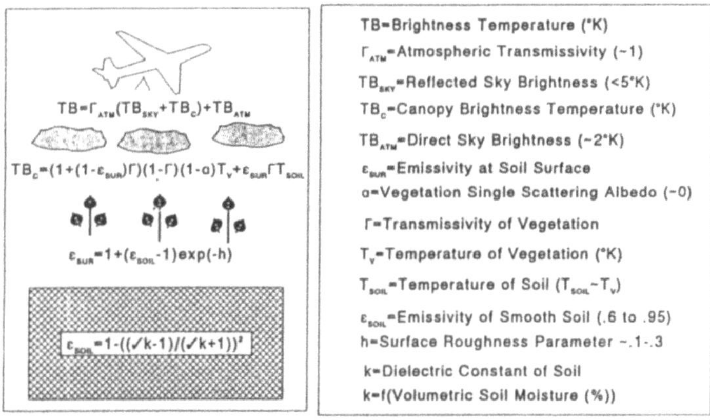

Figure 1. Schematic diagram of the factors affecting the microwave emission from land surfaces as observed by a radiometer on an aircraft or satellite. The angle of observation is assumed to be nadir.

3. Atmospheric Effects

One of the other major advantages for microwave remote sensing is the relative transparency of the atmosphere at these wavelengths. There are two interactions of microwave radiation with the atmosphere: absorption arising from the transition between the molecular energy states of the atmospheric gases and the scattering and absorption by hydrometeors in clouds and rain. The atmospheric gases which absorb microwave radiation are water vapor and oxygen. The H_2O absorption lines are at frequencies of 22.2 GHz(1.35 cm) and 183 GHz (0.164 cm), while the O_2 absorption is due to a band of lines centered about 60 GHz (0.5 cm) and a single line at 118 GHz (0.25 cm). The effect of these two gases can be calculated using their known absorption coefficients and their densities in the atmosphere. The results show that, for frequencies below 10 GHz (wavelengths > 3 cm), the effect of these gases is negligible with there being greater than 99% transmission through the entire atmosphere.

The effect of clouds will depend on the phase and size of the hydrometeors. An electromagnetic wave incident on a particle produces a motion of the charges contained within the particle. These moving charges will absorb some of the incident energy and reradiate some, producing a scattered field. Thus, some of the incident energy is scattered out of the beam and some is absorbed by the particle. As we noted earlier, the dielectric properties of water in the liquid state are much greater than those of the solid; therefore, the effects of the liquid hydrometeors will be greater than those of ice. The major effects of clouds and rain can be summarized as:

(1) Clouds of ice particles, i.e., cirrus clouds, have negligible effect on the microwave radiation because of the small particle sizes and the low dielectric constant of the ice;

(2) Because of the small particle sizes in non-raining clouds, e.g., stratus and fair weather cumulus, there is negligible effect for frequencies below about 15 GHz. The absorptivity for clouds of these types is proportional to the square to the frequency; and

(3) For rain, the scattering by the larger hydrometeors is significant for frequencies above 3 GHz, as seen by the use of weather radars at these frequencies.

Because the most effective frequencies for the microwave remote sensing of soil are between 1 and 5 GHz, the effects of the atmosphere will be small.

From Figure 1, we see that the intensity of the radiation observed by a radiometer is given by:

$$T_B = \Gamma_{atm} (T_{Bsky} + T_{Bcan}) + T_{atm} \tag{1}$$

The first term is the reflected sky brightness which depends on the atmospheric conditions and frequency and the surface or canopy reflectivity. For the frequencies of interest to us, the downwelling sky brightness under the normal range of atmospheric conditions is 5 to 10K. With a surface reflectivity, $R = 1 - \epsilon$, of at most 0.4 (for wet smooth soil), the downswelling sky brightness is generally less than 5K. The third term is the direct atmospheric contribution and as noted will be about 2K or less at the longer wavelengths. The atmospheric transmission, Γ_{atm} ,will typically be about 99%; hence, we are left with the emission from the surface, i.e., the second term as the main contributor to the observed T_B. The magnitude of error resulting from neglecting the atmospheric effect is less than 2%, i.e., 4 or 5K, and these effects are relatively constant for a non-raining atmosphere.

A more complete discussion of atmospheric effects can be found in Chapter 5 of the Manual of Remote Sensing (Colwell, 1983) and in Volume I of Microwave Remote Sensing by Ulaby et al. (1982).

4. Vegetation Effects

The simplest way to think of the effects of vegetation on the microwave emission from the soil is in terms of a cloud which can absorb and re-emit radiation, i.e., similar to the atmospheric situation. This assumes that scattering is negligible, which is true at the longer wavelengths. For a sufficiently thick layer of vegetation, only the radiation from the vegetation itself will be observed. The brightness temperature above the canopy is given by:

$$T_{Bcan} = (1 + (1 - \epsilon_{sur})\Gamma)(1 - \Gamma)(1 - a) T_v + \Gamma \epsilon_{sur} T_{soil} \tag{2}$$

For a = 0, i.e., no scattering, and $\Gamma = \exp(-\tau_p)$ this reduces to:

$$T_{Bcan} = T_v \left[1 + (\epsilon_{sur} - 1)\exp(-2\tau_p) + \epsilon_{sur}(T_{soil}/T_v - 1)\exp(-\tau_p) \right] \tag{3}$$

where $\tau_p = l\alpha \sec\theta$ is the one-way canopy absorption factor or optical depth, l and α are the thickness and absorptivity of the vegetation, and T_{soil} and T_V are the temperatures of the soil and vegetation. Because the temperatures are expressed in Kelvins, the ratio T_{soil}/T_V typically will be close to one, within 1 or 2%, and where the difference is large, e.g., a sparse canopy, the vegetation absorption will be small, i.e., $\exp(-2\tau_p) \approx 1$. Thus, Equation (3) reduces to:

$$\epsilon = T_B / T_{soil} = 1 + (\epsilon_{sur} - 1)\exp(-2\tau_p) \tag{4}$$

In an analysis of several data sets, Jackson and Schmugge (1991) found that:

$$\tau_p = bW \tag{5}$$

where W is the weight of water per unit area in kg/m^2, and b is a plant-dependent parameter. This relation holds for a variety of plants, at least at the longer wavelengths. The factor b is approximately proportional to the frequency; at 21-cm, it is about 0.1 ± 0.03 for crops ranging from small grains or grasses to broadleaf plants such as maize or soybeans. Figure 2 shows a plot of this parameter as a function of wavelength. It illustrates two important points: First, as the wavelength decreases, the vegetation factor b will increase. This means that the sensitivity to vegetation water content and the attenuation by the canopy will increase. The general trend of the data indicates that the rate of change is rather small down to 10 cm and then increases rapidly. The second point concerning Figure 2 is the scatter within any wavelength region. As λ decreases, the scatter increases. This is due to the increase in importance of vegetation structure (and scattering phenomena) at shorter wavelengths. If one wishes to extract soil moisture at these shorter wavelengths, accounting for vegetation will require the use of more sophisticated models and a great deal more information concerning the canopy that may not be available through remote sensing or data bases. Even if such information is available, the possibility of obtaining an accurate estimate is unlikely. Thus, to quantify the effect of vegetation on the microwave emission from soils, it is necessary to have some estimate of the plant above-ground biomass.

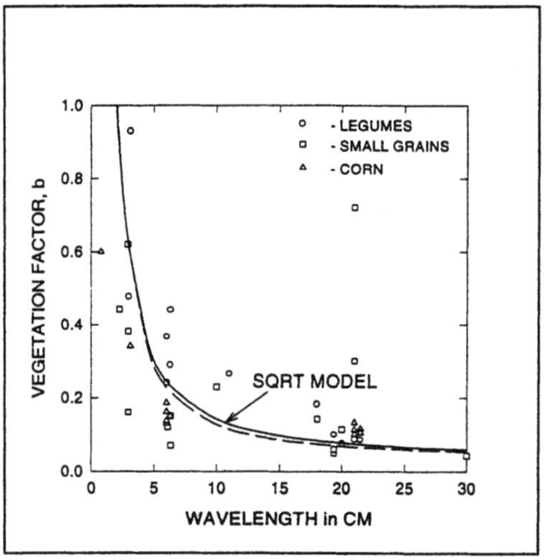

Figure 2. Observed values of b compared with model predictions. The solid curve is for a model of the dielectric behavior of vegetation, and the dashed curve is fitted to a $1/\lambda$ function.

A possibility for doing this lies with the use of visible and near infrared reflectance data which are strongly dependent on vegetation conditions. In one such analysis, Theis et al. (1984) used the perpendicular vegetation index calculated from visible and near IR as an estimator of plant biomass. The PVI, which is directly related to plant biomass, was used as a vegetation corrector in an algorithm to estimate soil moisture from 21-cm brightness temperature data. With this correction, the correlation between the microwave data and soil moisture improved from 0.09 to 0.75. This is an indication of the promise that the use of multispectral data can greatly improve our ability to estimate soil moisture remotely.

5. Surface Roughness

Another perturbing effect or noise factor on the microwave emission from soils is surface roughness. Qualitatively, roughness decreases the reflectivity and thus increases the emissivity. A rough surface can be thought of as having an increased area through which it can emit thermal energy. Using a simple model, Choudhury et al. (1979) showed that the increase in the emissivity can be expressed as:

$$\Delta\epsilon = R_o \, [1 - \exp(-h)] \tag{6}$$

where R_o is the smooth surface reflectivity and h is an empirical roughness parameter which increases as the surface height variance increases and the horizontal correlation length decreases. The bare soil emissivity is then:

$$\epsilon_{sur} = 1 - R_o \exp(-h)$$

or:

$$= 1 - (1 - \epsilon_{soil}) \exp(-h) \tag{7}$$

Both the vertical and horizontal scales of the roughness affect the emissivity. Values of h range from 0 for a smooth field to about 0.5 for a very rough field (Choudhury et al., 1979). For dry fields with $R_o < 0.1$, $\Delta\epsilon$ will be small, ≈ 0.02, while for wet fields with $R_o \approx 0.4$, the increase in emissivity will be larger, $\Delta\epsilon \approx 0.16$. The effect will be to decrease the sensitivity of the observed microwave emission to soil moisture. In practice, it has been found that $h \approx 0.1$ to 0.2 so that the decrease in sensitivity given by $\exp(-h)$ is on the order of 10 to 20%. This factor may be neglected and contribute to the noise or uncertainty in the soil moisture extraction. However, it may be possible to use supporting data from other sensors to correct for roughness effects, in this case from active microwave sensors, i.e. radars. In analyzing a data set containing both 1.6 GHz scatterometer and 1.4 GHz radiometer data, Theis et al. (1986) observed that the influence of surface roughness on the radar backscatter ($\sigma°$) was substantial at large incidence angles, while the sensitivity of $\sigma°$ to soil moisture remained about the same at 40° incidence angles. This fact enabled them to quantify the degree of roughness which was then used to correct for roughness in the radiometer data. Using this approach, they improved the correlation (R^2) of T_B with soil moisture from 0.69 to 0.95. Although conducted on only a limited data set, these encouraging results indicate the potential of multisensor techniques for improving microwave estimates of soil moisture.

By comparing the magnitude of the effects of surface roughness and vegetation, we concluded that the major perturbing factor affecting radiometer measurements for soil moisture detection is the amount of vegetation cover, because it can completely destroy the sensitivity to moisture variations for sufficiently thick covers. This will occur at the 21-cm wavelength for vegetation water contents in excess of 6 kg/m^2. The result of these two perturbing factors can be expressed as:

$$\epsilon_c = 1 + (\epsilon_{soil} - 1)\exp(-h - 2\tau) \tag{8}$$

where ϵ_{soil} is the smooth soil emissivity. Thus, without a priori knowledge of either roughness or vegetation amount, it will not be possible to separate the effects of these two factors in an experiment.

6. Dielectric Properties of Soils

As noted in the Introduction, the dielectric constant of soil increases from about 3 or 4 when dry to almost 30 when wet. While the range of variation is about the same for most soils, there are differences among the soils at the lower soil moistures. These arise from textural differences, i.e., the distribution of soil particle sizes. Wang and Schmugge (1980) observed that the initial water added to a soil is tightly bound to the particle surfaces and thus has dielectric properties somewhere between those of bound molecules in ice and those of freely rotating molecules in the liquid. Only after there are several layers of water on the particle surfaces does the water begin to behave like a liquid in its dielectric properties. Dobson et al. (1985) found that separation of the soil water into bound and bulk components depends on soil texture or, more directly, on the specific surface area of the soil. Their model treats the soil as a host medium of dry soil solids containing randomly distributed inclusions of bound water, bulk water, and air. With such models, it is possible to study the sensitivity of a soil's dielectric properties to such factors as density, texture, salinity, etc. (e.g., Jackson and O'Neill, 1987).

7. Emissivity and Reflectivity

Knowledge of the dielectric properties can be used in the Fresnel equations to calculate the reflectivity and emissivity for a soil with a smooth surface (Ulaby et al., 1982). In their simplest form they are:

$$R_V = \left| \frac{[\cos(\theta) - \sqrt{k - \sin^2(\theta)}]}{[\cos(\theta) + \sqrt{k - \sin^2(\theta)}]} \right|^2 \tag{9}$$

where θ is the incidence angle and k is the dielectric constant for the soil. For the horizontal polarization, the equation is:

$$R_H = \left| \frac{[k\cos(\theta) - \sqrt{k-\sin^2(\theta)}]}{[k\cos(\theta) + \sqrt{k-\sin^2(\theta)}]} \right|^2 \tag{10}$$

The horizontal polarization is the one with the electric field parallel to the soil surface, and the vertical polarization is where the electric field has a component perpendicular to the soil surface.

8. Soil Temperature

The intensity of radiation just above the soil surface is given by:

$$T_B = (1 - R)\,T_{soil} = \epsilon\, T_{soil} \tag{11}$$

from Equation (4). This is true for an isothermal soil; however, when there are variations of soil temperature, the factor T_{soil} is replaced by the integral:

$$T_{soil} = \int_{-\infty}^{0} T(z)\,\alpha(z)\exp\left[-\int_{z}^{0}\alpha(z')\,dz'\right]dz' \tag{12}$$

where:

$$\alpha(z) = 4\frac{\pi}{\lambda}\sqrt{k''(z)} \tag{13}$$

where T(z) and $\alpha(z)$ are the temperature and absorptivity at the depth z in the soil. The latter is seen to be proportional to the imaginary part of the refractive index and will be larger for wet soils. The integral essentially gives the intensity of the upwelling thermal radiation from the soil and is a weighted average of the soil temperature over the electromagnetic skin depth of the soil. In a study for a range of soil profiles (Choudhury et al., 1982), the value of T_{soil} was found to be given by:

$$T_{soil} = T_{deep} + c\cdot(T_{surf} - T_{deep}) \tag{14}$$

where T_{surf} and T_{deep} are the surface and deep soil temperatures. The former may be obtained from infrared surface temperature observations and the latter may be estimated from climatology. The constant, c, is a function of frequency and is larger at shorter wavelengths; it is 0.25 at the 21-cm wavelength. This result indicates that the effects of variations in surface temperature are reduced at the longer wavelengths.

9. Sampling Depth

The factor ϵ, the emissivity, gives the fraction of this upwelling radiation which is transmitted into the air, and it is this factor which shows the primary sensitivity to the soil moisture content. The critical factor is: What is the thickness of the layer at the surface whose dielectric properties determine e? It is this thickness which determines the soil moisture sampling depth for microwave sensors. An early theoretical study of the radiative transfer in a soil by Wilheit (1978) showed that this depth is only on the order of a few tenths of a wavelength thick, or about 2 to 5 cm at the 21-cm wavelength. This result has been confirmed by the experimental work of Newton et al. (1982) and Wang (1987).

10. Large-Scale Mapping Experiments

With this background, it is possible to apply this approach to map surface soil moisture in large-scale field experiments. Here we discuss results from one such experiment, Monsoon '90, as an example of the type of soil moisture maps the system described in the previous section can produce. The microwave brightness temperature (T_B) data presented here were acquired with an L-band ($\nu = 1.42$ GHz; $\lambda = 21$-cm) PushBroom Microwave Radiometer (PBMR) aboard the NASA C-130 aircraft. The PBMR has four horizontally polarized beams pointing at ±8° and ±24° from nadir. Each beam has a full width at half maximum power of about 16° so that the total swath is about 1.2 times the aircraft flight altitude (Harrington and Lawrence, 1985). Other good examples can be found in the PBMR results from the First ISLSCP Field Experiment (FIFE) (Wang et al., 1989, 1990) conducted in Kansas in 1987 and 1989 and HAPEX-MOBILHY conducted in France in 1986 (Nichols et al., 1993). In FIFE, the PBMR was successfully able to map surface soil moisture variations over a tall grass prairie.

The Monsoon '90 experiment was conducted during the summer of 1990 at an arid watershed in southcentral Arizona in the United States (Kustas et al., 1991). The purpose was to observe moisture fluxes in an arid climate during a drying period after a rain and to verify the role of remote sensing in determining these fluxes. The site selected is the well- instrumented Walnut Gulch Experimental Watershed operated by the Southwest Watershed Research Center of the USDA Agricultural Research

Service in southeastern Arizona. The region has 250-500 mm of annual precipitation with the majority falling during a summer "monsoon season" in July and August. This period was chosen for the experiment. Vegetation at the site is mostly sparse shrubs and desert grasses. There is a transition in the dominant vegetation types and density across the watershed. The eastern upland part of the watershed is classified as desert grassland, while the western part is classified as a desert steppe-shrub community. The biomass measured at the met sites ranged from 0.1 to 0.5 kg/m^2 at seven of the sites, with the eighth having slightly more than 1 kg/m^2. The near surface soils are coarse in texture, ranging from sandy loams to gravelly loamy sands.

A series of 6 PBMR flights was completed between 31 July (day 212) and 9 August (day 221) of 1990 at an altitude of 600 m which provides a spatial resolution of about 200 m. To provide complete coverage of the study area, seven 20-km flight lines were set up separated by 500 m. They lie roughly in an east-west direction. The flights were generally between 9:00 and 11:00 a.m. (local standard time, or 16 to 18 GMT). Volumetric soil moisture for the surface 5 cm of the soil was measured at 8 met sites on a daily basis (Schmugge et al., 1994). The measurements were made gravimetrically and converted to volumetric values using measured soil bulk densities.

The study period was characterized as very dry before the first flight on July 31 (day 212) with soil moisture <5%, followed by up to 5 cm of rain on August 1 over most of the study area. Thus, there was a significant decrease (50 to 60 K) in observed T_B during the next flight on August 2 (day 214) due to increases up to 20% in soil moisture. The succeeding flights on 4, 5, 8, and 9 of August (days 216, 217, 220, and 221, respectively) showed the effects of some lighter rains and the subsequent drydown of the study area. On day 221, the values of T_B rose to almost their pre-rain values. An example of the T_B variation is shown in Figure 3, where contour plots of T_B on days 212 and 214 and the rain amounts in mm on day 213 are presented. The top panel presents the T_B contours on the day before the rain, and the values are 280K or greater with the exception of the cool spot near the southwest corner of the map which is the town of Tombstone, Arizona with its many metal roofs. The locations of the 8 met stations are indicated by the triangles. The middle panel presents the rainfall contours for the rain on day 213 (August 1) with the + signs indicating the location of the rain gages. The amounts range from less 10 mm at the west side of the area to more that 50 mm just east of site 4. The bottom panel presents the T_B contours on day 214 after the rain and shows that T_B has decreased to less than 230 K in those areas where the rain was the heaviest. All of the area that received more than 10 mm of rain had T_B <240K, i.e., the 240K contour corresponds roughly to a 10-mm contour rain. When the change in T_B was compared to the rain amounts, there was a linear decrease up to rain amounts of 25 or 30 mm. Above this level, the surface soil apparently saturates and there is no further decrease in T_B. This relationship is pursued further in the recent paper by Jackson et al.(1993). In the T_B maps for all six flights, we observed no sensitivity to slope or topography which wasn't related to moisture differences.

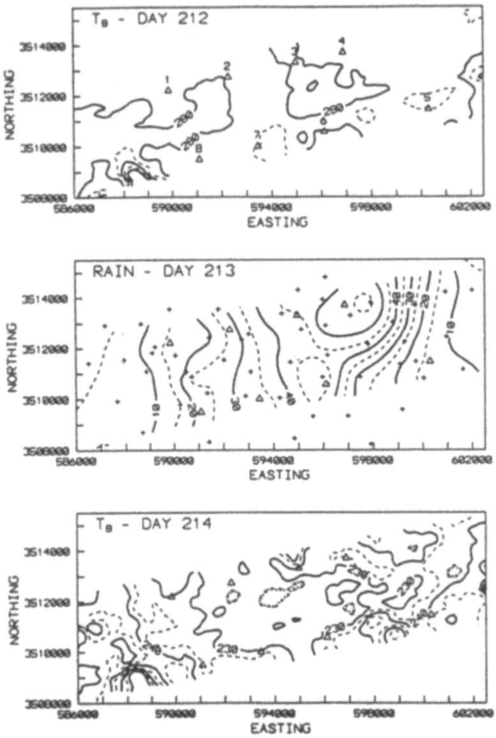

Figure 3. Contour maps of T_B and rain amounts in Monsoon 90. Top: T_B in K for day 212. Middle: rainfall in mm for day 213. Bottom: T_B in K for day 214. The Δ's are the location of the 8 met sites. The EASTING and NORTHING coordinates are in meters.

The comparison between the ground measurements of soil moisture in the 0 to 5-cm layer and T_B for the eight sites from the six flights is shown in Figure 4. The smooth curves are plots of Equation (8) for a soil with a texture similar to that found in Walnut Gulch with a range of surface parameters, $\alpha = h + 2\tau$, between 0 and 0.2. It is seen that the two curves roughly bracket the data. The lower curve for $\alpha = 0$ is the behavior expected for a bare smooth soil, and the upper curve for $\alpha = 0.2$ is the behavior expected for a slightly rough soil, $h = 0.1$, with a small amount of vegetation cover, $W = 0.5$

decrease up to rain amounts of 25 or 30 mm. Above this level, the surface soil apparently saturates and there is no further decrease in T_B. This relationship is pursued further in the recent paper by Jackson et al. (1993). In the T_B maps for all six flights, we observed no sensitivity to slope or topography which wasn't related to moisture differences.

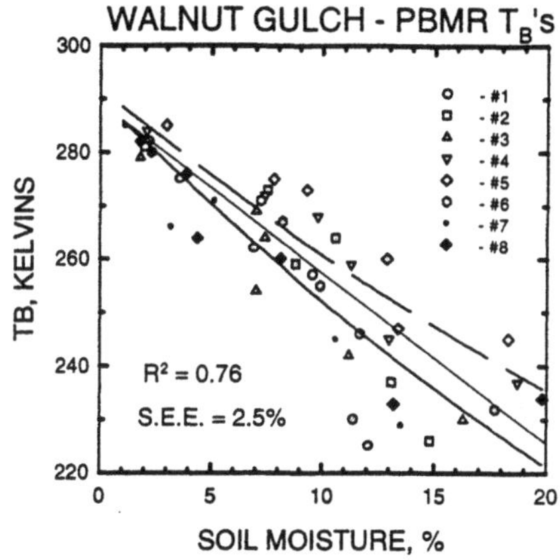

Figure 4. Comparison of PBMR T_B with ground measurements of volumetric soil moisture in the 0 to 5-cm layer. The linear regression results are shown.

The comparison between the ground measurements of soil moisture in the 0 to 5-cm layer and T_B for the eight sites from the six flights is shown in Figure 4. The smooth curves are plots of Equation (8) for a soil with a texture similar to that found in Walnut Gulch with a range of surface parameters, $\alpha = h + 2\tau$, between 0 and 0.2. It is seen that the two curves roughly bracket the data. The lower curve for $\alpha = 0$ is the behavior expected for a bare smooth soil, and the upper curve for $\alpha = 0.2$ is the behavior expected for a slightly rough soil, $h = 0.1$, with a small amount of vegetation cover, $W = 0.5$ kg/m^2, which is in reasonable agreement with the average value of the measured biomass. The straight line is the regression result for all 8 sites with an $R^2 = 0.76$. There was some difference in behavior among the eight sites with those at the western end, Nos. 1 and 8, having a slightly greater sensitivity of T_B to soil moisture which may be soils related.

Using these relationships, soil-moisture maps were derived from the T_B maps. Examples from the eastern portion of the watershed are given in Figure 5 for days 216 and 217. Recall that the cell size, PBMR resolution, used to generate these maps was 200m; thus, small-scale soil variations would not be seen. The spatial patterns qualitatively follow either the topography or soil patterns, mostly the former. For example, the main drainage running from the middle of the eastern edge toward the southwest is apparent in both images as is the increased wetness in the northeast corner, where the rain on day 213 was the heaviest. The availability of data on these successive days enabled us to evaluate the change in moisture storage and to compare it with estimates of the evapotranspiration flux at the met sites. This was done by multiplying volumetric moisture content by the thickness of the layer, 50 mm, to determine the amount of water stored in this layer. The difference or water loss is lower boundary for an estimate of ET. Figure 6 shows a contour plot of the "ET" flux derived using this approach. The values are generally between 3 and 4 mm, which is about 1 mm less than observed at the met sites. This discrepancy is probably due to our neglect of fluxes through the lower boundary of 0 to 5-cm soil layer, i.e., any recharge from below or extraction by the roots below this layer.

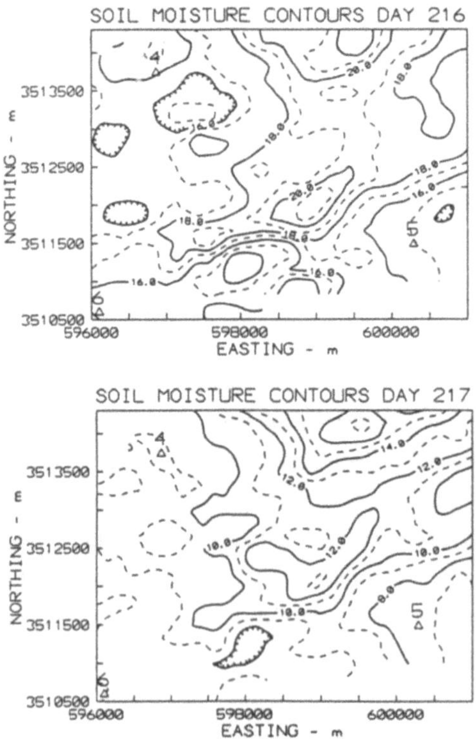

Figure 5. Contour plots of the volumetric soil moisture in percent for the 0 to 5-cm layer of the soil. They are for the eastern portion of the study area and include met sites 4, 5, and 6 which are indicated by the numbered Δ's. The upper figure is for day 216 and the lower for day 217.

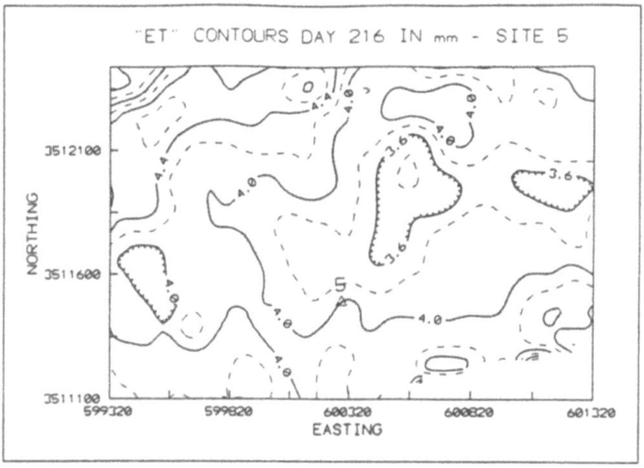

Figure 6. Estimates of the "ET" flux or water loss from the 0 to 5-cm layer of the soil for the same area as shown in Figure 5.

While it may not be possible to use this approach to estimate ET fluxes in general, it indicates that knowledge of surface soil moisture on a repetitive basis could be used to estimate the direct soil evaporation component of the ET flux (Chanzy and Kustas, 1993). Experiments such as Monsoon '90, FIFE, and the recently completed Washita '92 (Jackson et al., 1994) and HAPEX-Sahel have demonstrated the capabilities of long wavelength radiometers to monitor surface soil moisture from aircraft platforms. The next step will be to do it from space.

11. An Optimal System for Soil Moisture Sensing

Although there are several satellite-based microwave radiometers currently in operation or planned for the future, none of these is capable of providing reliable estimates of surface soil moisture. This is because they operate at short wavelengths (< 5 cm). Based upon the theory described in the previous sections and experimental results, an optimal system for soil moisture can be specified.

Of all the system design parameters, wavelength of observation is the most critical. There are several factors to consider:

(1) The dielectric constant of water decreases as wavelength decreases, and this is reflected in the dielectric properties of wet soils. The effects of this decrease in the dielectric constant on the brightness temperature-soil moisture relationship are illustrated in Figure 7 for the bare soil curve. Here the slope of the relationship between soil moisture and brightness temperature, the sensitivity, has been computed as a function of wavelength. The sensitivity is relatively constant down to a wavelength of 5 cm. Below this it drops off rapidly. A loss in sensitivity makes instrument error and other noise factors more important and will decrease our ability to accurately estimate moisture.

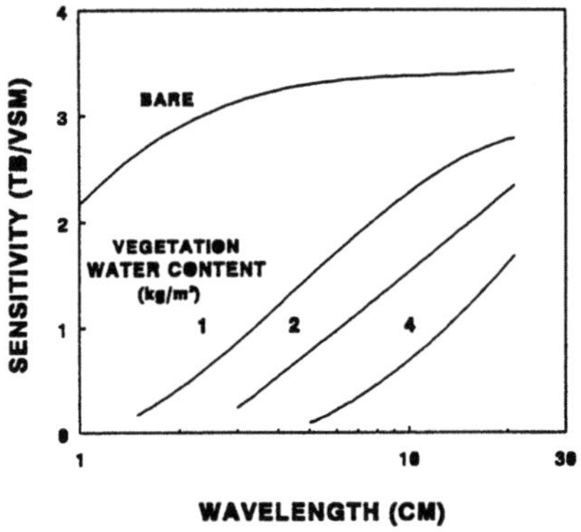

Figure 7. Sensitivity of T_B to soil moisture in units of kelvins/% soil moisture for bare and vegetated conditions.

(2) Attenuation by vegetation increases as wavelength decreases. As described previously, vegetation can mask the soil from the sensor. The correction algorithm presented in Equations (3) and (4) can be used to account for this effect. Figure 7 includes sensitivity curves for several values of the vegetation water content based on these equations and the work described in Jackson and Schmugge (1991). For the same canopy and vegetation water content, the sensitivity is reduced as wavelength decreases. While sensitivity to the soil moisture for bare soil remained about the same down to a wavelength of 5 cm, when vegetation is present the loss of sensitivity is more rapid. These results very clearly illustrate why longer wavelengths are more desirable for soil moisture estimation and the lack of information when shorter wavelengths (<5 cm) are used in the presence of vegetation.

(3) More ancillary vegetation information is necessary to estimate soil moisture at shorter wavelengths. Recent studies (Jackson and Schmugge, 1991) examined the effects of plant shape and wavelength on the vegetation parameter (*b*) used in Equation (5) and which were plotted in Figure 2. It illustrates two important points. First, that the attenuation is smaller at the wavelengths greater than 810 cm and, second, that scattering has a smaller effect at these wavelengths.

(4) The depth of the soil layer that contributes to the measured brightness temperature increases with wavelength. Radiative transfer theory shows that the brightness temperature observed by a sensor is the integrated result of the dielectric properties in a layer that is a few tenths of a wavelength thick.

Having said the above, the question arises: Why not go to wavelengths longer than the 21-cm? The reasons are practical because, at frequencies below 1 GHz, there is too much manmade interference for a radiometer to operate, and there is a radio astronomy quiet band at 1.4 GHz which allows the radiometer to operate relatively free of interference.

The implementation of this concept from satellite altitudes will require a large antenna to achieve useful spatial resolutions (~10 km). The spatial resolution in km is roughly determined by $\Delta x \sim H\lambda/D$, where H is the altitude in km and D is the linear dimensions of the antenna in m. Hence, to obtain 10 km resolution from a 500 km orbit will require an antenna size of at least 50λ or about 10 m. To reduce the mass of such a large antenna, various aperture synthesis techniques have been studied (Milman, 1988). A promising solution is the electronically scanned thinned array antenna (ESTAR) which consists of a partially filled array of linear antennae (Swift et al., 1991; Jackson et al., 1993). This affords about a 50% mass reduction, making a possible space flight more feasible.

12. Discussion and Conclusions

The accurate determination of soil moisture over large areas is a difficult problem using conventional in-situ approaches because of its small-scale variability. The large dielectric contrast between liquid water and dry soils at microwave frequencies enables remote sensing techniques at these frequencies to have the potential for making large area determinations of soil moisture. This paper reviewed the physical basis for the microwave remote sensing of soil and provides an assessment and examples of the current status of passive microwave techniques for soil moisture sensing. With microwave techniques, it is possible to estimate the soil moisture in the 0 to 5 cm surface layer under almost all non-raining cloud conditions, with an uncertainty of about 10% in relative terms. The sensitivity of the approach is effected by such factors as surface roughness, soil type, and most importantly vegetative cover, e.g., it is not possible to make a measurement through a heavy forest cover.

As described here, microwave remote sensing techniques can provide estimates of the soil moisture content for a surface layer about 5 cm thick. This depth is shallow compared to the 1 to 2 meter rooting depth of many crops. Estimating the root zone soil moisture from surface measurements has been studied using correlation techniques (Blanchard et al., 1981) and modeling studies (Jackson, 1980) which assume a moisture profile in hydraulic equilibrium. The conclusion from both approaches was that, if the water content of the surface 10 cm is known, the moisture content in the top meter could be calculated within acceptable limits, and that the lowest errors were obtained when the surface water contents were measured just before dawn.

Using a similar technique, Kondratyev et al. (1977) in the USSR obtained large area estimates of the pre-planting moisture stored in the top meter of the soil. In their study, they required knowledge of the moisture gradient for each soil type in the test area. This gradient describes the moisture profile assuming hydraulic equilibrium conditions. This approach had been implemented on a quasi-operational basis for irrigation scheduling, drainage, and structural surveys in the former Soviet Union (Reutov and Shutko, 1986, 1990) and in Bulgaria (Kostov, 1992). A review of these approaches is given by Kostov and Jackson (1992).

Repetitive observations of the surface soil moisture can also be used to estimate soil properties. Camillo et al. (1986) coupled moisture and energy balance models with remotely sensed data of surface soil moisture to estimate the soil's hydraulic conductivity, matric potential, and a soil texture parameter. The derived conductivities agreed with values measured with an infiltration ring. This is potentially a very useful technique for estimating the soil's average hydraulic properties over large areas.

Alternatively, knowledge of the surface layer moisture can be used to estimate moisture fluxes at the soil surface. These are of interest in themselves and can be used in moisture balance models to estimate the moisture in the profile. In Australia, Barton (1978) used soil moistures as estimated from an airborne radiometer in a model to estimate the evapotranspiration from a grassland. In France, Bernard et al. (1981) used a similar approach using simulated radar backscatter data. In a follow-up paper, they verified the technique using field measurements with a C-band radar system (Prevot et al., 1984). Both groups reported considerable success in estimating ET rates. In an analysis of the data from Monsoon '90, Kustas et al. (1993) found a correlation between T_B and evaporative fraction, EF = LE/(LE + H), where LE and H are the latent and sensible heat fluxes at the surface. This was true for EF > 0.5 and indicates that the microwave data are giving valuable information of the direct evaporation from the soil.

When these estimates of soil moistures were used to define pre-storm initial conditions for a distributed rainfall runoff model, the results were comparable to those obtained with conventional measurements (Goodrich et al., 1993). The spatial nature of the remotely sensed data makes it possible to incorporate the variations of the initial conditions in such models.

A remaining analysis task with these data sets is the development of a means to correct for vegetation using remotely sensed data. Approaches using various vegetation indices from visible and near infrared data will be considered. An example of this approach is given in the paper by van de Griend and Owe (1993). They used Normalized Difference Vegetation Index (NDVI) values from the NOAA AVHRR sensor to estimate the vegetation optical depth for the 6.6 GHz channel of the Nimbus-7 SMMR. The use of this vegetation correction improved the correlation (R^2) of satellite-derived surface emissivity and large-scale surface soil moisture from 0.45 to 0.72.

However, we recognize that these approaches may not provide vegetation water content for thick and dense vegetation conditions. In the future, methods involving microwave polarization difference (Choudhury et al., 1989; Paloscia and Pampaloni, 1988, 1992) at the higher microwave frequencies should be investigated. In this case, the absorption of the microwave emission from the soil by the vegetation at shorter wavelengths reduces the polarization difference at off-nadir observations. This reduction can be used to estimate vegetation water content. The first opportunity to test this will be with data acquired during the HAPEX-SAHEL experiment conducted in Niger during the late summer of 1992, where both the PBMR and the French multifrequency radiometer, PORTOS (Grosjean and Sand, 1994), were flown on several joint flights. The aircraft version of PORTOS has channels at 5, 10.7, 23.8, 36.6, and 90 GHz and should provide a wealth of data for studying the use of polarization difference for vegetation mapping. There was also a dual polarized 37 GHz operating at an angle of 50° on the same aircraft as the PBMR.

The microwave remote sensing of soil moisture is at a threshold at the present time. Theoretical and experimental research over the past 10 to 15 years has defined the capabilities of microwave approaches for soil moisture sensing. Briefly summarized, these capabilities are: (a) the ability to measure the moisture content of a surface layer about 5 cm thick to a relative accuracy of about 10%; (b) the measurement can be made under all weather conditions and for light to moderate vegetation conditions (the limiting case appears to be a mature corn crop); and (c) the factors of surface roughness and soil texture will introduce uncertainties into the soil moisture determinations which can be mostly accounted for with ancillary data. Before these systems are flown on spacecraft, it will be necessary to demonstrate the utility of such a remotely sensed soil moisture measurement with further experiments of the type described above. It is always important to realize that remote sensing measurements will not provide as accurate or as deep a measurement of soil moisture as can be obtained by conventional in-situ measurements, but they do provide information on its spatial variation and can provide this information on the long-term repetitive basis required for climate change observations.

References

Barton IJ (1978) A case study comparison of microwave radiometer measurements over bare and vegetated surface *Journal of Geophysical Research*, 83:3513-3517.

Bernard R, M Vauclin, D Vidal-Madjar (1981) Possible use of active microwave remote sensing data for prediction of regional evaporation by numerical simulation of soil water movement in the unsaturated zone, *Water Resources Research*, 17:1603-1610.

Blanchard BJ, MJ McFarland, TJ Schmugge, E Rhoades (1981) Estimation of soil moisture with API algorithms and microwave emission, *Water Resources Bulletin*, 17:767-774.

Camillo PJ, PE O'Neill (1986) Estimating soil hydraulic parameters using passive microwave data, *IEEE Transactions of Geosciences and Remote Sensing*, GE-24:930-936.

Chanzy A, Kustas W (1994) Evapotranspiration monitoring over land surfaces using microwave radiometry, *Proceedings of ESA/NASA International Workshop on Passive Microwave Remote Sensing Research Related Land-Atmosphere Interactions*, 11-15 January 1993, Saint Lary, France, VSP, Zeist, The Netherlands.

Choudhury BJ, TJ Schmugge, A Chang, RW Newton (1979) Effect of surface roughness on the microwave emission from soils, *Journal of Geophysical Research*, 84:5699-5706.

Choudhury BJ, TJ Schmugge, T Mo (1982) A parameterisation of the soil temperature for microwave emission, *Journal of Geophysical Research*, 87:1301-1309.

Choudhury B (1989) Monitoring global land surface using Nimbus-7 37 GHz data: theory and examples, *International Journal of Remote Sensing*, 10:1579-1605.

Colwell RN (ed) (1983) Manual of Remote Sensing, Second Edition, American Society of Photogrammetry, Falls Church, Virginia.

Dobson MC, FT Ulaby, MT Hallikainen, M Reyes (1985) Microwave dielectric behavior of wet soil: part II. dielectric mixing models, *IEEE Transactions on Geosciences and Remote Sensing*, Vol. GE-23:35-46.

Grosjean O, A Sand (1994) Etalonnage du radiometre hyperfrequence portion et methode de correction de mesures, *Proceedings of 6th International Symposium on Physical Measurements and Signature in Remote Sensing*, 17-21 January, 1994, Val d'Isere, France, pp. 513-518.

Harrington R, RW Lawrence (1985) An airborne multiple beam, 1.4 GHz pushbroom microwave radiometer, *Proceedings of IGARSS '85*, Amherst, Massachusetts, pp. 601-606.

Jackson TJ (1980) Profile soil moisture from surface measurements, *Journal of Irrigation Drainage Division*, IR-2, ASCE, pp. 81-92.

Jackson TJ, TJ Schmugge, JR Wang (1982) Passive microwave sensing of soil moisture vegetation canopies, *Water Resources Research*, 18:1137-1142.

Jackson TJ, PE O'Neill (1986) Microwave dielectric model for aggregated soils, *IEEE Transactions on Geosciences and Remote Sensing*, GE-24:920-929.

Jackson TJ, ME Hawley, J Shuie, PE O'Neill, M Owe, V Delnore, RW Lawrence (1986) Assessment of preplanting soil moisture using airborne microwave sensors, Hydrologic Applications of Space Technology, *Proceedings of Cocoa Beach Workshop*, Florida, August 1985, IAHS Publication Number 160.

Jackson TJ, PE O'Neill (1987) Salinity effects on the microwave emission of soil, *IEEE Transactions on Geosciences and Remote Sensing*, Vol. GE-25:214-220.

Jackson TJ, TJ Schmugge (1986) Passive microwave remote sensing of soil moisture, *Advances in Hydroscience*, 14:123-159.

Jackson TJ, TJ Schmugge (1991) Vegetation effects on the microwave emission of soils, *Remote Sensing of the Environment*, 36:203-212.

Jackson TJ, DM LeVine, A Griffis, DC Goodrich, TJ Schmugge, C Swift, PE O'Neill (1993) Soil moisture and rainfall estimation over a semi-arid environment with the ESTAR microwave radiometer, *IEEE Transactions on Geosciences and Remote Sensing*, 31:836-841.

Jackson TJ, DM LeVine, C Swift, TJ Schmugge (1994) Large area mapping of soil moisture using the ESTAR passive microwave radiometer, *Proceedings of 6th International Symposium on Physical Measurements and Signature in Remote Sensing*, 17-21 January, 1994, Val d'Isere, France, pp. 467-474.

Kondratyev KY, VV Malentyev, Y Rabinovich, EM Shulgina (1977) Passive microwave remote sensing of soil moisture, *Proceedings of 11th International Symposium on Remote Sensing of the Environment*, 2:164-1661.

Kostov KG (1992) Passive microwave remote sensing of soil moisture: experimental and modeling results, Paper A.3-S.1.07, Proceedings of XXIX COSPAR Meeting, Washington D.C., 28 August-5 September 1992, *Advances in Space Research*, 13(5):105-114.

Kostov KG, TJ Jackson (1993) Estimating profile soil moisture from surface layer measurements: a review, *Proceedings of SPIE Symposium on Ground Sensing*, 14 April 1993, Orlando, Florida, 1941:125-136.

Kustas WP, DC Goodrich, MS Moran, S Amer, LB Bach, J Blanford, A Chehbouni, H Classen, WE Clements, PC Doraiswamy, P Dubois, TR Clarke, CST Daughtry, D Gellman, TA Grant, LE Hipps, AR Huete, KS Humes, TJ Jackson, TO Keefer, WD Nichols, R Parry, E Perry, RT Pinker, PJ Pinter, J Qi, AC Riggs, TJ Schmugge, AM Shutko, D Stannard, E Swiatek, JD van Leeuwen, J van Zyl, A Vidal, J Washburne, MA Weltz, (1991) An interdisciplinary study of the energy and water fluxes in the atmosphere-biosphere system over semi-arid rangelands: description and some preliminary results, *Bulletin of the American Meteorological Society*, 72:1683-1705.

Kustas WP, TJ Schmugge, KS Humes, MA Weltz, TJ Jackson, R Parry (1993) Relationships between evaporative fraction and remotely sensed vegetation index and microwave brightness temperature for semi-arid rangelands, *Journal of Applied Meteorology*, 32:1781-1790.

Milman A (1988) Sparse-aperture microwave radiometers for earth remote sensing, *Radio Science*, 23:193-205.

Mkrtchjan FA, EA Reutov, AM Shutko, KG Kostov, MA Michalev, NM Nedeltchev, AY Spasov, BI Vichev (1988) Experiments in Bulgaria for determination of soil moisture in the top one-meter layer using microwave radiometry and a priori information, *Proceedings of IGARSS '88 Symposium*, ESA SP-284:665-666.

Mo T, BJ Choudhury, TJ Schmugge, JR Wang, TJ Jackson (1982) A model for microwave emission from vegetation covered fields, *Journal of Geophysical Research*, 87:11229-11237.

Mo T, TJ Schmugge, JR Wang, Calculation of the microwave brightness temperature of rough soil surfaces: bare field, *IEEE Transactions on Geosciences and Remote Sensing*, GE-25:47-54.

Newton RW, QR Black, S Makanvand, AJ Blanchard, BR Jean (1982) Soil moisture information and thermal microwave emission, *IEEE Transactions on Geosciences and Remote Sensing*, GE-20:275-281.

Nichols WE, RH Cuenca, TJ Schmugge, JR Wang (1993) Pushbroom microwave radiometer results from HAPEX-MOBILHY, *Remote Sensing of the Environment*, 46:119-128.

Owe M, EB Jones, TJ Schmugge (1982) Soil moisture patterns observed in Hand County, South Dakota, *Water Resources Bulletin*, 18:949-954.

Owe M, AA Van de Griend, ATC Chang (1992) Surface soil moisture and satellite passive microwave observation in semi-arid southern Africa, *Water Resources Research*, 28:829-839.

Paloscia S, P Pampaloni (1988) Microwave polarization index for monitoring vegetation growth, *IEEE Transactions on Geosciences and Remote Sensing*, GE-26:617-621.

Paloscia S, P Pampaloni (1992) Microwave vegetation indexes for detecting biomass and water conditions of agricultural crops, *Remote Sensing of the Environment*, 40:15-26.

Prevot L, R Bernard, O Taconet, D Vidal-Madjar (1984) Evaporation from a bare soil evaluated from a soil water transfer model using remotely sensed surface soil moisture data, *Water Resources Research*, 20:311-316.

Reutov EA, AM Shutko (1986) Prior knowledge-based soil moisture determination by microwave radiometry, *Soviet Journal of Remote Sensing*, 5:100-125 (Harwood Academic Publishers GmbH).

Reutov EA, AM Shutko (1990) Microwave spectroradiometry of water content of nonuniformly moistened soil with a surface transition layer, *Soviet Journal of Remote Sensing*, 6:72-79 (Harwood Academic Publishers GmbH).

Schmugge T (1989) Microwave Remote Sensing of Soil Moisture, in F Toselli (ed) Applications of Remote Sensing to Agrometeorology, 257-284 ECSC, EEC, EAEC, Brussels, Belgium.

Schmugge TJ, TJ Jackson, WP Kustas, R Roberts, R Parry, D Goodrich, S Amer, MA Weltz (1993) PBMR observations of surface soil moisture in Monsoon '90, *Water Resources Research*, 30:1321-1327.

Spencer RW, JR Christy (1990) Precise monitoring of global temperature trends from satellites, *Science*, 247:1558-1562.

Swift CT, DM LeVine, CS Ruf (1991) Aperture synthesis concepts in microwave remote sensing of the Earth, *IEEE Transactions on Antennas and Propagation*, 39:1931-1935.

Theis SW, BJ Blanchard, AJ Blanchard (1986) Utilization of active microwave roughness measurements to improve passive microwave soil moisture estimates over bare fields, *IEEE Transactions on Geosciences and Remote Sensing*, GE-24:334-339.

Theis SW, BJ Blanchard, RW Newton (1984) Utilization of vegetation indices to improve microwave soil moisture estimates over agricultural lands, *IEEE Transactions on Geosciences and Remote Sensing*, GE-22:490-496.

Tučker CJ, HE Dregne, WW Newcomb (1991) Expansion and contraction of the Sahara Desert, *Science*, 256:299-301.

Ulaby FT, RK Moore, AK Fung (1982-86) Microwave Remote Sensing: Volumes I, II, and III, ARTECH HOUSE, Inc., Dedham, Massachusetts.

van de Griend A, M Owe (1993) Microwave vegetation optical depth and inverse modeling of soil emissivity using Nimbus/SMMR satellite observations, To be published in *Meteorology and Atmospheric Physics* Special Issue on Space-Based Microwave Remote Sensing of Hydrological Variables.

Wang JR, TJ Schmugge (1980) An empirical model for the complex dielectric permissivity of soils as a function of water content, *IEEE Transactions on Geosciences and Remote Sensing*, GE-18:288-295.

Wang JR (1987) Passive microwave sensing of soil moisture: the frequency dependence of microwave penetration depth, *IEEE Transactions on Geosciences and Remote Sensing*, GE-25:616-622.

Wang JR, TJ Schmugge, JC Shiue, ET Engman (1989) Mapping surface soil moisture with L-band radiometric measurements, *Remote Sensing of the Environment*, 27:305-312.

Wang JR, TJ Schmugge, JC Shiue, ET Engman (1990) The L-band PBMR measurements of surface soil moisture in FIFE, *IEEE Transactions on Geosciences and Remote Sensing*, GE-28(5):906-914.

Wilheit TT (1978) Radiative transfer in a plane stratified dielectric, *IEEE Transactions on Geosciences and Electronics*, GE-16:138-143.

Remote Sensing of Soil Moisture and Moisture Related Parameters by Means of Microwave Radiometry: Instruments, Data, and Examples of Application in Hydrology

A.M. Shutko

Chief Lab., Institute of Radioengineering and Electronics
Russian Academy of Sciences
Head, Institute of Geoinformatics, Ltd.
Moscow, RUSSIA

1. Introduction

The seven-year period of perestroika in the former Soviet Union has clearly shown that there are no borders between the environmental sciences developed in different countries. That is why it is important for me, as a representative of Soviet High-Technology in the applied sciences of Radio Waves Propagation and Remote Sensing by Radiophysical Methods and Means (designed in particular on the basis of military electronics), to participate in the NATO Advanced Research Workshop. Therefore, I am honored to present this paper on the state-of-the-art, based on the results of more than 20 years of activity in passive microwave remote sensing, to contribute through these data and future visions to our common understanding of global change and land surface processes in hydrology and thereby stimulate further work and contacts in this field. The main content of this presentation will be the role of passive microwaves in these studies.

2. Physical Background

Microwave radiometry is based on measurements of the natural electromagnetic radiation of environmental objects in the ranges of millimeter, centimeter, and decimeter wavelengths. Clouds and rain are the main sources of interference at wavelengths shorter than 5-8 mm, and galactic and ionospheric radiation as well as radar and TV installations are the main sources of interference at wavelengths longer than 20-30 cm. Between these bands, radiation is mostly sensitive to the free water content in soil. The sensitivity to other parameters, such as depth to a water table, salinity, temperature, and

NATO ASI Series, Vol. I 46
Land Surface Processes in Hydrology
Trials and Tribulations of Modeling and Measuring
Edited by S. Sorooshian, H. V. Gupta, and S. C. Rodda

biomass of vegetation, is dependent on wavelength (Shutko, 1982; Shutko, 1986; Shutko, 1987; Shutko, 1992).

3. Microwave Radiometers

Currently available instruments designed by the Special Construction Bureau at the Institute of Radioengineering and Electronics (IRE), formerly the Academy of Sciences of the USSR, and recently designed by the Institute of Geoinformatics (IGI) Ltd., consist of several portable and sensitive radiometers operating at different electromagnetic wavelengths ranging from 0.5 to 30 cm (frequency range from 1 to 60 GHz) (Shutko, 1985; Shutko, 1986; Reutov and Shutko, 1990; Shutko et al., 1991; Otte, 1992). A typical set consisting of two to three instruments may be packed into a briefcase (Shutko et al., 1991). The characteristic antenna size for radiometers operating at decimeter wavelengths is (60-80) x (60-80) x 3 cm. The instruments may be used in laboratory conditions or on mobile or aircraft platforms.

4. Feasibility of Partial Parameter Determination by Means of Microwave Radiometry

Software for the remote sensing of physical and chemical parameters of land and water surfaces by measuring the natural radiation at microwave frequencies has been developed by the IRE during the last 20 years (and during the last 3 years by the IGI) through extensive scientific research, studies, experimental tests, and practical applications (Shutko, 1982; Shutko, 1986; Shutko, 1987; Shutko et al., 1991; Shutko, 1991; Shutko et al., 1991; Reutov and Shutko, 1992; Shutko, 1992; Otte, 1992). The instrument measurements provide quantitative estimates for the detection of parameters. They are practically independent of illumination and meteorological conditions, are sensitive to physical and chemical parameters of land surface and water areas, and provide soil/canopy sensing through vegetation of agricultural types. Currently available procedures allow the measurement of the following parameters:

- soil moisture (volumetric free water) content within the top 20 cm without prior information and within the top 100 cm if prior knowledge is available and used;
- depth to a shallow water table down to 1.5-4 m, depending on soil type and soil-climatic zone;
- counters of water seepage from canals, etc.;
- crop biomass, conditions of agricultural and wetland vegetation;
- water quality in terms of general mineralization (conductivity) or salinity of the water surface (top 1-2 cm);
- thermal phenomena (physical temperature contrasts) at ground in volcano (fumarol) areas and at the sea surface (warm waste water discharge, etc.); and
- snow and ground melting-freezing conditions.

5. Feasibility of Generalized Parameter Determination by Means of Microwave Radiometry

It has been shown that passive microwave instruments carried by satellites can be used for operational measurements of some generalized parameters of soil/vegetation and ocean/atmosphere systems (Shutko, 1986; Shutko and Reutov, 1987; Grankov and Shutko, 1992; Shutko, 1992).

The dryness index is one of those parameters. It indicates the ratio between real incident solar energy and the hypothetic energy required to be incident on the surface to totally evaporate the moisture contained in the soil. The dryness index is an indicator of plant productivity. Variations of either temperature or moisture content from certain optimum values result in a decrease of potential yield. The expressions for brightness temperature (which indicates the intensity of microwave radiation and for dryness index expansions) are rather similar (within accuracy of 10-20%) and are functions of the average soil temperature and moisture content. It follows from this explanation that the brightness temperature maps of the Earth's land surface being obtained from satellites may provide access to operational inventories of areas with different potential plant productivity and may help in yield forecasting. In a similar fashion, passive microwave measurements can be used for determining the elements of energy budget studies.

Satellite microwave data can also be used for determining the heat flows in energy-active zones of the ocean to estimate effects of these on the seasonal variability of the weather of continental areas (Grankov and Shutko, 1992). Estimates of heat flows between ocean and atmosphere are available from radiometric information. The annual energy balance in the climatic machine ocean/atmosphere may be represented in the form of two-dimensional diagrams in terms of brightness temperature data measured at two different wavelengths. By the weather of continental areas, we mean the variations of temperature and soil moisture (precipitation) and thus the changes in the indices of dryness described above.

Other generalized parameters that may be mentioned include the polarization (or microwave vegetation) index which provides information dependent on the stress index and the type and state of the vegetation cover on the Earth.

6. Scientific and Operational Programs and Campaigns in Which Our Team Has Participated

The instruments and retrieval algorithms have been successfully tested since 1975 in the semiarid area of Crimea in the USSR. The main goals of these studies were the measurement of soil moisture, determination of depth to a shallow water table, and detection of the contours of water seepage from irrigation canals. Extensive and representative ground truth measurements were collected during these tests. Airplanes of different types, such as the AN-2 (one-engine), IL-12, IL-14 (two-engine), and IL-18 (four-engine) and helicopters of the MI-2 and MI-8 types were used as microwave instruments carriers. Nearly every year, experiments were conducted during all seasons, flying 300 to 700 hours annually and thereby observing the microwave behavior of various natural phenomena from the beginning of snow melt, through the moistening and drying cycles and until freezing and snow covering.

Between 1975 and the end of the 1980s, over 7 million hectares in 10 republics of the USSR have been covered, working in collaboration with the Ministries of Agriculture, Land Reclamation, Hydrometeorology, Ecology, Geology, Education, Defense, and the Academy of Sciences (Shutko, 1982, 1985, 1986; Shutko and Reutov, 1987; etc.--see List of References). Since 1983, scientific studies and applicational flights have been conducted in several Eastern European countries including Bulgaria, Hungary, Poland, and East Germany and also in Vietnam and Cuba. These studies were conducted either in the form of scientific programs (including those in the framework of the Intercosmos Council) or in the form of contracts. In 1990, we participated in the MONSOON '90 energy budget experiment at the USDA-ARS Walnut Gulch experimental watershed in Arizona at the invitation of Dr. T.J. Jackson of USDA, Beltsville Agricultural Research Center, Hydrology Lab. In these experiments, strong relationships between the remotely sensed data and in situ measurements of soil moisture related to meteorological conditions and surface relief were observed (Jackson et al., 1992).

Other scientific international programs in which our team has participated during the last 3 years include:

- a Russian-Ukrainian program on application of remote sensing in agriculture, land reclamation, hydrology, and ecology in a framework of Center of Geoecology activity;
- a similar Russian-Bulgarian program in a framework of the Association Interwaterautomatics activity;
- a Russian-Ukrainian-Bulgarian program on ecological studies of water bodies; and
- a Russian-Bulgarian-USA multidisciplinary program on application of remote sensing from satellites and aircraft in river deltas studies, wetland observation, epidemiology, etc. (Louisiana State University is a partner from the U.S. side).

In addition, several cooperative scientific programs in the framework of the space project Priroda (Nature) have been signed with a number of universities, institutes, and laboratories in USA, UK, France, Italy, Switzerland, and Canada. In addition, in keeping with the challenges of selling Soviet technology in the United States (Otte, 1992), we are attempting to develop commercial cooperation with American organizations in the area of application of passive microwave technology (Otte, 1992) and experiments are currently in preparation in the framework of commercial contracts.

7. Typical Areas of Application

Typical areas of application include (Shutko, 1982, 1985, 1986; Shutko and Reutov, 1987; etc.--see List of References):

- soil moisture mapping for determining sowing dates, filterization, optimization of rates, and dates of irrigation;
- shallow water table determination and detection of areas of water losses from canals, dams, and destroyed drainage systems;
- monitoring of marshy, wetland areas;
- mapping the biomass of rice and other crops over the water surface or wet grounds for harvest forecasting, determining the extent of flooded land overgrown with cane and other plants;
- control of inland waters and coastal zones in relation to salination, freshening and contamination due to river outflows and industrial urban water flows;
- detecting talic zones in permafrost areas during the winter season; and
- detecting areas of snow melting.

The characteristic areas where these data may be usefully applied are in:

- land reclamation
- water management
- agriculture
- agrometeorology

- hydrometeorology
- rice crop cultivation
- hydrology
- wetland and flood-land studies
- biocenology
- ecology
- forestry

8. User Services We Can Provide

These include (Shutko et al., 1991; Shutko, 1991; Shutko et al., 1991; Otte, 1992):

- aircraft observations, processing and thematic mapping of agricultural fields, irrigation systems, off-shore zones, outflow areas, lakes, and other surfaces (airplane or helicopter may be given by users; owner installs instrumentations on board of carrier during 1-2 days);
- software for retrieving above-mentioned parameters from measured microwave radiation;
- organization of joint aircraft laboratories and services for remote sensing and monitoring;
- organization of geoinformation monitoring systems (GIMS) on the base of GIS, remotely sensed data, in situ measurements and modeling;
- hardware (microwave radiometers at the wavelengths from 0.5 to 40 cm) and supporting services; and
- conducting joint scientific investigations.

9. Synergism

In order to increase the effectiveness of remote sensing of the environment (i.e., to increase the accuracy and productivity of mapping and to provide a wider range of measured parameters, etc.), multiband and multichannel measurements may be recommended to the user. In doing these, problems of synergism must be taken for consideration. In our investigations, we use the following definitions of synergism.

(1) *Parametric Synergism*: This type (or level) of synergism assumes application of one or more spectral bands in addition to the main one selected as the most informative for determination of a certain physical (geophysical) parameter. The main goals of this type of synergism based on multichannel measurements are an increase in accuracy, range, and conditions of determination of certain parameters, etc. A very important problem here is in optimization (including minimization) of the number of bands and wavelengths inside the bands. Examples are: (a) soil moisture determination using passive microwaves as a main band with the aid of a thermal infrared, and (b) water temperature determination using the thermal infrared band as the main one with due regard for atmospheric conditions using passive microwaves, etc.

(2) *Technical Synergism*. This type/level of synergism assumes solving problems of multiband and multichannel calibration, taking into consideration data of signal to noise ratio, footprint/pixel size, time of integration, rate/frequency of measurements, etc. The main goal of this type of synergism, based on adjusting technical parameters and operational characteristics of multiband and multichannel systems, consists in providing equal opportunities for all sensors to manifest their feasibilities with equal weighting functions.

(3) *Problematic Synergism*. This type/level of synergism assumes a much higher level of generalization of remotely sensed and a priori information than with levels 1 and 2. It includes not only multiband and multichannel measurements but also a multidisciplinary approach. The main goal of this type of synergism consists in focusing the scientific and technical advantages of spaceborne and airborne systems along with many types

of multidisciplinary prior knowledge-based information, on solving regional or global problems relevant to the customers, users (actual or potential), and the funding agencies. Examples are: (a) global changes: water content in soil and vegetation biomass, deforestation, warming-cooling, etc., (b) wetland studies: contours, dynamics, etc., and (c) weather forecasting, etc.

10. Conclusion

The results of scientific research and operational application presented in this paper show that the microwave radiometry methods for determining soil moisture, moisture related parameters, and some other physical and chemical characteristics of environmental objects are highly effective. Hence, they may play a valuable role in studies related to global environmental change and land surface processes in hydrology.

11. Acknowledgements

It is my sincere pleasure to express my thanks to my colleagues, Dr. E.A. Reutov, Dr. A.G. Grankov, Dr. B.M. Liberman, Dr. G.I. Chukhray, Dr. E.P. Novichikhin, Eng. Z.G. Yazerian, Eng. A.A. Milshin, Eng. A.A. Haldin, Eng. S.P. Golovachev, Eng. L.A. Parshina, and Eng. E.A. Shutko for their involvement in the research and operational application of the methods of microwave radiometry. I am also grateful to Dr. T. Jackson for his support in conducting joint experiments in the USA; to Dr. T. Schmugge for fruitful discussions; and to Profs. Y. Kerr and B. Choudhury for goodwill in having our team in the ESA/NASA field of view. My cordial thanks are also addressed to the organizers of this research workshop, especially to Prof. Soroosh Sorooshian of The University of Arizona, for the invitation to participate in the workshop, and the financial support which gave me the opportunity to realize the invitation and present this paper.

References

Shutko AM (1982) Microwave radiometry of lands under natural and artificial moistening, *IEEE Transactions on Geosciences and Remote Sensing*, GE-20:1:18-26.

Shutko AM (1985) (Interview) Radiometry for farmers, *Science in the USSR*, 6:97-102.

Shutko AM (1986) Microwave radiometry of water surface and grounds, Moscow, Nauka (Science) Publ. House (in Russian; English translation is available), 190 p.

Reutov EA, AM Shutko (1992) Prior knowledge-based soil moisture determination by microwave radiometry. *Soviet Journal of Remote Sensing* 5:100-125.

Shutko AM (1987) Remote sensing of the waters and lands via microwave radiometry (The principles of the method, problems feasible for solving, economic use). In: *Proceedings of Study Week on Remote Sensing and Its Impact on Developing Countries*, Pontifical Academy of Science, Vatican City, pp. 413-441.

Shutko AM, EA Reutov (1987) On interconnection between radiation field and generalized parameters of natural objects. In Abstracts of International Symposium Commission F URSI on Natural Signature in Remote Sensing, Goteborg, Sweden, Chalmers University of Technology, pp. 120-121.

Reutov EA, AM Shutko (1990) Microwave spectrometry of water content of nonuniformly moistened soil with a surface transition layer, *Soviet Journal of Remote Sensing*, 6:72-79.

Shutko AM, EA Reutov, BM Liberman, AG Grankov, AA Milshin, AG Yazerian, AA Haldin, SP Golovachev (1991) Airborne passive microwave remote sensing of land and water surfaces, *Proceedings of the 5th International Colloquium on Physical Measurement and Signature in Remote Sensing*, Courchevel, France, ESA-SP319, pp. 487-489.

Shutko AM (1991) Present and future applications of land surface remote sensing by microwave radiometry, *Preventive Medicine*, 11:289-292.

Shutko AM, EA Reutov, AA Haldin (1991) Airborne passive microwave remote sensing of land and water, Report at American Hydrological Society Meeting, Orlando, Florida, 2 p.

Reutov EA, AM Shutko (1992) Estimation of the depth to shallow water table using microwave radiometry, *International Journal of Remote Sensing*, 13(12)2223-2232.

Grankov AG, AM Shutko (1992) Use of remote radiophysical methods to determine the role of energy-active zones of the ocean in the development of the weather on the continent, *Soviet Journal of Remote Sensing*, 9(6):926-941.

Jackson TJ, TJ Schmugge, R Parry, WP Kustas, JC Ritchie, AM Shutko, AA Haldin, EA Reutov, EP Novichikhin, BM Liberman, JC Shue, MR Davis, DC Goodrich, SB Amer,

LB Bach (1992) Multifrequency passive microwave observations of soil moisture in an arid rangeland environment, *International Journal of Remote Sensing*, 13(3) 573-580.

Shutko AM (1992) Soil/vegetation characteristics at microwave wavelengths. In: Proceedings, Terra-1, International Symposium, Nottingham University, United Kingdom, pp. 53-66.

Otte J (1992) (Economics Editor) Challenges in selling Soviet technology in the United States, Wallaces Farmer, March 28, p. 18.

The Challenges We Face: Panel Discussion on Soil Moisture

Edited by Hoshin Gupta and Soroosh Sorooshian

1. Introduction

This panel discussion was moderated by *Jens Refsgaard* and *Dara Entekhabi*. The following is a summary of the major points of discussion, based primarily on notes prepared by participants *Maria Costa-Cabral, Paul Houser, Joseph Mas-Pla,* and *Jim Washburne*. Their assistance in documenting the discussions is greatly appreciated. The moderators focused the discussion mainly on (a) the interaction between vegetation and soil parameters, (b) problems related to scale and heterogeneity, and (c) the calibration and validation of hydrologic models.

2. We need to understand the interaction between vegetation and soil parameters

The present understanding of the dependence of soil parameters upon vegetation is somewhat limited and, as pointed out by *Beven*, the appropriate scale for representation of soil and vegetation parameters depends on the nature of the problem at hand. However, it was generally agreed that soil water-holding capacity is the main parameter of importance in soil moisture models. *Becker* pointed out that the spatial variability of soil water-holding capacity is strongly dependent on the natural habitat, especially on vegetation type and distribution. By plotting normalized areas of different vegetation cover against water-holding capacity, a distribution function of holding capacity can be obtained that can be incorporated in soil moisture models. *Calder* suggested that we may only need to consider vegetation cover at large scales, using information such as tree density obtained from remote sensing. However, more work is needed to study the relationships between larger-scale vegetation cover and hydrology. It is possible that at smaller scales, the assumption of homogeneous water-holding capacity values may be appropriate because lateral root spread may offset the spatial variability of vegetation cover observed at the surface.

NATO ASI Series, Vol. I 46
Land Surface Processes in Hydrology
Trials and Tribulations of Modeling and Measuring
Edited by S. Sorooshian, H. V. Gupta, and S. C. Rodda

3. Should we be looking for aggregated responses instead of homogeneous scales?

A great deal of discussion centered on the application of fundamental laws to describe physical processes in heterogeneous media. *Entekhabi* presented the view that all hydrologic processes are fundamentally due to heterogeneity and that there may not exist any homogeneous scales at which to apply fundamental laws such as the Richards equation. Furthermore, not only do the soil parameters have complex and mixed distribution but the initial and boundary conditions do so as well. He asked "Shouldn't we be looking at aggregated responses instead of searching for homogeneous scales?" *Dozier* suggested that a main difficulty in applying fundamental laws resides in being able to represent the interaction between physics and geometry. While there are appropriate physical laws at any heterogeneous scale, the relevant geometry will be different for different scales. *Entekhabi* responded that many so-called physical laws can also be viewed as parameterizations of phenomena taking place at even smaller scales, for example, Darcy's law. In fact, as pointed out by *Dozier,* even the concept of a wetting front may be viewed as a parameterization or "empiricism".

On the issue of measurement, *Goodrich* pointed out that there is often a serious incompatibility between the scale of the data sources and the scale(s) at which processes are represented in the model. For example, if there is only one rain gage over a basin, the partitioning of the basin into small areal segments may be inappropriate. While there is always within-grid variability in either modeling, observing, or validating (*Becker*), the scale of representation of soil heterogeneity must be consistent with the scale of the model. Smaller-scale heterogeneity may be overcome by larger-scale heterogeneity, so that we may need only consider the latter. *Goodrich* suggested that we should simplify our models accordingly so that the smaller scales are disregarded without any loss of accuracy. *Refsgaard* mentioned that in his experience, the requirements on the spatial resolution of remotely sensed soil moisture measurements are dictated partially by the requirements of distributed hydrologic models. In particular, intergrid spatial variability of soil moisture information has been found to be useful for distributed hydrologic models, but intragrid variability has not been found to be very useful.

4. How complex do our models need to be?

The discussion on model calibration centered on the issue of optimal degree of model complexity. *Entekhabi* started the discussion by suggesting that the following distinction between simple and complex models be considered: simple models (e.g., lumped parameter models) are those which must be calibrated with model input and output fluxes, and complex models are those that contain more physically realistic parameters that must be specified using field data. *Kabat* and *Collier* disagreed with the equating of "simple" and "complex" with "black box" and "physically based", because there is no reason that a physically based model cannot also be simple while also having physically meaningful parameters that need not be calibrated with system input/output data. *Sorooshian* and *Goodrich* emphasized that it is important to bear in mind that a model should not be more complex than the available data support. For example, the use of a complex physically based model requiring estimation of several parameters is not justifiable if only output data are available for its calibration. Further, if a simple model provides good predictions, there is no reason to use a more complex one. However, *Smith* reminded the panel that so-called "black-box" models (having little or no basis in physics) offer no hope for use in ungaged basins, while physically based models can at least be used with some rough estimate of parameters. *Michaud* agreed, having found in her research on semi-arid watersheds that complex models perform better than simple models when uncalibrated; however, her findings also indicate that when calibration data are available, simple models may provide better results.

The issue of appropriate model complexity is a longstanding one. *Gupta* remarked that the whole issue of appropriate model complexity is really related to the intended use. If the primary objective is prediction, a simple model with few parameter requirements may be most appropriate, while if the objective is to improve our understanding of physical processes, a more complex model may be more useful. Along these lines, *Shutko* pointed out that the inherent paradox is that the best models for "direct" problems are the ones that represent as much detail of our understanding as possible, while the best models for inverse problems are the simplest ones that match the data.

Beven brought up the issue of model validation and proposed various levels of validation: a very weak validation being one where parameters are estimated based on calibration and split-sample tests, and a strong validation being one where parameters are estimated "a priori" without reference to system output. *Gupta* pointed out that the issue of validation is complicated by the fact that validation is not possible for those parameters to which the model has very little sensitivity.

5. Challenges for the hydrological community

Panel moderator *Refsgaard* reminded the participants that the predominant problems with process representation lie in several areas of basic science and in the applications of hydrochemistry, ecohydrology, and the spatial variability of energy and water fluxes. However, in the near future, the following challenges need immediate attention:

(a) Remote sensing may not provide soil moisture data at the spatial and temporal scales that present models are prepared to use. Conversely, present models are not written to use available data. For example, most General Circulation Models contain a soil moisture parameter which is not physically based and, thus, cannot be obtained from soil moisture measurements. *Engman* suggested that models need to be developed or modified to make the best use of data in their available form. In particular, models which make use of SAR soil moisture data should be developed.

(b) Remote sensing will show us the existence of patterns of soil moisture. *Beven* suggested that the attempt to understand the structure of those patterns will provide insights which current models do not predict, thus presenting us with a new challenge.

Part IV - Evapotranspiration

Areal Evaporation from Satellite Thermal Infrared Data

Bernard Seguin
Station de Bioclimatologie INRA
Centre d'Avignon - BP 91
F. 84143 Montfavet Cedex
FRANCE

1. Introduction

Evaporation is a major component of land surface processes in hydrology, both in terms of water cycle and energy balance. At the continental scale, roughly 2/3 of precipitation is evaporated, and slightly more than 1/2 of the available net radiation is converted into latent heat (Tables 1 and 2).

Table 1. Estimates of World Water Balance Since 1970 (m y^{-1}) (from Brutsaert, 1982).

Reference	Land		Oceans		Global
	(1.49 x 10^8 km^2)		(3.61 x 10^8 km^2)		
	P	E	P	E	P = E
Budyko (1970, 1974)	0.73	0.42	1.14	1.26	1.02
Lvovitch (1970)	0.73	0.47	1.14	1.24	1.02
Lvovitch (1973)	0.83	0.54	---	---	---
Baumgartner and Reichel (1975)	0.75	0.48	1.07	1.18	0.97
Korzun et al. (1978)	0.80	0.485	1.27	1.40	1.13

These values evidently correspond only to the order of magnitude on a large global scale and need to be precise according to the large possible variations across spatial and temporal scales.

Global circulation models and watershed hydrologic models are roughly characterized by the same scales (daily time scale and 100 km horizontal spatial scale). However, they

NATO ASI Series, Vol. I 46
Land Surface Processes in Hydrology
Trials and Tribulations of Modeling and Measuring
Edited by S. Sorooshian, H. V. Gupta, and S. C. Rodda

could require shorter time description of evaporation during drying processes after rainfall episodes, and smaller scale sizes (10 km and even 1 km) could be required for small watersheds. Topography and land use then become important factors, while these factors are only approximately assessed in GCM. Nevertheless, the main challenge in the two modeling approaches lies in a correct assessment of daily evaporation from land surfaces at a spatial scale between 10 and 100 km.

Table 2. Estimates of Mean Global Heat Budget at the Earth's Surface in kcal cm^{-2} y^{-1} (from Brutsaert, 1982).

Reference	Land			Oceans			Global		
	Rn	LoE	H	Rn	LoE	H	Rn	LoE	H
Budyko (1974)	49	25	24	82	74	8	72	60	12
Baumgartner and Reichel (1975)	50	28	22	81	69	12	72	57	15
Korzun et al. (1978)	49	27	22	81	82	8	79	67	12

1. Estimation of Evaporation by Usual Methods

1.2 At a Local Scale

Most of the enormous amount of knowledge about evaporation (or evapotranspiration) measurement and modeling concerns the local scale, covering some square meters and a well-identified surface type.

Evaporation may be described as the water vapor flux in the lower atmosphere (surface layer) in terms of micrometeorological turbulent exchanges, or the latent heat flux component in the surface energy balance, or the water loss by the soil reservoir as a liquid and gas flow in a porous medium. When soil and plants are not considered as a single layer ("big leaf" assumption), soil evaporation and plant transpiration need to be separated, which implies the analysis of water transport between roots and stomata using analogies with hydraulic or electrical transports and possible measurements of sap flow.

These various approaches, which have to be combined in modeling (the latent heat flux is considered as the boundary condition when the approach is based soil water transfer, for instance), are summarized in Figure 1.

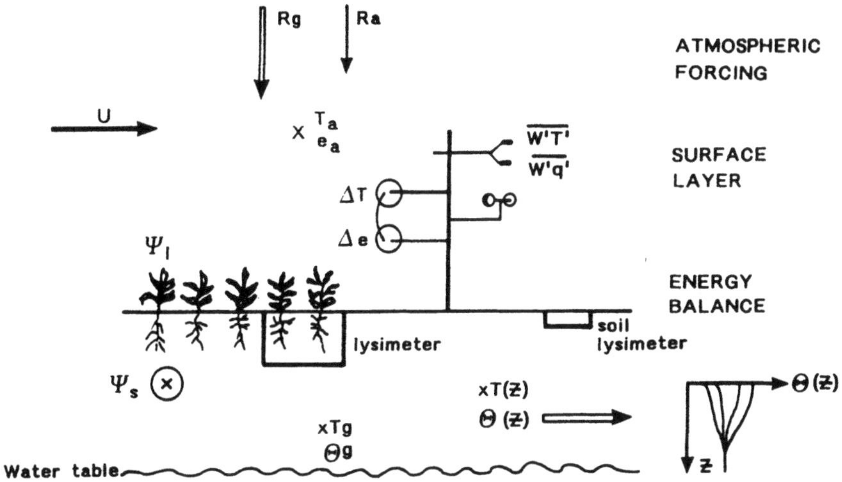

Figure 1. Schematic representation of the soil-plant-atmosphere continuum, summarizing the main processes and measurement techniques.

Figure 1 assumes a sufficiently large homogeneous surface (in order to ensure the vertical conservation of convective fluxes in the lower atmosphere surface layer adapted to this surface). It may be complicated by the occurrence of horizontal exchanges in the case of surface heterogeneities (advection). Measurements on a local scale may be based on:

- micrometeorological systems (eddy-correlation, aerodynamic method, Bowen-ratio);
- mass budget (weighing lysimeter for evapotranspiration, or soil lysimeter plus sap flow for separately accounting for evaporation and transpiration); and

♦ water balance from soil moisture profile evolution with time.

At this local scale, modeling approaches generally combine heat and mass transfers between surface and soil on one hand, and atmosphere on the other hand. The surface is considered as one layer (soil or vegetation assimilated to a big leaf), or may be treated as two-layer (vegetation and soil), or even multiple layers where vegetation is considered as a pile of vegetation layers. Most of these models use an analog electrical circuit, where transfers are described between successive layers by means of individual resistances (soil resistance for bare ground, crop resistance for a one-layer vegetation including an internal resistance due to the effect of vegetation structure and an equivalent stomatal resistance, aerodynamic resistance between surface and the lower atmosphere, or the addition of aerodynamic and stomatal resistances at leaf level for each vegetation layer in the case of multiple layers).

At a larger scale, both measurements and models need to consider the emission of fluxes by the various contributing surfaces more or less totally mixed in an integrated large scale flux, which is possible inside the surface layer when the horizontal extension of contributing surfaces is small (some hundred meters) or, more currently, leads to consider the whole atmospheric boundary layer (Figure 2).

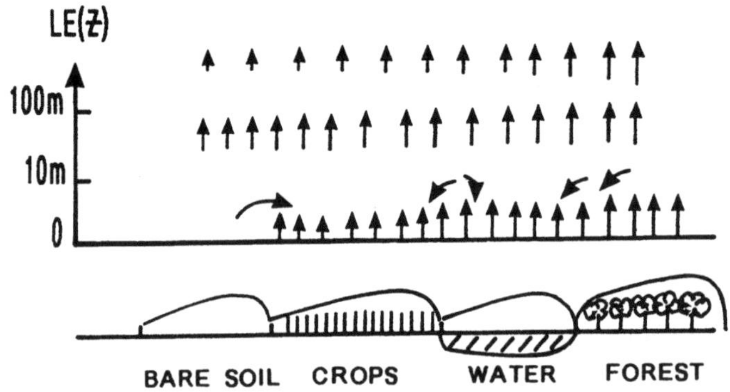

Figure 2. Large scale evaporative flux from various contributing surfaces.

1.2 Extension to Larger Scales

At this large scale, several approaches may be considered, as follows:

(a) Measure the individual contribution by each elementary surface (i) and derive the total flux E by E = Σ Si X Ei, where Ei is weighed by the surface Si of each element.

(b) Directly measure the spatially integrated fluxes by airborne flux measuring systems (Desjardins and MacPherson, 1991; Desjardins et al., 1992) or by the atmospheric profile derivation using atmospheric boundary layer laws (Brutsaert and Mawdsley, 1976; Sugita and Brutsaert, 1991; Betts, 1992). The same concept may be applied for the derivation of evaporation flux from the atmospheric water budget method for a finite-size mesh in the atmosphere; reasonable results have been obtained for parts of continents, but the method seems more suited to areas larger than 10^6 km^2 (Brutsaert 1982).

These two approaches need specific measurement facilities which are not available on a routine basis. They may be used in the specific context of large field experiments, designed to allow the testing and validating of other approaches (Hapex-MOBILHY, FIFE, EFEDA, Hapex-Sahel, etc., already effective and the forthcoming campaigns within IGBP or GEWEX programs). However, operational procedures need to only rely on flux estimation methods using regularly available data (mainly meteorological from the standard network).

(c) Directly derive ET from climatological data. Empirical formulas have been adjusted, like for PET calculations, between ET and the main climatic elements. Among others (see Brutsaert, 1982; Dooge, 1982; Eagleson, 1982), Schreiber (1904), Budyko (1974), and Turc (1954) proposed climatological formulas relating ET to rainfall P and a radiation index of dryness (Rn/P) for Budyko or a thermal factor (LT) for Turc. These formulas have been adjusted on a large number of water balance components from watersheds in various climate areas. They allow to obtain reliable values for annual ET estimates on a large (more than 100 km) scale, but may hardly be used for more precise scales. They give, at least, global reference values, which have been used to check ET values derived from large-scale models.

A rather different approach is based on the development of the "complementary relationship", first derived by Bouchet (1963) and reappraised, among others, by Morton (1976). For a regional scale, ET is assured to be linked to the estimate of potential evaporation PET on a central site by the relationship:

$$ET + PET = 2PET_o \tag{1}$$

where PET_o is the value of PET in humid conditions. This relationship is based on the argument that, when the regional surface dries and ET diminishes, the advective effect due to the heating and drying of the air mass will simultaneously increase PET. The approximation of PET by the Priestley-Taylor equation (Priestley and Taylor, 1972) has allowed Brutsaert and Stricker (1979) to derive satisfactory values of ET when compared to local measurements. In addition, several derivations by Morton (1975) have proved to be acceptable. However, the difficulty in precisely identify regional surfaces to be considered, as well as less satisfying results (MacNaughton and Spriggs, 1989) presently limits the common use of this interesting approach.

(d) Compute ET by a single water budget procedure, where soil is considered as a single reservoir (possibly improved by a separation into a top-layer reservoir and a deeper part, which allows a better description of the short-term drying process and deep infiltration). ET equals the value of potential evapotranspiration PET (or eventually to maximum evapotranspiration ETM for each type of crop or vegetation) when water is not limiting and prescribed as a fraction of PET when the reservoir diminishes and soil moisture falls to values smaller than field capacity θ_{fc} (Figure 3).

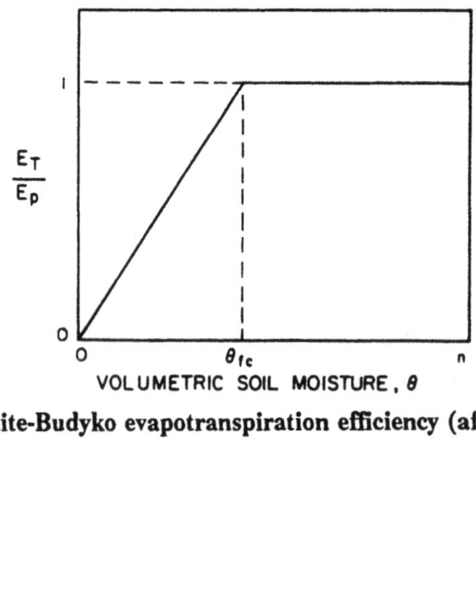

Figure 3. Thornthwaite-Budyko evapotranspiration efficiency (after Eagleson, 1982).

PET may be approached by the Penman formula (Penman, 1956) or by the simplified Priestley-Taylor equation (Priestley-Taylor, 1972), ETM needing to add a specific crop coefficient Kc or to be expressed by the Penman-Monteith approach (Thom, 1973).

The same general approach is often used in GCMs by assuming a value of ET/PET = β related to the available soil moisture, PET being specified by the flux equation for saturated surface conditions:

$$PET = \rho C_E u(z)\ (q(z) - q_s) \qquad (2)$$

where q(z) and q_s refer to air and surface specific humidity (assumed to be saturated at the surface temperature Ts), ρ is the air density, C_E is the bulk water vapor transfer coefficient, and u(z) is the wind velocity at the level Z.

These methods are based on a rather simple approach from the physical point of view and could be improved by introducing more sophisticated processes (Carson, 1982; Eagleson, 1982). They also must consider the grid cell as a uniform surface (in terms of soils and land use) more or less representative of the various contributing areas which compose it. However, they give a correct first-order appraisal of evaporation at the spatial scales needed for both GCMs and hydrologic watershed models, so that they are often used in operational systems.

(e) Following the progress of research on a local scale, more sophisticated models have been specifically developed for larger scales using coupled energy and mass transfers with a one- or two-layer approach (SIB, developed by Sellers et al., 1986; BATS by Dickinson, 1984). They are now commonly used in GCMs and could be incorporated in hydrologic models. Some of the needed inputs (for instance, the leaf area index LAI, available at 1 km resolution scale with NOAA-AVHRR satellites) are derivable from remote sensing, so that a significant increase of precision is expected compared to the simple approaches previously referred. These models allow for obtaining a much more precise assessment of ET fluxes in terms of space and time scales (daily, and even more up to the hourly basis; better than 100 km, down to 10 or even 1 km).

Obviously, remote sensing is now able to offer an alternative way for estimating ET at the required scales by the recent availability of operational observing systems covering the global scale with permanent acquisitions (essentially meteorological satellites like geostationary Meteosat, GOES, and/or the polar orbiting NOAA-AVHRR).

The use of NDVI from NOAA is one possibility, as mentioned above. Computations from soil moisture estimation using microwaves also offer interesting perspectives (see the other presentations within this workshop), but radar measurements are still not usable on a global scale, and passive microwave data from satellite presently suffer from an insufficient resolution (100 km and possibly 10 km only). The use of thermal IR data to estimate ET is an alternative way, which will be analyzed in the second part of this presentation.

2. Estimation of ET from Satellite Thermal Infrared Data

2.1 The Basic Procedure on a Local Scale

It is based on the well-known energy balance for usual daytime conditions:

$$Rn = G + H + LE \tag{3}$$

where Rn, G, and H are the flux densities (W/m^2) of net radiation, soil heat flux, and sensible heat flux, respectively, LE being the latent heat flux equivalent to the evaporation rate ET.

When it is instantaneously applied at the time of measurement of Ts by the thermal infrared, the usual laws of convective exchanges in the lower atmosphere give:

$$H = \rho Cp\ h\ (T_s - Ta) \tag{4}$$

or:

$$H = \rho Cp CHu(z)(T_s - Ta) \tag{4a}$$

or:

$$H = \rho Cp \frac{T_s - Ta}{ra_h} \tag{4b}$$

where the sensible heat flux H is related to the deviation T_s - Ta by an exchange coefficient h, or a bulk heat transfer coefficient C_H, or the resistance to heat transfer exchange ra_h (ρ and Cp being the air density and specific heat capacity, u(z) the wind velocity at the standard level).

The use of Equation (4) allows the computation of H, and then after to derive LE as the residual term of the energy balance LE = Rn - G - H , so that LE may be estimated from the knowledge of Rn and G. Soil heat flux G may be neglected at the daily time scale (except at some periods of regular heating, like spring or inversely during autumn). On an instantaneous basis, it is lower than some percent of Rn for a dense and actively transpiring vegetation, but may amount to 20 and even 30% for bare or scarcely covered soil. A reasonable approach relates G/Rn to a spectral estimate of soil cover (see Moran and Jackson, 1991).

There has been a large amount of experimental studies on this subject, most of them having been successful (i.e., an agreement between measured and calculated values within say $\pm$20%) (Bartholic et al., 1972; Brown, 1974; Stone and Horton, 1974, Idso et al., 1975; Heilman and Kanemasu, 1976; Hatfield et al., 1984; Choudhury et al., 1986; Kalma and Jupp, 1990; etc.).

The experimental approach also has been completed by a large number of modeling studies (based on the one-layer or several-layer assumptions already referenced in the first part of this chapter), the measured radiative temperature Ts being compared to the surface temperature as computed by the model (Soer, 1980; Gurney and Camillo, 1984; Lagouarde, 1991; etc.)

The correct estimation of ra_h is evidently a crucial point for this approach, with two main points:

(1) ra_h is first derived from the aerodynamic resistance ra, with the basic expression in neutral conditions:

$$ra = \frac{[\mathrm{Log}(z-d)/Zo]^2}{k^2 u(z)} \qquad (5)$$

being corrected for the stability effects by using either the Richardson number (Mahrt and Ek, 1984) or the Monin Obukhov length L by an iterative procedure (Paulson, 1970). Estimates of Zo and d from the canopy height Zh are currently obtained by simple relations, like Zo = 0.13 Zh and d = 0.65 Zh (Thom, 1973), which, however, do not fit all types of canopy structure. The knowledge of Zh is a difficult challenge without ground-based observations. Attempts exist for deriving it from spectral measurements like NDVI (see Moran and Jackson, 1991), but it is still a matter of research.

(2) A second problem is that heat (and vapor) transfers near the surface may significantly differ from the momentum transfer (see Thom, 1975; Brutsaert, 1982), so that a supplementary resistance (named kB^{-1} for historical reasons) needs to be added to ra as a result of the "bluff-body" effect. This effect leads to values of the roughness length for heat (and vapor) Zoh significantly smaller than the aerodynamic roughness Zom:

$$ra_h = ra + B^{-1}/u* \qquad (6)$$

so that:

$$kB^{-1} = \mathrm{Log}\,\frac{Zom}{Zoh}$$

It also corresponds to a difference between the canopy aerodynamic surface temperature To (defined by the extrapolation of temperature profile down to the level Zo + d) and the canopy radiative surface temperature Ts as measured by infrared thermometry.

The definition of To by:

$$H = \rho Cp \frac{To - Ta}{ra} \tag{7}$$

when compared to Equation (4) leads to the expression:

$$Ts - To = (To - Ta) \frac{ra_h}{ra} \tag{8}$$

The difference Ts - To will not only depend on the micrometeorological parameters but also on the factors governing the radiometric measurements of Ts (as canopy structure, LAI, solar elevation, and view angle, especially for row crops and sparse canopies) (see Thom, 1975; Brutsaert, 1981, for a general discussion). Currently reported values of kB^{-1} range between 3 and 6, depending upon the type of canopy and the geometrical conditions (Garratt, 1978; Kustas et al., 1989; Kohsiek et al., 1993).

In terms of surface temperature, they correspond to differences between aerodynamic and radiative values of the order of 6° (as computed by the modeling approach by Brunet et al., 1991).

If this effect is ignored (i.e., if kB^{-1} is neglected), errors amounting to 50 and even 100% may be observed in the estimation of H and LE (Figure 4).

This fact has been reported by several authors (Soer, 1980; Leckie et al., 1981; Huband and Monteith, 1986; Vining and Blad, 1992; Hall et al., 1992) (see Figure 5). The last results, obtained during the FIFE campaign, may lead to the conclusion that the estimation of H (and then LE) using radiative measurements of Ts is not feasible in operational procedures. However, a proper assessment of kB^{-1} reduces the error to $\pm$20% in the estimation of H, and there exists a reasonable hope to derive correct sets of values both from experimental results and modeling approaches, which could be used for operational assessment of H and LE with a reasonable accuracy in the near future.

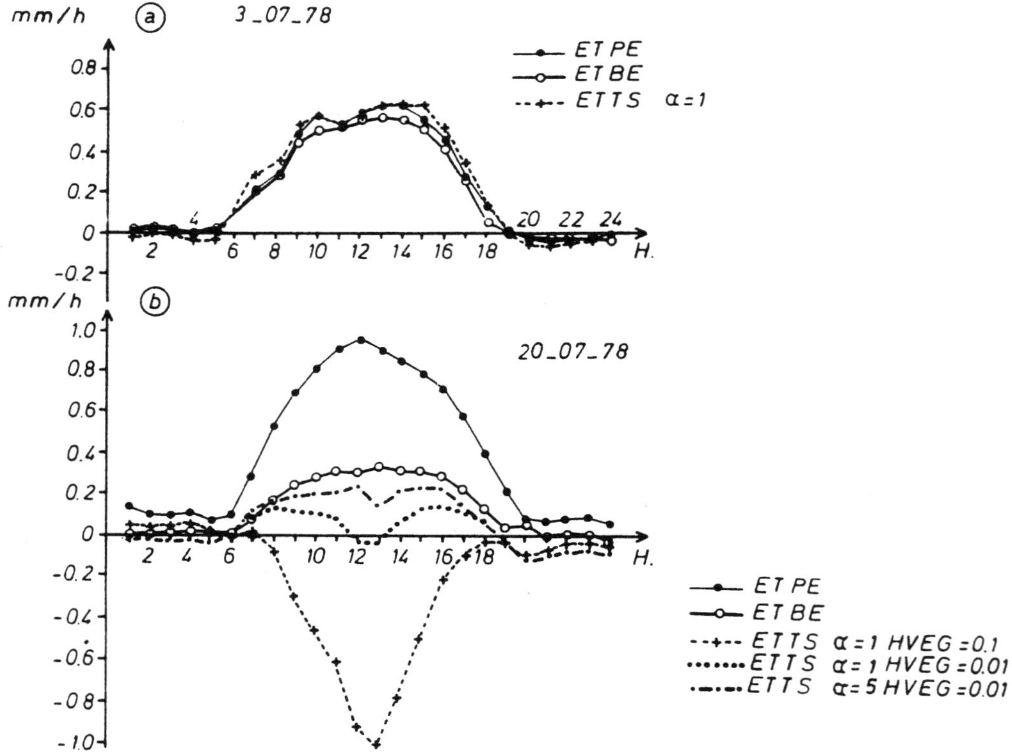

Figure 4. Hourly variations of latent heat fluxes for: (a(irrigated zones, (b) dry area. PET is computed by the Penman Formula, ET BE is the reference measurement of ET by energy balance (Bowen ratio); ETTS corresponds to the estimation by Equation (4), using two values of the canopy height (HVEG) and of the ratio Zom/Zoh = α (from Seguin et al., 1982).

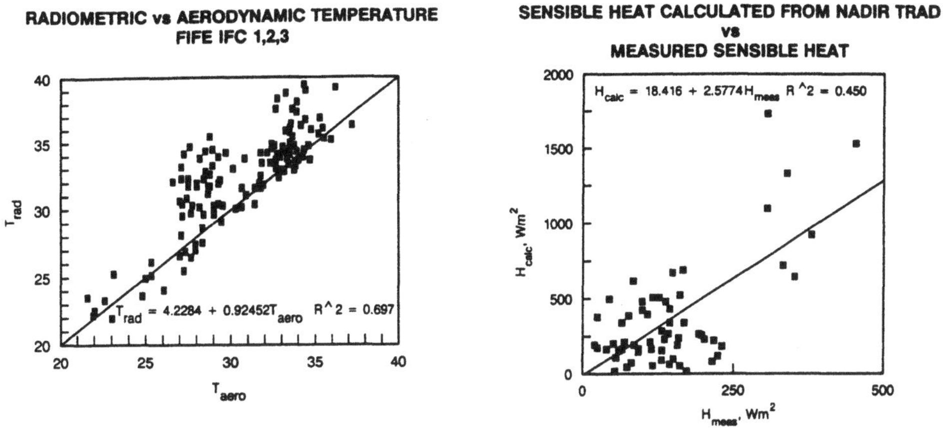

Figure 5. (a) Radiometric temperature from the nadir-viewing, helicopter-mounted MMR acquired for 6 June, 11 July, and 15-17 August 1987, versus aerodynamic temperature as calculated from flux station measurements and (3) and (4); (b) Sensible heat calculated using radiometric temperature from the nadir-viewing, helicopter-mounted MMR acquired for 6 June, 11 July, and 15-17 August 1987, versus measured sensible heat (from Hall et al., 1992).

2.2 Instantaneous Flux Estimates on a Local Scale Using Remote Sensing Data

The described procedure may be applied to remote sensing data using aircraft-mounted thermal instruments or high-resolution satellites (like HCMM or TM) for mapping the spatial variations of instantaneous values of LE. When proper atmospheric corrections are applied, good results are obtainable using aircraft or TM (Nieuwenhuis and Menenti, 1986; Moran and Jackson, 1991). Jackson et al. (1987) were able to map LE with an agreement better than 10% when compared to ground measurements, which confirms the feasibility of the method for describing its spatial variations (Figure 6). However, for operational purposes, the need of frequent data limits its potential use to the case of surface temperature derived from meteorogical satellites. Here, the limiting factor comes from the precision of Ts estimation from these satellites, which depends upon the accuracy of satellite thermal IR channels calibration and atmospheric corrections.

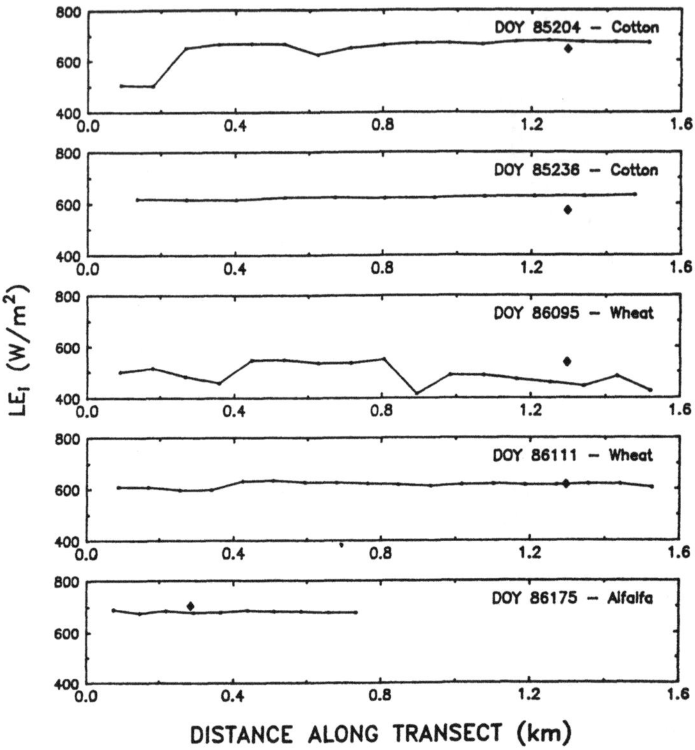

Figure 6. Spatial distribution of instantaneous evaporation (LE_i) over the 1.6 km length of fields (0.8 km for alfalfa). The dots connected by lines indicate a remote measurement. The diamond (♦) indicates the point measurement of LE_i (average of 12 minutes) with the Bowen ratio apparatus (from Jackson et al., 1987).

Ts may be derived from the thermal infrared measurements, knowing that the measured radiance (in terms of radiative brightness TB) corresponds to:

$$\sigma TB^4 = \epsilon \sigma TS^4 + (1-\epsilon)\ Ra \qquad (9)$$

The occurrence of ϵ smaller than 1.0 leads to an underestimation of Ts by TB (ϵ = 1.0) of the order of 0,6° per 0.01 error on ϵ (this range depends upon the value of the atmospheric radiation Ra, within the used spectral band in thermal IR). For the spectral bands between 10 and 12 m, the emissivities of natural surfaces range between 0,96 and 0,98 (for vegetation and soils with some smaller values for bare soils), so that the order of the underestimation lies between 1 and 2° (Becker, 1987; Labed and Stoll, 1991). This effect is noticeable for ground measurements on a local scale, but hardly separable from atmospheric effects in the case of satellite data. Atmospheric absorption acts in the opposite sense by reducing the measured TB (generally by 6 to 8°, and even more for geostationary satellites). This effect may be corrected by applying an atmospheric correction model (Price, 1984; Kneizys et al., 1988), but this procedure requires the availability of simultaneous radiosoundings, so that the use of a split window (combining in a linear way two wavelengths) is generally preferred. The original coefficients, proposed by Deschamps and Phulpin (1980), have been adapted for land surfaces (Kerr, 1991; Ottle and Vidal-Madjar, 1992) so that an overall precision of $\pm$2°C (including emissivity and atmospheric effects) appears reasonable in the case of NOAA-AVHRR data. This 2°C precision of Ts roughly corresponds (depending upon the ra values) to a precision of the order of 20 to 60 W/m^{-2} for H, which may be acceptable compared to the other sources of errors. Hence, positive results have been obtained when using satellite thermal IR data as an input in surface models (Gurney and Hall, 1983; Taconet et al., 1986). The instantaneous time scale, however, is of limited interest, except for testing methods or illustrating the potential of spatial variability assessment. Thus, most of the applications aim at determining fluxes on a daily time scale.

2.3 Daily Evaporation on a Regional Scale

As we have mentioned in the introduction section, daily time scale appears to be best suited both for meteorological and hydrologic models. How is it attainable from satellite measurements?

The best opportunity could be obtained with the quasi-permanent observation by geostationary satellites. The analysis of thermal cycles has opened the way some time ago (Favard, 1976), and methods have been developed for a quantitative assessment based on the physical modeling of land surface exchanges (Carlson et al., 1981; Abdellaoui et al., 1987). However, the poor precision on Ts estimates by these geostationary satellites, both due to the insufficient calibration and the existence of only one broad thermal channel (which prevents the use of the split-window procedure), has limited up to now an effective development of these methods.

The use of at least two passes per day (day-night) allowing for obtaining the surface thermal inertia has supported the conception of the HCMM satellite. Specific models have been developed to relate daily ET to thermal inertia (Rosema et al., 1978), and interesting results have been obtained during its lifetime from 1977 to 1979 (Price, 1982), but HCMM was only operational during two years. The same procedure may be applied to NOAA data, but the processing of both day and night images leads to practical limitations. Because the weight of night information appears low compared to that of daytime data, most of the studies aiming at the operational derivation of daily ET values presently rely on the use of one (generally near midday) value obtained from satellite (mainly NOAA-AVHRR).

The combination of these data with simulation models (see the previous paragraph) appears as the most precise solution for computing daily ET. Instantaneous values are computed using the single surface temperature data and the model runs for extrapolating fluxes in time. However, the procedure is rather complex; it supposes to use the model for each pixel on a quasi-permanent time scale (basically an hourly computation), with the availability of meteorological parameters like solar radiation, air temperature, and wind velocity at this time scale. At present, this procedure is by far too heavy for GCM or hydrologic models (even if SiB or BATS models could be adapted for that purpose).

The need for less-sophisticated methods has led to elaborated simplified procedures, directly linking daily values of H and LE to one single midday Ts value. In the case of well-established clear weather, energy balance components are known to follow a regular quasisinusoidal temporal evolution during daytime, so that it seems feasible to use an instantaneous value as representative of the whole day. Jackson et al. (1977) then proposed such an adjustment:

$$ETd - Rnd = a - b(Ts - Ta)_i \qquad (10)$$

based on an experiment with wheat irrigation treatments in the clear sky sunny conditions of Phoenix, Arizona. Similar linear relationships have been obtained in our own experiment in the Crau site (near Marseille, in southeastern France), grouping values for a dry area and large pasture irrigated fields. However, the coefficients a and b were rather different from those obtained in Phoenix (Seguin et al., 1982) (Figure 7), which appears quite logical because the coefficients b (and also a as a residual) will depend mainly upon the surface roughness and the wind velocity (Seguin and Itier, 1983).

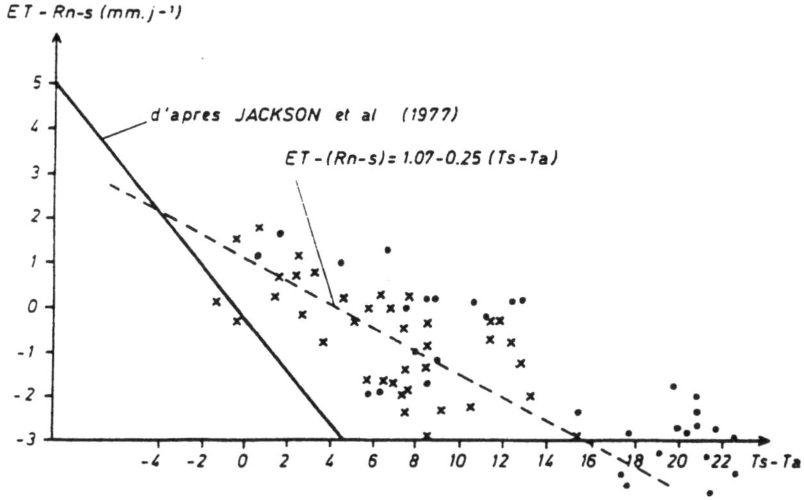

Figure 7. Relationship between ET - (Rn - S) and Ts - Ta for 32 days during the summer of 1978 on the Crau site (x = irrigated Crau; • = dry Crau) (from Seguin et al., 1982).

Both analytical (Riou et al., 1988; Vidal and Perrier, 1989) and experimental (Lagouarde and MacAneney, 1992; Sandholt and Andersen, 1993) studies allow for proposing sets of a and b coefficients which may be used for the mapping of daily ET in case of a clear sky. In these studies, air temperature Ta at the time of satellite passing is generally replaced by the air maximum temperature Tam, more easily derivable from climatological data. Slightly different approaches have been proposed by several authors:

♦ Nieuwenhuis et al. (1985) related the differences between ET and potential ET on a daily basis to the difference between the instantaneous surface temperatures for the two surfaces. This method promises more accuracy, but needs the identification of a zone evaporating at the potential rate, which is rather difficult on an operational basis.

♦ Carlson and Buffum (1989) related LEd - Rnd to the time evolution of surface temperature (during the morning, for instance). This method does not require the knowledge of air temperature, but is presently limited by the insufficient precision of geostationary satellite data (as the previously cited methods based on thermal cycles or inertia).

♦ Kustas et al. (1993) proposed using the evaporative fraction concept (temporal conservation of the ratio LE/LE + H) for extrapolating instantaneous to daily values (see Brutsaert and Sugita, 1992).

The simplified approach has been applied for ET mapping on a regional scale using satellite measurements (Meteosat and NOAA-AVHR) and mean values for a and b coefficients in a first-order approach (Seguin et al., 1985, for the Rhone Valley; Seguin et al., 1989, for Senegal) or adjusting these coefficients to the vegetation cover for a finer analysis (Lagouarde and Brunet, 1989, for the Hapex-MOBILHY site, Dabrowska-Zielinska et al., 1991, for Poland; Dunkel et al., 1993, for Hungary, Kustas et al., 1993, for the Walnut Gulch watershed, etc.).

These applications demonstrate that daily ET mapping for scales of 10 to 100 km (with a basic pixel of 1 km for NOAA) is feasible for clear day conditions. The precision is not very high and may be assumed within ± 1 mm/day. Some improvements in the precision may be gained by a more detailed analysis of the effects of factor like Zo, the conservation of evaporative fraction etc., but the statistical origin of Equation (9) induces a necessary dispersion for one single day. Nevertheless, obtained mappings express the main features of ET spatial variations with a reasonable accuracy and may be considered as adding a real new source of information to be incorporated in regional scale hydrologic studies.

2.4 ET Estimates at Larger Time and Space Scales

The computation of ET for one isolated day is one thing, but its derivation for long periods is another challenge, especially when it is necessary to deal with the presence of clouds. The combination of daily estimates for cloud-free conditions and energy balance-water budget models for time interpolation is one possibility (Vidal et al., 1987). As already noted in the previous paragraph for a finer scale, this combination is rather heavy for operational perspectives; therefore, the extension of the simplified procedure appears as a preferable solution. It is based on the integration of the daily linear relationship for n days:

$$\sum_n ET - \sum_n Rn = \sum_n a - b\sum_n (Ts - Ta)_i \qquad (11)$$

This extended simplified mean relationship is valid for a longer-term basis (typically one month), Ts being computed as a selection of one representative day for a 5- or 10-day period, in a similar way as used for deriving NDVI from the maximum composite value MVC with NOAA data. According to the generally admitted correspondence between time and space scales, the method will apply to typical spatial scales ranging between 10 and 100 km. At this scale, the mixing of individual surfaces appears to result in a weighted-equivalent homogeneous surface, with values of b ranging between 0.20 and 0.30, the contribution of the "a" term becoming negligible (Seguin et al., 1991).

The obtained values of Σ (Ts - Ta) allow for describing the general feature of water availability amount in Sahelian regions (Seguin et al., 1989) or European countries (Seguin et al., 1991) and explain the potential of thermal IR data from satellites for drought characterization (Gutman, 1990; Seguin et al., 1992). They also may be used for mapping of large-scale ET, when calibrated with a water balance model (Seguin et al., 1991, using Meteosat data; see Figure 8, Courault et al., 1994 using NOAA data).

Obtained values appear in good agreement with those derived by the complementary approach between ET and PET as described in the first part of this paper (Figure 9). They also compare favorably with statistical formulas like that of Turc (1964) computed for some selected sites (Figure 10). Although it is not possible to qualify this agreement as a validation test (because of the uncertainities for each approach), it confirms that some link may be established between available conventional methods and the remote sensing approach for this large time-space scale domain.

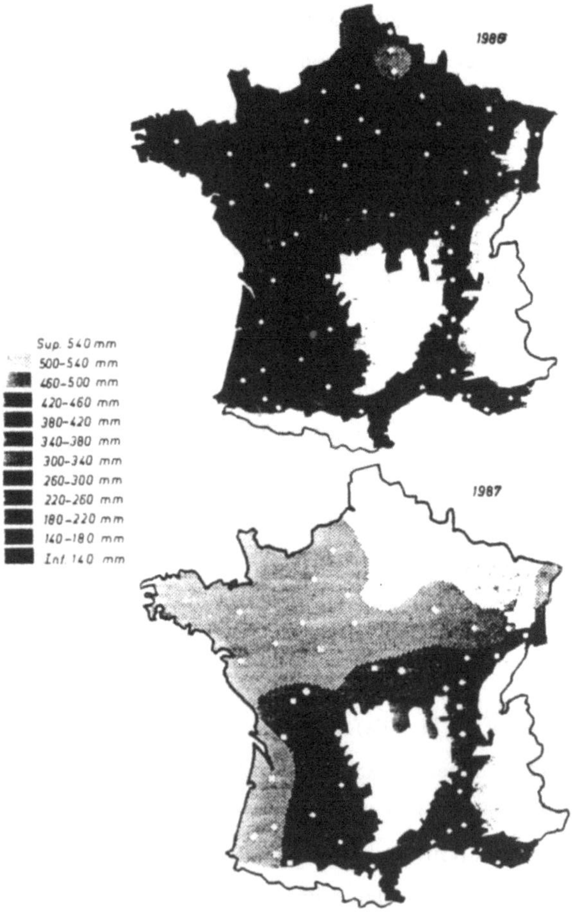

Figure 8. Maps of ET for six months (April to October) in France during the years 1986 (a) and 1987 (b) (from Seguin et al., 1991).

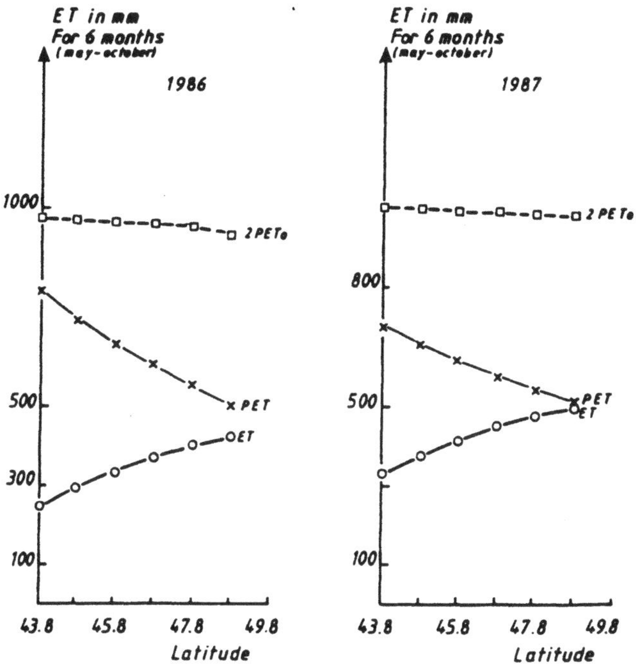

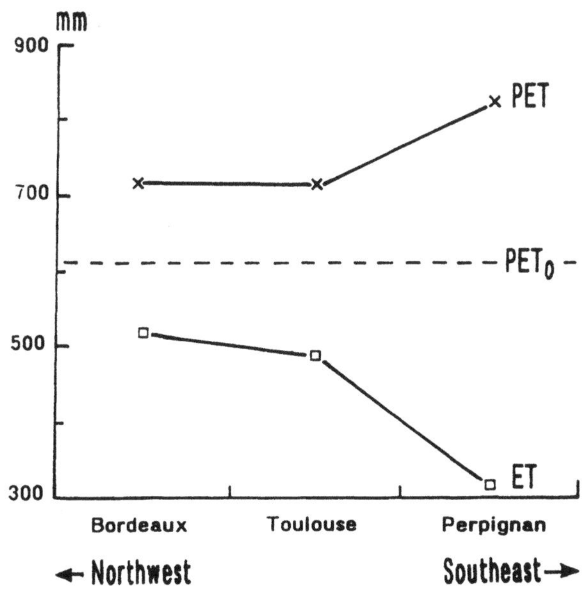

Figure 9. Complementary evolution of ET and PET for transects along the Rhone (top) and Garonne (bottom) rivers. (Adapted from Seguin et al., 1991, and Courault et al., 1993).

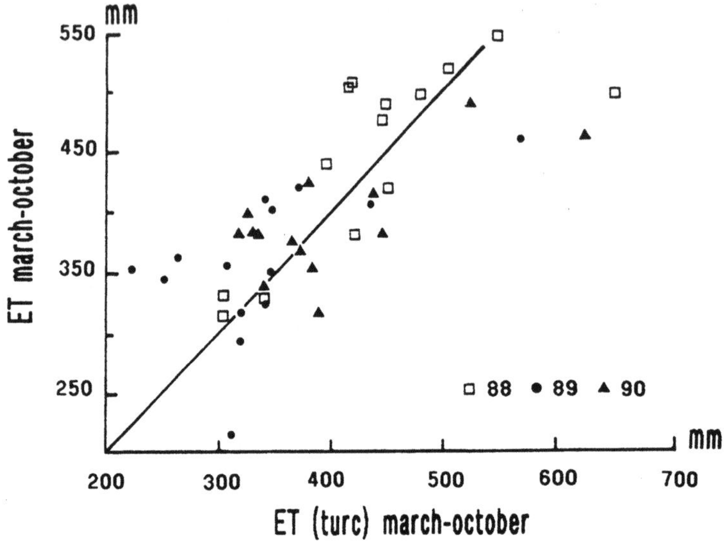

Figure 10. Comparison between ET (March-October for three years: 1988-1990) derived for the combination of NOAA data/agrometeorological MAGRET model and computed by the TURC formula (from Courault et al., 1993).

3. Conclusion

In spite of clear limitations (mainly due to the impossibility of continuous record with the presence of clouds), the approach using satellite thermal IR data from satellites at various scales (ranging from local to regional scales) is able to derive reasonable estimates of daily ET. From the presently available results, it appears that progress is needed in several directions for improving its reliability :

(1) On a local scale, values of kB^{-1} need to be established for a large range of natural surfaces, as well as methods for giving rules to be applied for remote sensing use (ways to obtain correct estimates of Zo are also a subject of research);

(2) The combination with numerical models at various scales, based on the description of energy balance and water budget. These models, including the more

difficult case of partially covered vegetation (Kustas et al., 1993; Chehbouni et al., 1993), are now available. Ways to research an effective assimilation of remote sensing data need to be explored;

(3) The scaling problem (how to compute equivalent effective parameters for a large surface including heterogeneous land covers) must also be adressed;

(4) The present approach, using only one datum per clear day, evidently suffers from a limited precision. The availability of the diurnal cycle of surface temperature offers a large perspective improvement, presently hindered by the lack of precision of geostationary satellites. Projects like those of the Next Generation Meteosat (with a better calibration of the thermal sensors and the presence of at least two thermal channels allowing the use of the split window method) will open new ways to approach ET; and

(5) The combination with other sources of informations from different wavelengths (especially microwaves, which allow for obtaining reliable values even in the presence of clouds) must also be explored in the near future.

References

Abdellaoui A, F Becker, E Olory-Hechinger (1987) Use of Meteosat for mapping thermal inertia and evapotranspiration over a limited region of Mali. *Journal of Applied Meteorology*, 25, 1489-1506.

Bartholic JF, LN Namken, CL Wiegand (1972) Aerial thermal scanner to determine temperatures of soils and crop canopies differing in water stress, *Agronomy Journal*, 64, 603.

Becker F (1987) The impact of spectral emissivity on the measurement of land surface temperature from satellite. *International Journal of Remote Sensing*, 8:1509-1522.

Betts AK (1992) FIFE atmospheric boundary layer budget methods. *Journal of Geophyscial Research*, 97, D17, 18523-18532.

Bouchet RJ (1963) Evapotranspiration reelle, evapotranspiration potentielle et production agricole. *Annals of Agronomy*, 14, 743-824.

Brutsaert W, JA Mawdsley (1976) The applicability of planetary boundary layer theory to calculate regional evapotranspiration. *Water Resources Research*, 13, 637-643.

Brutsaert W, H Stricker (1979) An advection aridity approach to estimate actual regional evapotranspiration. *Water Resources Research*, 15, 443-450.

Brutsaert W (1982) Evaporation into the Atmosphere. Reidel Publishing Company, 299 p.

Brutsaert W, M Sugita (1992) Application of self preservation in the diurnal evolution of the surface energy budget to determine daily evaporation. *Journal of Geophysical Research*, 97, D17, 18377-18382.

Brown KW (1974) Calculations of evapotranspiration from crop surface temperature. *Agricultural Meteorology*, 14, 199.

Brunet Y, UKT Paw, L Prevot (1991) Using the radiative surface temperature in energy budget studies over plant canopies. Proceedings of 5th Colloquium on Physical Measurements and Signatures in Remote Sensing. ESA, SP 319, 557-560.

Budyko MI (1974) Climate and Life, Academic Press, 508 p.

Carlson TN, JK Dodd SG Benjamin, JN Cooper (1981) Satellite estimation of the surface energy balance, moisture availability, and thermal inertia. *Journal of Applied Meteorology*, 20, 76-87.

Carlson TN, MJ Buffum (1989) On estimating total daily evapotranspiration from remote surface temperature measurements. *Remote Sensing of the Environment*, 29, 197-207.

Carson DJ (1982) Current parametrizations of land-surface processes in atmospheric general circulation models. In Land Surfaces in Atmospheric General Circulation Models, PS Eagelson (Editor) Cambridge University Press, pp. 67-108.

Chehbouni A, JP Lhomme, EG Njoku, B Monteny, YH Kerr, D Troufleau (1993). Model for sensible heat flux estimation over a heterogeneous surface during the HAPEX Sahel experiment. Submitted to *Remote Sensing of the Environment*.

Choudhury BJ, RJ Reginato, SB Idso (1986) An analysis of infrared temperature observations over wheat and calculation of the latent heat flux. *Agricultural and Forest Meteorology*, 37, 75-88.

Courault D, PH Clastre, JP Guinot, B Seguin (1994) Analyse des secheresses de 1988 a 1990 en France a partir des donnees satellitaires NOAA-AVHRR. *Agronomie*, 147, 41-56.

Dabrowska-Zielinska K, M Gruzczynska, S Levinski (1991) From evapotranspiration to soil moisture, investigation based on NOAA thermal infrared data: a case study, Poland. Proceedings of 5th International Colloquium on Physical Measurements and Signatures in Remote Sensing. ESA, SP 319, 675-678.

Deschamps PY, T Phulpin (1980) Atmospheric correction of infrared measurements of sea-surface temperature using channels at 3.7 m, 11 m, and 12 m. *Boundary Layer Meteorology*, 18, 131-143.

Desjardins RL, JI MacPherson (1991) Water vapour flux measurements from aircraft. In Land Surface Evaporation Measurements and Parameterization. T Schmugge, JC Andre (Editors) Springer-Verlag, 245-260.

Desjardins RL, RL Hart, JI MacPherson, PH Schwepp, SB Verma (1992) Aircraft and tower-based fluxes of carbon dioxide, latent and sensible heat. *Journal of Geophysical Research*, 97, D17, 18477-18485.

Dickinson RE (1984) Modeling evapotranspiration for three-dimensional global climate models. In "Climate Processes and Climate Sensitivity", Geoph. Monograph 29, AGU, 58-72.

Dooge JCI (1982) Parametrization of hydrologic processes. In Land Surface Processes in Atmospheric General Circulation Models, P Eagleson (Editor) Cambridge University Press, 243-288.

Dunkel Z, V Vadasz (1993) Estimation of regional evapotranspiration over Hungary combining standard and satellite data. COSPAR-IAF World Space Congress, Washington, August 1992, Symposium A9M, Adv. Space Res.

Eagleson PS (1982) Dynamic hydrothermal balances at macroscale. In Land Surface Processes in Atmospheric General Circulation Models. PS Eagleson (Editor), Cambridge University Press, 289-360.

Favard JC (1978) Determination de cycles thermiques a partir de satellites meteorologiques geostationnaires. Proceedings of Conference on Earth Observation from Space, ESA-SP 134, 141-143.

Garrat JR (1978) Transfer characteristics for a heterogeneous surface of large aerodynamic routhness. *Quarterly Journal of Royal Meteorological Society*, 104, 491-502.

Gurney RJ, PJ Camillo (1984) Modeling daily evapotranspiration using remotely sensed data. *Journal of Hydrology*, 69, 304-324.

Gurney RJ, DK Hall (1983) Satellite-derived surface energy balance estimates in the Alaskan subarctic. *Journal of Climate and Applied Meteorology*, 22, 115-125.

Gutman GG (1990) Towards monitoring droughts from space. *Journal of Climate*, 3:282-295.

Hall FC, KF Huemmrich, SJ Goetz, PJ Sellers, JE Nickeson (1992) Satellite remote sensing of surface energy balance: success, failures, and unresolved issues in FIFE, *Journal of Geophysical Research*, 97, D17, 19061-19090.

Hatfield JL, RJ Reginato, SB Idso (1984) Evaluation of canopy temperature-evapotranspiration models over various crops. *Agricultural Meteorology*, 32, 41-53.

Heilman JL, ET Kanemasu (1976) An evaluation of a resistance form of the energy balance to estimate evapotranspiration. *Agronomy Journal*, 68, 607-611.

Huband NDS, JL Monteith (1986) Radiative surface temperature and energy balance of a wheat canopy. *Boundary Layer Meteorology*, 36, 1-18 (I), 107-117 (II).

Idso SB, T Schmugge, RD Jackson, RJ Reginato (1975) The utility of surface temperature measurements for the remote sensing of water status. *Journal of Geophysical Research*, 80, 3044.

Jackson RD, RJ Reginato, SB Idso (1977) Wheat canopy temperature: a practical tool for evaluation of water requirements. *Water Resources Research*, 13: 651-656.

Jackson RD, MS Moran, LW Gay, IH Raymond (1987) Evaluating evaporation from field crops using airborne radiometry and ground-based meteorological data. *Irrigation Science*, 8, 81-90.

Kerr YH (1991) Corrections thermiques dans l'infrarouge thermique: cas de l'AVHRR. Proceedings of 5th Colloquium on Physical Measurements and Signatures in Remote Sensing. ESA, SP 319:29-34.

Kalma JD, DLB Jupp (1990) Estimating evaporation from pasture using infrared thermometry: evaluation of a one-layer resistance model. *Agricultural and Forest Meteorology*, 51, 223-246.

Kneizys FX, EP Shettle, WO Gallery, JH Chetwynd, LW Abreu, JEA Selby, SA Clough, RW Fenn (1988) Atmospheric transmittance/radiance: computer code LOWTRAN-7, Air Force Geophysics Lab, Hanscomb Air Force Base, Massachusetts.

Kohsiek W, HAR de Bruin, H The, B Van den Hurk (1993) Estimation of the sensible heat flux of a semi-arid area using radiative temperature measurements. *Boundary Layer Meteorology*, 63, 213-230.

Kustas WP, BJ Choudhury, MS Moran, RJ Reginato, RD Jackson, LW Gay, HL Weaver (1989) Determination of sensible heat flux over sparse canopy using thermal infrared data. *Agricultural and Forest Meteorology*, 44, 197-216.

Kustas WP, EM Perry, PG Doraiswarmy, MS Moran (1993) Using satellite remote sensing od extrapolate evapotranspiration estimates in time and space over a semi-arid rangeland basin. Submitted to *Remote Sensing of the Environment*.

Labed J, MP Stoll (1991) Spatial variability of land surface emissivity in the thermal infrared band: spectral signature and effective surface temperature. *Remote Sensing of the Environment*, 38, 1-18.

Lagouarde JP (1991) Use of NOAA-AVHRR data combined with an agrometeorological model for evaporation mapping. *International Journal of Remote Sensing*, 12, 1853-1864.

Lagouarde JP, Y Brunet (1991) Suivi de l'evapotranspiration reelle journaliere a partir des donnees NOAA-AVHRR lors de la campagne HAPEX-MOBILHY. Proceedings of 5th Colloquium on Physical Measurements and Signatures in Remote Sensing. ESA, SP 319:569-572.

Lagouarde JP, KJ MacAnneney (1992) Daily sensible heat flux estimation from a single measurement of surface temperature and maximum air temperature. *Boundary Layer Meteorology*, 59:341-362.

Leckie DGT, TA Black, PA Murtha (1981) Development and testing of a method of estimating sensible heat from natural surfaces using remotely sensed surface temperatures. *Canadian Journal of Chemical Engineering*, 59, 189-193.

MacNaughton KG, TW Spriggs (1989) An evaluation of the Priestley and Taylor equation and the complementary relationship using the results from a mixed layer model of the convective boundary layer. IAHS Publication 177. Estimation of Areal Evapotranspiration, 89-104.

Mahrt L, M Ek (1984) The influence of atmospheric stability on potential evaporation. *Journal of Climate and Applied Meteorology*, 23, 222-234.

Moran MS, RD Jackson (1991) Assessing the spatial distribution of evapotranspiration using remotely sensed inputs. *Journal of Environmental Quality*, 20, 725-737.

Morton FI (1975) Estimating evaporation and transpiration from climatological observations. *Journal of Applied Meteorology*, 14, 488-497.

Nieuwenhuis GJA, EH Schmidt, HAM Tunnissen (1985) Estimation of regional evapotranspiration of arable crops from thermal infrared images. *International Journal of Remote Sensing*, 6:1319-1334.

Nieuwenhuis GJA, M Menenti (1986) Application of thermal infrared remote sensing in water management of humid and arid areas. *Geocarto Int.*, 1, 35-46.

Ottle C, D Vidal-Madjar (1992) Estimation of land surface temperature with NOAA9 data. *Remote Sensing of the Environment*, 40, 27-41.

Paulson CA (1970) The mathematical representation of wind speed and temperature profiles in the instable atmospheric surface layer. *Journal of Applied Meteorology*, 9, 857-861.

Penman HL (1956) Evaporation: an introductory survey. *Neth. Journal of Agricultural Science*, 4, 9-29.

Price JC (1983). Estimating surface temperatures from satellite thermal infrared data: a simple formulation of the atmospheric effect. *Remote Sensing of the Environment*, 13, 353-361.

Price JC (1982) Estimation of regional scale evapotranspiration through analysis of satellite thermal infrared data. *IEEE Transactions of Geology and Remote Sensing*, GE20, 286-292.

Priestley CHB, RJ Taylor (1972) On the assessment of surface flux and evaporation using large-scale parameters. *Monthly Weather Review*, 100, 81-92.

Rambal S, B Lacaze, H Mazurek, G Debussche (1985) Comparison of hydrologically simulated and remotely sensed actual evapotranspiration from some Mediterranean vegetation formations. *International Journal of Remote Sensing*, 6:1475-1481.

Riou CH, B Itier, B Seguin (1988) The influence of surface roughness on the simplified relationship between daily evaporation and surface temperature. *International Journal of Remote Sensing*, 9:1475-1481.

Rosema A, JH Bigleveld, P Reiniger, G Tassone, K Blyth, RJ Gurney (1978) TELL-US, a combined surface temperature, soil moisture, and evaporation mapping approach. *Proceedings of 12th ERIM Symposium*, 2267-2276.

Sandholt I, HS Andersen (1993) Derivation of actual evapotranspiration in the Senegalese Sahel using NOAA-AVHRR during the 1987 growing season. *Remote Sensing of the Environment*, 46, 164-172.

Schreiber P (1904) Cited in Brutsaert (1982).

Seguin B, S Baelz, JM Monget, V Petit (1982) Utilisation de la thermographie IR pour l'estimation de l'evaporation regionale. *Agronomie*, 2, 7-16.

Seguin B, B Itier (1983) Using midday surface temperature to estimate daily evaporation from satellite thermal IR data. *International Journal of Remote Sensing*, 4: 371-383.

Seguin B, JC Mandeville, Y Kerr, JP Guinot (1986) A proposed methodology for daily ET mapping using thermal IR satellite imagery. ASAE Conference on Evaporation. ASAE Ed., 385-392.

Seguin B, E Assad, JP Freteaud, J Imbernon, YH Kerr, JP Lagouarde (1989) Use of meteorological satellite for water balance monitoring in Sahelian regions. *International Journal of Remote Sensing*, 10:1001-1017.

Seguin B, JP Lagouarde, M Savane (1991) The assessment of regional crop water conditions from meteorological satellite thermal infrared data. *Remote Sensing of the Environment*, 35:141-148.

Sellers PJ, Y Mintz, YC Sud, A Dalcher (1986) A simple biosphere model (SiB) for use within General Circulation Models. *Journal of Atmospheric Science*, 43, 505-531.

Sellers PJ, FG Hall (1992) FIFE in 1992: results, scientific gain, and future research directions. *Journal of Geophysical Research*, 97, D17, 19091-19109.

Soer GJR (1980) Estimation of regional evapotranspiration and soil moisture conditions using remotely sensed crop surface temperature. *Remote Sensing of the Environment*, 9, 27-45.

Stone LR, ML Horton (1974) Estimating evapotranspiration canopy temperatures: field evaluation. *Agronomy Journal*, 66, 450-454.

Sugita M, W Brutsaert (1991) Daily evaporation over a region from lower boundary layer profiles measured with radiosondes. *Water Resources Research*, 27, 747-752.

Taconet O, R Bernard, D Vidal-Madjar (1980) Evapotranspiration over an agricultural region using a surface flux/temperature model based on NOAA-AVHRR data. *Journal of Climate and Applied Meteorology*, 25, 284-307.

Thom AS (1975) Momenture, mass, and heat exchange of plant communities. In JL Monteith (Editor), Vegetation and the Atmosphere, Academic Press, I, 57-109.

Turc L (1955) Le bilan d'eau des sols: relations entre les precipitations, l'evaporation et l'ecoulement. *Annals of Agronomy*, 6, 5-131.

Vidal A, YH Kerr, JP Lagouarde, B Seguin (1987) Teledetection et bilan hydrique. Utilisation combinee d'un modele agrometeorologique et des donnees de l'IR thermique du satellite NOAA-AVHRR. *Agricultural and Forest Meteorology*, 39, 155-175.

Vidal A, A Perrier (1989) Analysis of a simplified relationship for estimating daily evapotranspiration from satellite thermal IR data. *International Journal of Remote Sensing*, 10:1327-1337.

Vining RC, BL Blad (1992) Estimation of sensible heat flux from remotely sensed canopy temperatures. *Journal of Geophysical Research*, 97, D17, 18951-18954.

Hydrological Models, Regional Evaporation, and Remote Sensing: Let's Start Simple and Maintain Perspective

W. James Shuttleworth
Department of Hydrology and Water Resources
The University of Arizona
Tucson, Arizona, USA

1. Introduction

This paper seeks to be provocative in the sense that it is intended to stimulate a line of thought and perhaps a climate of opinion in which the global modeling of terrestrial interactions, evaporation in particular, is predisposed towards simplicity. The justification for adopting this intention is an opinion that, having been faced with the new challenge to provide an acceptably realistic description of hydrological process at a global scale, but having simultaneously been faced with powerful visual impact of remote sensing images to remind them of the heterogeneity of terrestrial surfaces, hydrologists may have suffered some loss of confidence and an associated lack of perspective on the scale of the problem to be addressed. Perhaps this loss of confidence is reflected in the phrase "The Trials and Tribulations of Modeling and Measuring" which is part of the title of this workshop.

In addressing our new and exciting challenge, we should draw courage and motivation from the fact that hydrological models are already routinely applied with considerable success to practical problems at catchment scale. Often the point-to-point variability in soil characteristics across an ostensibly uniform 'patch' of vegetation or hill slope is huge, and the plot-to-plot variability in vegetation cover across a catchment can itself be substantial. Yet hydrologists continue to triumphantly build and apply models to catchments. In fact, much of the land surface heterogeneity which gives concern is already present at subcatchment scale, whether visible in remotely sensed images or otherwise.

Within scientific hydrology, as we have acquired more detailed understanding of hydrological processes over the last 25 years and as computer resources became more

NATO ASI Series, Vol. I 46
Land Surface Processes in Hydrology
Trials and Tribulations of Modeling and Measuring
Edited by S. Sorooshian, H. V. Gupta, and S. C. Rodda

available, our first instinct was to seek to build more and more complex models. With time, we learned that our ability to provide proper initiation and calibration of the parameters used in such models was a severe and fundamental limitation. We learned to strike a balance between detail in model description and the justifiable data input we could make to them. Shouldn't we transfer that acquired common sense and experience when approaching this new challenge? Here I hope to speed progress towards reaching the equivalent compromise when hydrologists address global change by providing a positive counterbalance to any overestimation of the scale of the challenge we face in providing adequate hydrological models and in providing representation of the evaporation process in particular.

This I do, first, in Section 2 by distinguishing several distinct objectives behind seeking improved representations of hydrological processes in the context of better modeling Global Environmental Change (GEC). The point is that, if there are distinct objectives, then there may well be distinction in the complexity of the modeling required to achieve each objective. Explicit representation of slopes and surface features is presumably necessary in the hydrological models which give interpretation of predicted global change in terms of modified surface and subsurface flows, but these models can be locally specific. On the other hand, feedbacks and mixing in a dynamic, fluid atmosphere might mean that there is little need for *explicit, geographically accurate* description of aspects of land surface heterogeneity within atmospheric General Circulation Models (GCMs) to improve climate change prediction. For instance, there is now substantial evidence that heterogeneity in land cover, providing it has appropriate size, can generate enhanced vertical transport in the atmosphere by stimulating mesoscale circulations (e.g., Dalu and Pielke, 1993; Avissar and Chen, 1993). In practice, perhaps, this enhancement can be parameterized by defining semi-empirical functions involving the statistics of land cover distribution (Zeng and Pielke, 1995). Equally, the complexity required when describing area-average surface exchanges in GCMs is arguably less than that required for interpretative hydrological modeling, because land surface hydrology is just one of several influences on climate, and GCMs must retain an appropriate balance in the way they include these influences. In this case, an appropriately defined averaging procedure applied to component land covers in each grid square might suffice to define the effective values of relevant land surface parameters (Shuttleworth, 1991, 1993; Noilhan and Lacarrere, 1994).

Over the last 5-10 years, there has been an explosive proliferation in the range of submodels [Soil-Vegetation-Atmosphere Transfer Schemes (SVATS)] which seek to provide a description of the energy-water interactions of terrestrial surfaces for the purposes of global environmental change prediction. There has also been a gradual but ongoing increase in the complexity with which they represent the physical and physiological processes involved. This proliferation gives more insight into human ingenuity and the now ready availability of computers than it does into the complexity which is truly required in SVATS. Just how broad a range of distinguishable land covers do we really need? What characteristics of these land covers really matter? We will look at these issues in Section 3 with an emphasis on seeking simplicity and on seeking greater relevance to satellite data, assuming that such data are a likely mechanism to affect global coverage of model parameters. Having this emphasis, we then consider recent experimental and modeling evidence which suggests that perhaps quite simple averaging schemes might provide acceptable aggregate representation of heterogeneous vegetation cover.

Part of the pressure from hydrologists to include hydrological complexity in GCMs (beyond that justified to improve climate prediction) is generated by the parallel need to allow more ready and more transparent hydrological interpretation of any changes they predict. The final part of this paper seeks to address this problem by advocating a computational mechanism which will allow GCM-generated, globally defined, vector fields of atmospheric humidity to be expressed in terms of the net input of water to irregularly shaped catchments. In this way, the persistently distinctive aspects of meteorological and hydrological models, namely their differing requirements for representational complexity and their optimum modeling elements (rectangular on the one hand, irregular topographically determined catchments on the other) can be reconciled, leaving each type of model free to develop in its own optimum and balanced way.

2. The Objectives of Hydrometeorological Modeling

Traditionally, the term hydrological models is used to describe a combination of mathematical formulae which provide a description of the movement of surface and subsurface waters within a catchment, from information on the precipitation input and other information which allows estimation of the evaporative loss from the catchment. Therefore, such models necessarily describe both horizontal and vertical water movement, with at least as much emphasis placed on modeling the former as the latter.

Until recently, the 'hydrological models' used in GCMs have been profoundly different to this. Their main objective has been to describe the vertical movement of that portion of the precipitation which is returned to the atmosphere by evaporation from the surface of leaves and soil or by transpiration through plants from the shallow layer of soil accessible to the roots. Here we broaden this narrow definition of 'GCM hydrology' to include the representation of both horizontal and vertical flows and, with this re-definition, there are at least four main purposes for carrying out hydrometeorological modeling, as follows:

(1) To allow predictions of global environmental change to be better interpreted in terms of its impact on hydrological and agrohydrological systems.

(2) To allow validation of the description of hydrology provided by General Circulation Models in present-day climates by making comparisons against historical hydrological records of rainfall and runoff.

(3) To enhance the accuracy and credibility of predictions of global environmental change made with global models by improving the description of the land surface energy and water exchanges.

(4) To enhance the potential for improved prediction of global environmental change by allowing better description of other surface exchanges controlled or significantly influenced by hydrological flow or soil moisture.

The following subsections, 2.1 to 2.4, separately discuss modeling needs in relation to the above objectives, along with opinions on the activity required to address them with *minimum complexity*. The alternative approach to satisfying these objectives is to seek to write hydrological components into models of global environmental change which address these four needs simultaneously. This might be assumed to be the best and most complete approach. However, it requires modeling with maximum complexity and modeling which rapidly becomes intractable (or at best inefficient) and which requires that large amounts of calibration data are available worldwide. A consistent argument in this paper is that the above distinct objectives are best considered separately and that, in each case, the approach should be to address the needs with a predisposition to start with the simplest assumptions consistent with physical and biophysical insight.

2.1 Hydrometeorological Models and Improved Interpretation

To allow interpretation of global change predictions in terms of hydrological and agrohydrological impacts, the primary need is for an interfacing procedure which can interpret the global fields of weather variables provided by GCMs in terms of the weather variables required by traditional hydrological models. These hydrological models can then operate independently in the local region of interest. Indeed, they may well already exist and may even have a relevant calibration based on existing data if the predicted climate change lies within the range of past climate variability.

Section 4 outlines a novel approach to providing the required interface between GCMs and local hydrological models. However, there are alternative approaches, sometimes called 'Weather Generators' (IGBP, 1993), although these tend to be more complicated and more location specific. They include the possible use of mesoscale meteorological models, perhaps operating off-line from a GCM at less than GCM grid scale, but driven by GCM-gridded data. Alternatively, they may access existing weather data, if available, to provide the statistical linkage between mean and distributed weather variables for the region of interest, under the assumption that such statistical linkages will not change substantially when the climate alters. However, the key point made here is that *none of these potential linkage procedures requires that the hydrological models used for interpretation are directly coupled to the GCM providing the source data*, thereby removing the need to compromise the operational efficiency of the GCM just to facilitate the hydrological interpretation of its output.

2.2 Hydrometeorological Models and Improved Validation

Existing hydrological records are never in an appropriate form for direct validation of the simplified hydrological schemes in GCMs because there is a basic inconsistency between both the spatial and temporal characteristics of the hydrological data on the one hand and the computed GCM outputs on the other. The easiest way to deal with this problem is to write some form of hydrological model which can provide credible routing of the GCM runoff computed for individual grid points through simplified river systems (e.g., Liston, 1993). In this way, derived runoff data can be synthesized which has appropriately delayed river flow timing for the few large rivers that can be represented at the grid mesh scale of GCMs.

It is clearly necessary that the representation of river routing is sufficiently credible and well calibrated that it is, in fact, the GCM--rather than the routing scheme--which is validated. Further, bearing in mind the probability that the grid mesh length used in GCMs will reduce in size over the next decades as computer technology advances, it is sensible to seek a robust, grid mesh size independent procedure to define the form of the river system represented and to calibrate the constants used to calculate the river flow against available hydrological records. Such a procedure will likely now be based on Geographical Information System (GIS) techniques (Moore et al., 1992). In the context of this paper, the relevant point is that, *for the purposes of validating a GCM against runoff data, the required hydrological modeling is essentially simple and large scale* and that there is no real need to couple even this simple routing model into the GCM. There may, however, be other reasons for doing this coupling, as described in Sections 2.3 and 2.4.

A further mode of validation might exist in which an advanced hydrological model is used to synthesize the area average values of some of the model-calculated variables which the hydrological model shares with the GCM. Such variables might include soil moisture, evaporation, or perhaps surface temperature. In this case, the primary requirement is that the model is realistic and credible, which might well mean that it is itself validated at sample points against patch-scale data and that it is used as an interpolation mechanism. To be realistic, the hydrological model used for interpolation may need to be complex but, as in the case of validation, it can operate independently of the GCM in the local region where validation is to be made. In this case, the hydrological modeling methods are not necessarily new--they are merely being applied over a somewhat bigger area than previously--and the most exacting novel requirement is not the model method; rather it is for distributed near-surface weather data with which to drive the model across the now larger validation area.

2.3 Hydrometeorological Modeling and Improved Prediction

Most of the considerable effort which has so far been directed towards improving the representation of hydrology in GCMs has focused on providing a more realistic description of the vertical movement of water. This is because it is these vertical flows which are most important in determining the net return and the rate of return of water to the atmosphere by evaporation. However, there has already been substantial activity in this subject area and there are, in the author's opinion, too many alternative SVATS, some with arguably too much complexity. However, we postpone further consideration of this last point until Section 3 and here focus attention on the as yet unfilled need to include at least some simple representation of horizontal water flows in GCMs.

Accessibility is the key word when addressing the need for an adequate representation of hydrological processes in atmospheric models (see Shuttleworth, 1991 and Section 3). It is the extent to which water is made accessible to the atmosphere and the ease with which it is accessible that are responsible for the sensitivity of weather and climate predictions to the manner in which hydrology is described in GCMs. Therefore, it is only those aspects of horizontal flow processes that change the atmosphere's access to surface and surface waters which need to be included (along with existing SVATS) to improve the predictive ability of GCMs.

2.3.1 Modeling the Consequences of Horizontal Flows in GCMs

Within an individual grid square, water is accessible to the atmosphere only if it is on the surface of vegetation or soil, in moving or stagnant bodies of water, or in the near surface soil to depths which can be reached by the roots of actively transpiring vegetation. Once water drains from the rooting zone or enters the body of a deep river, it is no longer available for evaporation. However, it is possible for soil water lost to the rooting zone at the top of slopes to become accessible to the atmosphere again if hillslope-induced flow gives concentration of water lower down, so that water once again rises into the plants' rooting zone (see Figure 1).

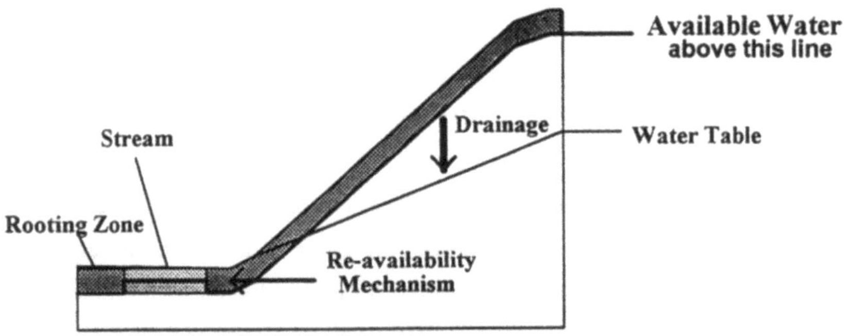

Figure 1. Schematic diagram illustrating the surface water 'available' for return to the atmosphere and the effect of horizontal, subterranean flow in making water which has drained from the plants' rooting zone at the top of a slope 're-available' at the bottom of the slope by returning it to the rooting zone.

Similarly, water which has been 'lost' to rivers or lakes can in time become accessible to the atmosphere again if the level of the water body drops by evaporation to reveal successively deeper layers of water. This process could be important, for instance, in the case of many shallow, stagnant lakes in high latitude North America. Otherwise, stream water drains through the bed of the river or lake or it runs off down river out of the grid square and, for all practical purposes, it is no longer available to support evaporation from the grid square in question.

In principle, it is possible for water lost as river runoff or by deep drainage in a particular grid area to become re-accessible to the atmosphere elsewhere in the same grid square, if the rates of surface or subsurface flow and the underlying topography of the streams or aquifers are appropriately combined to give flooding or springs. In practice, the area of the free water surface so generated is unlikely to be more than a few percent of that of the grid square; this phenomenon can probably be neglected from the standpoint of improving GCM predictions. On the other hand, exactly the same phenomenon might be more significant to regional weather in the case of very large rivers, such as the Amazon and the Nile, which are prone to extensive seasonal flooding and which are sufficiently large to be acknowledged in the routing algorithms described in Section 2.2. In this case, water which has become inaccessible to the atmosphere in one grid area might become re-accessible again elsewhere in another grid area (see Figure 2).

From the restricted standpoint of providing better climate change prediction, it could be argued on the basis of the above discussion that the most important, and perhaps the only worthwhile, horizontal water flow related improvements which could be made in GCMs are the following:

(a) In the case of grid squares with a marked topography, description of the topography-generated, intragrid, horizontal flow mechanisms which allow water that has drained from the rooting zone at the top of hill slopes to become re-accessible to the atmosphere again at the bottom of hill slopes, as illustrated in Figure 1.

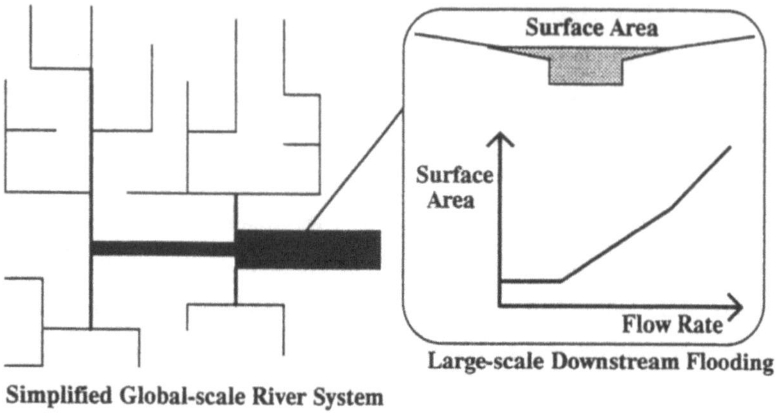

Figure 2. Schematic diagram showing a simple river routing scheme appropriate for incorporation in a GCM and how water which has been lost to the atmosphere by becoming stream flow in one grid square might become 're-available' to the atmosphere if it results in large-scale seasonal flooding in grid areas downstream in the simplified river system.

(b) In the case of very large, flood-prone rivers, description of the intergrid flow mechanisms which allow water lost to river flow in one grid area to become re-accessible to the atmosphere in other grid areas because the surface area of the river is substantially increased as it spreads across the flood plain, as illustrated in Figure 2.

In line with the philosophy of this paper, we should seek to accommodate description of the above two processes with minimum complexity. In this respect, the second, intergrid process is easier than the first. It merely requires the real-time coupling into the GCM of the simplified river systems discussed in Section 2.2 and then including the contribution of a river flow dependent, free water source area to the (vertical) return of water vapor to the atmosphere. Remote sensing measurements of river area and conventional measurements of river flow for the few relevant large rivers might provide calibration for this model feature.

Providing description of the first process, the intragrid process (a) above, is more difficult, but some useful, relevant work has already been done using fairly simple distributed hydrological models, particular TOPMODEL (Beven and Kirkby, 1979). The work of Beven and Wood (1983), Band and Wood (1988), Wood et al., (1990) and most relevant, that of Famiglietti and Wood (1994a) is valuable in this regard. Such research is starting to show that intragrid horizontal flows may impact the calculated area-average evaporation rate, at least in certain conditions (Famiglietti and Wood, 1994b), even though the area of the downslope wetted zones may, in practice, be only about 10% of the total surface area.

Such distributed hydrological models still retain a level of representational complexity for horizontal flows which, were it suggested they be explicitly incorporated into GCMs, might be considered inconsistent with the relative importance to global change prediction of this within grid, horizontal flow generated re-accessibility of water. This is yet to be determined. Could such models be used to define the effective values of a few topography-determined parameters in a simpler modeling approach which involves including a few 'representative slopes', perhaps one for each allowed vegetation cover, for which hillslope flow is modeled in each GCM grid square? Progress in this branch of research is rapid, and it is likely these questions will be answered in the next few years.

In any case, the suggestion made here again is that, for the purposes of improved climate and weather prediction, *the additional modeling in GCMs to acknowledge the relevant impacts of horizontal flow on the atmospheric return of water and energy should, and probably could, be simple. Simple description should at least be attempted first.*

2.4 Hydrometeorological Modeling and Other Surface Exchanges

The hydrometeorological modeling of river flows and soil moisture is a necessary component of the modeling of certain other surface atmosphere exchanges which are of significance in improving the prediction of global environmental change. The most significant hydrology-related processes in this regard are perhaps changes in ocean salinity and the exchange with the atmosphere of biogeochemicals.

Accommodating the impact of (large) river flows on ocean salinity should be comparatively simple and is already being addressed in so-called "Climate System Models". The need is to incorporate the simplified river routing schemes discussed in Section 2.2 into the GCM and then to formulate an algorithm to describe the mixing of salt and fresh waters in those grid squares which subtend the river mouth of the world's major rivers.

The carbon cycle is arguably the most important global cycle after those of energy and water; providing a worthwhile description of carbon dioxide exchange over continental surfaces is currently a high priority (IGBP, 1993). It is extremely helpful that the release of water from plants (as transpiration) and the photosynthetic absorption of carbon dioxide are both strongly and jointly controlled by the same stomatal mechanism. It is also helpful that some aspects of this stomatal control seem to be related to measurements which can be made from space (see Hall et al., 1992; Sellers et al., 1992; and Section 3). In the context of our present discussion, the relevant point to make is that, because these important, largely vegetation-determined exchanges share a common dominant controlling mechanism, it is possible (indeed essential) that the models used to provide their global scale description should share a common level of complexity. If, as we argue in Section 3, the SVATS used to describe the energy-water exchanges of vegetation stands are to remain simple, then those used to describe large-scale carbon dioxide exchanges should be similarly simple.

Providing adequate hydrometeorological models to serve as components in the description of other biogeochemical releases will be altogether a more complex matter. This is primarily because some of the processes involved in such release are not likely to be linear functions of soil water content and because a significant proportion of those biogeochemical releases with most global significance, such as that of methane from swamps or rice paddies, may occur from areas so small that they are, in global terms, essentially point sources. These two factors argue for the use of detailed high-resolution hydrological modeling to address this issue, but it is not clear that implementing models with this complexity across the globe is justified, especially if it were also proposed to include such a submodel in (GCM) real-time. Perhaps some compromise will emerge in the course of the next decade involving detailed submodeling for key source areas, but using a model time step which is less than that used in the atmospheric component.

In the specific case of biogeochemical release then, it is still too early to advocate a predisposition towards a simple modeling approach as for other processes; indeed, the best first step forward may be to write complex locally specific models. In doing this, it is essential not to lose sight of the true numerical significance of the calculated biogeochemical releases when viewed from the perspective of global balances. This is easily done.

In summary, when modeling changes in ocean salinity and net carbon photosynthesis at global scale, the recommendation is again to use models with simplicity consistent with that used elsewhere in predictive global scale models and which is consistent with the relative sensitivity of prediction to these processes. For other, essentially point source biogeochemical exchanges, the current recommendation is to "act local, but think global".

3. The Complexity of Soil Vegetation Atmosphere Transfer Schemes

Earlier, the opinion was expressed that there are now too many SVATS and that they are probably too complicated. Here I return to this point by again asking what about such submodels is most important from the point of view of modeling the surface-atmosphere exchange of energy and water vapor, this being currently the primary purpose of SVATS. In pursuing this argument, I draw heavily on ideas and opinions previously expressed in Shuttleworth (1991) and at an earlier NATO Advanced Study Institute (Shuttleworth, 1993).

3.1 What Matters?

It is first necessary to consider which aspects of the energy-water interactions of vegetation-covered continental surfaces are truly essential to include in SVAT submodels for the purposes of adequate climate and weather prediction. As stated earlier, the key characteristic of such aspects (and the parameters which determine them) is that they must strongly influence the extent to which energy and water is available to the atmosphere. Again, the key word is *availability*.

To become available, energy and water must be first captured at the land surface. The area-average value of albedo must be important, because this directly influences the amount of short-wavelength radiant energy captured at the ground that is available for return to the atmosphere. It is well known that land cover type can alter the value of albedo significantly. The surface emissivity is also important because this directly determines long-wavelength radiation exchange, but defining its effective, area-average value is less difficult than for albedo because the variation from one land cover type to the next is less. Clearly, the area-average maximum rate at which water can infiltrate the soil is important too, because water which runs off into rivers and lakes is generally more concentrated and less available to the atmosphere, as discussed in Section 2.3.

To remain available once captured, energy and water must remain in stores near the surface which are accessible to the air above, and the area-average magnitude of such accessible stores are clearly key parameters because they determine the net amount of energy and water available for return. The atmosphere is inherently a 'fast' system: it can respond to changes in surface inputs at the daily time scale--witness the daily cycle of convective rain. However, it also responds to slower changes if there is 'memory' in the stores of energy and water to which it has access. Therefore, it seems appropriate to acknowledge and include two types of stores in models of the land-atmosphere interaction. The first is the energy and water stored in comparatively easily accessible form in, or in the case of water on, the vegetation and soil surfaces. The second store is the energy and water held in less readily accessible form in the soil, but only to the depth where they are still (ultimately) available for return to the atmosphere.

Because the timing of the release of available energy and water is important, the control exerted by the vegetation and soil in determining the rate of atmospheric access to the stores of available energy and water is also important. Commonly, the mechanisms involved in this control are expressed in terms of a resistance network, which includes at least two resistances, one in the atmosphere and one associated with the surface (the latter controlling the rate of diffusion of water vapor through the stomatal apertures from inside plant leaves into the canopy air stream outside).

In summary then, the primary land surface characteristics which influence the magnitude and rate of access of the atmosphere to energy and water available at the ground are the albedo, a; the surface emissivity, e; the maximum surface infiltration rate, I; the aerodynamic roughness, z_0, because this determines the efficiency of the return of energy and water near the ground; the surface resistance, or its reciprocal the surface conductance, g_s, because this determines the efficiency of the return of soil water to the air adjacent to the plant leaves or bare ground; and the size of two types of stores for energy and water, one in the vegetation which is easily accessed, one less accessible in the soil.

Most of the many SVATS now available for use in GCMs include calculations which involve these key characteristics, although some (Shuttleworth, 1992) still follow the approach of many practicing hydrologists in not including explicit description of the stomatal control and rainfall interception mechanisms when calculating evaporation rate. It is not yet clear what level of mathematical complexity in SVATS is justified by an associated improvement in the predictive ability of the GCMs in which they reside. It is likely that the Project on Intercomparison of Land-surface Parameterization Schemes (PILPS) will shed light on this issue and provide insight into the accuracy required for adequate representation of land-atmosphere exchanges in GCMs, because this project will in due course evaluate their comparative performance against observational data for selected sites (Henderson-Sellers and Dickinson, 1993).

Pending categorical evidence from PILPS, simple physical insight suggests that many of the SVATS currently available have little substantial difference in the way they represent the component processes. Assuming this is the case, it is the author's opinion that the creative energy and skill currently being wasted in creating additional SVATS of increasing complexity would be better directed towards providing credible calibration of the few (common) parameters which do matter and, in particular, in seeking ways to rewrite SVATS in (probably simpler) ways, such that these important parameters have enhanced relevance to globally available satellite measurements.

3.2 The Area-Average Values of Parameters

If, as suggested above, there are only a few key processes (and associated parameters) which matter in the description of energy and water exchanges between continental surfaces and the atmosphere, the question arises as to how to define the effective, area-average values of these parameters. This is clearly a significant issue for continental, as opposed to ocean surfaces, because the heterogeneity of the land surface is very obvious and occurs at all scales, from local to global. Given that we can separate out the problem of providing parameterization of "mesoscale" circulations in the lower atmosphere induced by surface heterogeneity (Zeng and Pielke, 1995), the primary requirement of a SVAT is to give a simple description of the aggregate, area-average exchanges of a heterogeneous mix of different land covers for each grid square, which is adequate for calculating their influence on the atmosphere in GCMs.

This issue is addressed here head-on and in a way which is consistent with the message of this paper, by advocating that the first model to try is the simplest. That is, adopt (or perhaps simplify) one of the currently available SVATS--providing it includes the key processes and parameters described above--and then define effective area-average values of model parameters to be applied to the vegetation-covered portions of each grid area using a set of simple *aggregation rules* for each parameter (Shuttleworth, 1991, 1992). The philosophy is that, although such simple aggregation rules may not stand the test of time, once in place any critical evaluation of them based on observation or modeling can only result in their replacement by superior alternatives, and progress will automatically follow.

SVATS are required to describe area-average surface fluxes, primarily of energy and water, although some attempt is now being made to include description of the exchange of CO_2. Hence, the required area-average parameters are those which most nearly compute these area-average fluxes, either their integrated value over a given period in the case of stores, or the rate of flow into and out of the stores. In general, the simplest aggregation rule, and the one which is advocated here pending categorical evidence to the contrary, is that *the effective area-average value of land surface parameters is estimated as a weighted average over the component cover types in each grid through that function involving the parameter which most succinctly expresses its relationship with the associated surface flux.*

Let us take the example of momentum flux, τ, and the controlling parameter, z_o, the aerodynamic roughness. In this case, the relevant relationship between momentum flux and aerodynamic resistance (neglecting other surface parameters) is:

$$\tau \propto \ln^{-2}(z_b/z_o) \qquad (1)$$

where z_b is a hypothetical 'blending height' at some level in the air above the ground where turbulent mixing is sufficient for it to be assumed that the atmosphere has, in effect, lost its knowledge of the differing types of land cover on the ground below. Mason (1988) estimated z_b to be of the order L/100, where L is the characteristic horizontal scale of the vegetation patches with different surface roughness covering the surface. In this case, for a grid square with i component covers of fractional area w_i (perhaps determined from satellite), the relevant aggregation rule for z_o follows from Equation (1) as:

$$\ln^{-2}\left[z_b/\bar{z}_o\right] = \sum_i w_i \ln^{-2}\left[z_b/(z_o)_i\right] \qquad (2)$$

Table 1 interprets the general aggregation rule given above more generally for this and other key parameters in the land-atmosphere interaction. Certain new symbols are listed in Table 1 in addition to those already defined above. These are specified as follows: C_c is the amount of thermal energy stored in unit area of biomass for each degree change in temperature; S_c is the amount of precipitation which can be stored on the surface of vegetation (and soil) during rain; K_t is the thermal conductivity of the soil; C_s

Table 1. 'Strawman' Aggregation Rules for Key Parameters in Soil-Vegetation-Atmosphere Transfer Schemes (taken from Shuttleworth, 1991; 1993)

PARAMETER		'STRAWMAN' AGGREGATION RULE
Albedo		$\bar{\alpha} = \sum_i w_i \alpha_i$
Maximum Infiltration Rate [Note (i)]		$\bar{I} = \sum_i w_i I_i$
Emissivity		$\bar{\varepsilon} = \sum_i w_i \varepsilon_i$
Aerodynamic Roughness [Note (ii)]		$\ln^{-2}\left[z_b / \overline{z_o}\right] = \sum_i w_i \ln^{-2}\left[z_b / \left(\overline{z_o}\right)_i\right]$
Surface Conductance	[Note (iii)]	$\overline{g_s} = \sum_i w_i (g_s)_i$
or	[Note (iv)]	$\left(\overline{g_s^{-1}}\right) = \sum_i w_i (g_s)^{-1}$
or	[Note (iv)]	Average of above two alternatives
Canopy Stores:		
Heat	[Note (vi)]	$\overline{C_c} = \sum_i w_i (C_c)_i$
Water	[Note (vii)]	$\overline{S_c} = \sum_i w_i (S_c)_i$
Soil Stores:		
Heat	[Note (viii)]	$\overline{(K_t C_s)^{1/2}} = \sum_i w_i (K_t C_s)_i^{1/2}$
Water	[Note (ix)]	$\overline{S_s} = \sum_i w_i (S_s)_i$

Notes:

(i) Neglects possible effect of topography on infiltration
(ii) Averages momentum flux; z_b is an assumed 'blending height'
(iii) Assumes very efficient aerodynamic mixing in the atmospheric boundary layer
(iv) Assumes very inefficient aerodynamic resistance in the atmospheric boundary layer
(v) Some evidence of adequacy of this assumption (eg. Blyth et al., 1993)
(vi) Neglects any subgrid variation in radiation (cloud)
(vii) Neglects any subgrid variation in precipitation
(viii) Assumes cyclic conduction of soil heat flux into the soil
(ix) Neglects any impact of topography on water stored in rooting depth (see Section 2.3)

is the specific heat of the soil; and S_s is the amount of water which can be stored per unit area within the vegetation's rooting zone.

Although the 'strawman' aggregation rules given in Table 1 are strictly plausible only in the case of overhead conditions which are constant over a grid square, there is some evidence from modeling studies that they may be more generally valid. It seems wise to consider land surface heterogeneity as two general types according to the length scale of the organization in cover type they contain (Shuttleworth, 1989), because this can influence the atmosphere's response. For Class 'A' land cover, i.e., with vegetation in large patches of about 10 km or greater, conventional mesoscale meteorological models (with intergrid distances of 5-15 km) are a convenient mechanism to test hypothetical aggregation rules. For Class 'B' land cover, which has smaller average patch size (and no obvious surface-linked response in the atmosphere), coupled surface-boundary layer models with intergrid distances of 10-1000 m are appropriate.

Model-based tests of simple, hypothetical aggregation rules, such as those in Table 1, have been made using models of the above two types. Noihlan and Lacarrere (1994) used a mesoscale meteorological model with a 10-km grid to calculate the area-average surface flux exchanges for the Class 'A' mixed cover of the 100 x 100 km, HAPEX-MOBILHY (Andre et al., 1988) experimental square. They regarded these calculated fluxes as the true area-average values, against which they compared estimates of the area-average fluxes made assuming a single one-dimensional SVAT with parameters estimated from simple aggression rules similar to those defined in Table 1. The results suggest that such rules may be adequate, at least in fine weather. Figure 3 shows a comparison of the two estimates of latent heat which they calculated for one day.

Some similar preliminary tests of aggregation rules have been made for Class 'B' type land cover (actually the most common type) using an atmospheric boundary layer model with short intergrid distances of 100 m. Blyth and Dolman (1993) investigated aggregation rules for aerodynamic resistance and surface conductance using the Penman-Monteith equation as a simple SVAT, with alternate small patches of vegetation with different values assigned to these two parameters. Again, simple aggregation rules work well in dry conditions, but they found additional complications during and immediately after convective rain storms, when some of the vegetation may be covered by surface water. They found, for instance, that the average of the linear mean and reciprocal mean values for g_s is a more robust aggregation rule in conditions of mixed surface wetness, but that the aerodynamic transfer resistance for sensible heat becomes undefined in such conditions because of the plot-to-plot advection (Blyth and Dolman, 1993).

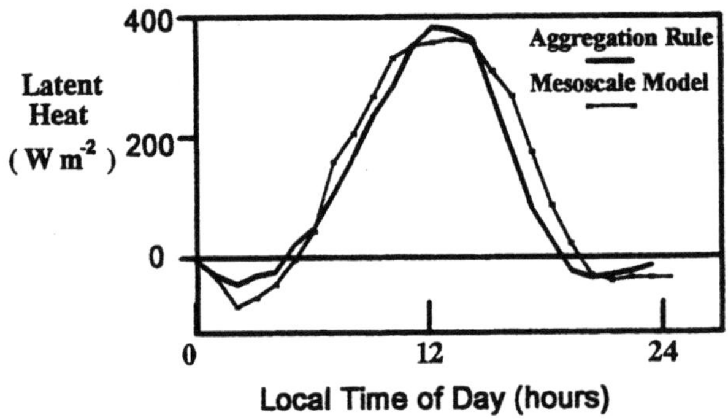

Figure 3. Comparison between area-average evaporation rates calculated for the 100 x 100 km HAPEX-MOBILHY study site for 16 June 1986 using a nesoscale meteorological nodel to synthesize the hourly values as the sum of the surface fluxes separately calculated for points on a 10-km grid and the value calculated by assuming a single one-dimensional model in which the effective model parameters have been calculated using simple aggregation rules similar to those given in Table 1.

Similar modeling studies have been made by Sellers et al. (1992) using data from the First International Satellite Land Surface Climatology Project Field Experiment (FIFE). In this case, however, the atmospheric part of the (otherwise coupled) model was, in effect, substituted by measurements; the area-average surface conductance was calculated using remote sensing measurement to allow for spatial variability in the light dependence of this parameter. The dependency of g_s on other variables was assumed to be the same or, in the case of soil moisture, measured. This study gives similar good agreement between calculated area-average fluxes in fine weather but, in this case, the result is given additional credence by the availability of independent aircraft measurements.

Although the modeling studies described above have provided pointers, there is a clear need for more methodical modeling studies of this type, preferably with better focus on the relevant aggregation rules for the more advanced and realistic SVATS. Recent studies of the aggregation rule approach applied within the framework of BATS (Dickinson et al., 1986), using plausible mixes of vegetation covers in a coupled model initiated at selected times with field data from the FIFE study (Arain, 1994), confirm their usefulness in all but artificial situations involving continuously irrigated patches of vegetation in an otherwise dry landscape.

3.3 Space-Data-Relevance

There is increasing evidence that the some of the important vegetation-determined parameters used in SVATS may have reasonably robust relationships to variables which can be measured from space. Further, it might be that these relationships are more strongly determined by plant metabolism (whether the plants have a C_3 or C_4 metabolism) rather than the vegetation class to which they are conventionally assigned in advanced SVATS.

Results from the First International Satellite Land Surface Climatology Project Field Experiment (FIFE), for instance, provide encouragement that there may be relationships between one of the parameters which determine canopy resistance and the so-called "Simple Ratio" vegetation index which can be measured from satellite. Following Sellers et al. (1992), the surface conductance, g_s, can be written in the form:

$$g_s = \nabla_F \cdot F_o \cdot [f(\delta e) \cdot f(T) \cdot f(\varphi)] \qquad (3)$$

where f( ) indicates a 'stress function' for vapor pressure deficit, de, temperature, T, and soil moisture deficit, j, respectively; F_o is the incident flux of Photosynthetically Active

Radiation (PAR), while v_F is a factor relating g_s to F_o. Analysis of the FIFE data, which involves first removing the effect of the stress functions to determine v_F, reveals that the value of this last factor bears an approximately linear relationship to the Simple Ratio, SR, with the general form:

$$\nabla_F = (SR - a_1)\cdot(a_4 \cdot V_3 + a_3) + a_2 \tag{4}$$

where V_3 is the fractional cover of vegetation with C_3 (as opposed to C_4) metabolism, and a_1, a_2, a_3, and a_4 ($a_4 = 5.668 \times 10^{-5}$) are empirical constants which were determined for the prairie grass at the FIFE site.

It is implied that the constants a_1, a_2, a_3 and especially a_4 are strongly determined by the (C_3 or C_4) metabolism of the plants and that the form of Equation (4) and perhaps the value of some of the constants can be more widely applied. It would, however, be naive not to acknowledge that these constants may also in part be determined by the optical characteristics of the leaves. However, because there are fewer metabolic pathways than readily distinguishable biomes, this FIFE result may provide a pointer to the future and suggest that the number of vegetation classes acknowledged in SVATS might reduce over the next few years. Some morphology-related distinctions surely will still need to be made between plants even if they have the same metabolism; perhaps this reduction might just be from the current 15-20 classes to (say) 6-12 classes of land cover.

4. Interfacing Atmospheric and Hydrologic Fields

In Section 2, the case was made for separating the objectives of hydrological modeling in global environmental research as a means of retaining simplicity in the required models. One of the main arguments against making such a separation is that it may inhibit the direct, credible hydrological interpretation of predicted changes in global climate. In this section, it is argued that a more profitable approach is to seek an interface which allows the retention of separation between the predictive and interpretative objectives.

4.1 Nature of the Problem

There can be little doubt that it is through the use of hydrological models that observed variations or predicted changes in global meteorological fields will find most relevant local interpretation in terms of hydrological impact, but the nature of hydrological models is such that they are inherently best framed around geomorphologically determined, irregular river basins. Atmospheric General Circulation Models (GCMs) are primarily required to describe the (largely unrestricted) flow of a fluid medium and, as such, most naturally operate with a rectangular grid which is difficult to relate to underlying irregular watersheds. At first sight, therefore, the information provided by atmospheric models is inherently inconsistent with that required as input to hydrological models.

For computational reasons and at the present time, GCMs operate with a grid mesh vastly greater than that consistent with describing the processes which determine surface and subsurface flows. This complicates use of the meteorological information they provide in the context of worthwhile predictions of the hydrological flows for real catchments. Of more fundamental and lasting significance, however, is the fact that GCMs only require that hydrologically relevant exchanges are described with a level of realism sufficient to allow them to maintain adequate description of globally distributed atmospheric fields of energy, mass, and momentum, because this is their purpose. Therefore, it is cost-effective to formulate hydrologically relevant processes in GCMs only with a level of computational complexity which is broadly consistent with that used elsewhere in the model.

Thus, although the fact that GCMs operate at a vastly different scale to hydrological models is a problem in seeking to provide realistic hydrological interpretation of the meteorological fields, this is primarily a technical limitation which will no doubt reduce in importance with time. The more fundamental and probably more permanent inconsistency between these two types of models arises because of the difference in the relative importance they *a priori* must assign to the continental precipitation and evaporation fields and the inherently different natural geographical elements on which the two types of models operate with most efficiency and realism--rectangles in one case and watersheds in the other.

Some means is required to allow the ongoing development and enhancement of GCMs to proceed in an internally balanced way, but with the need to give them hydrological interpretation still catered for. In this way, GCMs are only required to further one of their own primary objectives, namely that of providing the best possible calculation of the vector field of atmospheric water vapor. Meanwhile, the attention for hydrological modelers need not be on 'cluttering' GCMs with hydrological detail purely for hydrological purposes; rather, it might be on deriving an interface to exploit GCM improvement. The required interface should respond to improving source data, but allow a time series of GCM-generated, global fields of atmospheric water vapor to be interpreted in terms of regional maps of soil moisture and point maps of stream flow discharge at selected locations.

4.2 Reynolds Transport Theorem (RTT)-Based Linkage

For small watersheds with dimensions on the order of 10 km, within-catchment recycling of water is a small component in the water budget and, at this scale, the input required for hydrological modeling is the net input of water across the small catchment. Given such an input for each individual small basin in a larger watershed, conventional and now well-established routing procedures can be applied. In many cases, these can exploit an existing locally relevant calibration. Surface observations are the normal source data for calculating the area-average net water input to small catchments, but the vector fields of atmospheric water vapor generated by GCMs are an alternative source of net water input, at least for the purpose of calculating monthly or annual flows. The Reynolds Transport Theorem can be applied to that portion of the atmosphere which overlies each catchment to provide the required interface.

Discharge is only defined for irregularly shaped watersheds (Figure 4a). It is possible to define a vector field of atmospheric moisture on the GCM grid by taking the product

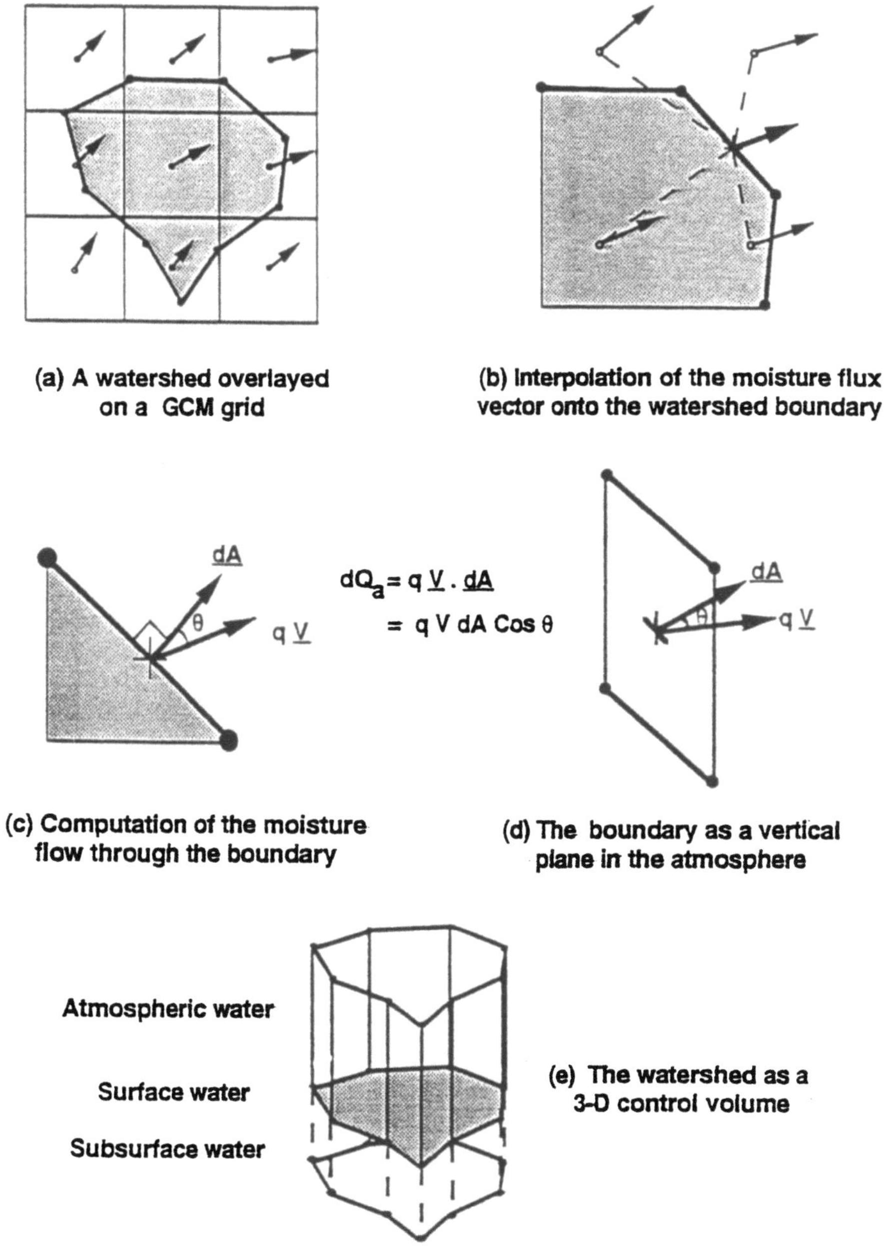

Figure 4. Illustrating steps in the development of a Reynolds Transport Theorem based interface between the vector field of atmospheric humidity provided on a regular rectangular grid by a GCM (or NWP model), and the net input of water to an underlying array of irregularly shaped watersheds for which traditional hydrological modeling is required.

of the specific humidity, q, and the velocity components u and v for each position in the grid box, integrating the values qu and qv vertically through the atmosphere, and then averaging them through time if necessary. The values of qu and qv together form the moisture flux vector, $q\underline{V}$, and can be defined anywhere within the domain of the vector field by spatial interpolation between the four most adjacent point values. In particular, they can be defined at the center of each line segment in the watershed boundary, as in Figure 4b. (Note: To simplify the explanation here, the linear dimension of the rectangular grid in Figure 4a has the same order of magnitude as that of the schematic watershed. With present day GCMs, this is not the case, but this fact does not alter the proposed procedure in any fundamental way. It is, however, likely to degrade the quality of the product provided by an RTT-based interface pending a reduction in the GCM grid size on which meteorological fields are calculated. This could give rise to the need for grid-size dependent corrections in the short term, as discussed in Section 4.3).

If the line segment of the watershed boundary is projected vertically from the ground to the maximum height used in the original GCM field, it defines a surface area which we represent as a vector $\underline{dA}$, which is defined to be normal to the area always pointing in the outward direction from the watershed, as in Figure 4c and Figure 4d. The contribution to atmospheric moisture flow above the catchment through this surface, dQ_a, is given by:

$$dQ_a = q\underline{V} \cdot \underline{dA}$$

or:

$$dQ_a = qVdA\cos\theta \tag{5}$$

where θ is the angle between the vectors $\underline{V}$ and $\underline{dA}$. Because the length and orientation of all line segments are known, θ can be determined and dQ_a calculated.

In this way, the watershed can be visualized as a three-dimensional control volume by similarly projecting vertically all the line segments along its boundary and then enclosing the volume with horizontal planes at the top and bottom, as shown in Figure 4e. The net atmospheric moisture flow out of the watershed is then found by integrating Equation (5) over all the planes making up the control surface, thus:

$$Q_a = \oiint_{c.s.} q\underline{V} \cdot d\underline{A} \tag{6}$$

Once this method has been successfully implemented for a single watershed, its application could be automated using a Geographical Information System, and it could then be applied to a landscape which may comprise hundreds or even thousands of watersheds if necessary. In this way, it is possible to conceive the watershed as a fundamental spatial primitive and to use it to describe the hydrological systems in a landscape in much the same way that the rectangular grid cell is the fundamental spatial primitive is used to describe atmospheric flow over the Earth's surface.

The net atmospheric moisture flux upwards from the watershed surface is easily found as:

$$q_a = Q_a / A_w$$

where A_w is the watershed area, so that:

$$q_a = \frac{1}{A_w} \oiint_{c.s.} q\underline{V} \cdot d\underline{A} \tag{7}$$

and q_a = E - P where E is evaporation and P is precipitation. The water balance on the watershed is hence given by:

$$\frac{dS}{dt} = -Q_a - Q_s - Q_g \tag{8}$$

$$\frac{dS}{dt} = -(E - P)\, A_w - Q_s - Q_g$$

where Q_a, Q_s, and Q_g represent the outflows of atmospheric water, surface water, and groundwater across the control surface, respectively, and S is the total of all water storage components in the watershed.

The above logic, the Gauss theorem, can be used to relate flux divergence within a control volume to the integral of flux over the control surface. However, if we wish to examine the time rate of change of a fluid passing through a control volume, it is necessary to apply the Reynolds Transport Theorem (Aris, 1962), which incorporates the Gauss theorem along with basic physical laws and applies them to the fluid flowing through a control volume at a fixed location. This can be done as follows.

If B is the amount of a fluid property, b is the amount of this property per unit mass of fluid, ρ is the fluid density, and dv is an element of volume within the control volume, then the Reynolds Transport Theorem, when applied to a control volume of fixed size such as that shown in Figure 4e, takes the form:

$$\frac{dB}{dt} = \frac{d\left[\iiint_{c.v.} (\beta\rho)\, dv\right]}{dt} + \oiint_{c.s.} (\beta\rho)\, \underline{V} \cdot \underline{dA} \qquad (9)$$

where the term dB/dt is the material derivative of the fluid property (that is, its total rate of change with respect to time); while the first term on the righthand side is the time rate of change of the amount of property stored in the volume, and the second term is the net outflow of the property across the control surface. When this theorem is applied to water vapor transport in air, B becomes the mass of water vapor, β the mass of water vapor per unit mass of moist air, (i.e., the specific humidity q), ρ the density of moist air, and dB/dt becomes the sum of the rate of extraction of vapor from the air stream by precipitation and the rate of addition to the air stream by evaporation, respectively.

If the flow is assumed to be in steady state, an assumption which can be argued to be valid for monthly average calculations, the first term on the righthand side of Equation (9) disappears and the physics in the equation reduces to that used to derive Equation (6). Alternatively, when studying short-term atmospheric moisture motion, such as might occur during thunderstorm development over a small watershed, the moisture content of the air changes rapidly. It is then necessary to take account of all three terms in the Reynolds Transport Theorem. In this case, the first term on the righthand side of

Equation (9) can be accounted for by interpolating the vertically integrated values of specific humidity into a subgrid across the watershed, areally integrating these values at each point in time, and then approximating the time derivative of atmospheric water vapor from the successive differences between these integrated quantities. Exactly the same procedure can, of course, be used in the case of monthly average calculations to account for longer term changes in water vapor concentration.

4.3 Short-Term Weakness and Upcoming Opportunity

GCMs are used in two main ways. They can be run as free-standing, self-regulating, predictive models, and it is in this form that they are used as the basis of global change prediction. However, in the form of Numerical Weather Prediction (NWP) models, they are routinely run at regularly repeated intervals, typically every few hours, to simulate the evolution of global fields for hours and days ahead, and they are then the basis of weather prediction. Prior to each short run, meteorological data from around the world in the form of surface and balloon borne measurements and, if feasible, satellite measurements are *assimilated* using advanced error balancing techniques into the last predicted global field calculated by the model. This field is thus modified in a model-consistent way so that the initial state from which the next prediction is then made is as close as possible to reality, within the limits of the data available. The, albeit in part model-determined, global fields created by assimilating all the available meteorological data to hand at the time of re-calibration are in themselves a source of meteorological data. These data are by their nature available worldwide and, being model-generated, have the advantage of being consistent with the constraints of energy and mass imposed in the model.

The time sequence of evolving global meteorological fields generated by the means just outlined is sometimes referred to as *four-dimensional (4-D) model assimilation data* and is routinely stored by the major numerical weather prediction centers several times each day as a byproduct of their operation. As stated above, although substantially determined by the assimilated input data, they are also significantly influenced by the form and assumptions employed in the model used in their synthesis. They need to be used with understanding and care because of this.

A point of particular relevance to the interface outlined above is that the information on net water exchange with the ground derived by the Reynolds Transport Theorem calculations can only be no more than that contained in the original meteorological source data. These data provide average values of this surface exchange sampled at the spatial interval of the grid used in the GCM (or NWP) model, and the surface fluxes calculated by the interfacing method we derive can, therefore, be no more than smoothed grid-average values. Thus, it is probable that there will be differences between the RTT-calculated net input of water and the measured runoff for individual catchments whose size is less than that of the grid used in the atmospheric source data.

The fact that an RTT-based interface can at present provide no more than smoothed (GCM or NWP) grid-average values of surface exchange over small watersheds will be of most significance in regions with marked topography. With atmospheric humidity fields derived from large grid mesh GCMs, it is likely that there will be substantial differences between the RTT-calculated net input of water and measured runoff for individual hilly catchments. Further, although these discrepancies will likely decrease with catchment size, it is possible that some systematic discrepancy will persist.

On the basis of the above comments, it is of interest to determine whether discrepancies in RTT-based calculations with present-day GCM fields originate from:

(a) inherent regional weaknesses in the GCM-calculated source data with regard to its description of large-scale humidity fields;

(b) scale-dependent inconsistencies between the meteorological source data and the catchment scale to which it is applied; or

(c) shortcoming in the GCMs physics associated with poor recognition of topography-generated vapor extraction arising from processes which occur at subgrid scales.

However, there is opportunity to address evaluation of these potential discrepancies in the course of the next few years.

Activity under the GEWEX Continental-scale International Project (GCIP) will provide upgraded global fields of meteorological data as part of the enhanced observational program which will run for five years from 1995. Numerical Weather Prediction models at the National Meteorological Center (NMC) in Washington and at the European Centre for Medium Range Weather Forecasting (ECMWF) in Reading, UK will be among the operational models used primarily for this purpose. These centers already have model-calculated data records for an extended period which were generated in the course of their day-to-day operational activities and which are available for research purposes. Both centers have plans to upgrade the continuity of these global fields by re-analyzing, with a consistent model applied to the assimilation of the original source data used in past assimilations.

In parallel with this improvement in meteorological data under GCIP, there will be an equivalent synthesis of hydrological data (WMO, 1992). Specifically, it is proposed to define a substantial set of watersheds sampled from across the Mississippi River basin and to prepare a high-quality data set of rainfall and runoff for these in the near future. These data could be used for initial comparison against calculated net water inputs made using the Reynolds Transport Theorem interface for these basins. The extensive nature, range of climates, and number of available catchments will ensure the provision of a worthwhile initial test of the interfacing methods and will provide an ample source of information for investigating the relationships between catchment characteristics and any differences in calculated net water input to the catchment and measured runoff.

Presumably, the area average values at larger scales ought to agree better and, in the context of GCIP, an investigation could be made to establish the scaling characteristics of the RTT-based interface. This would involve applying it to progressively larger catchments in the Mississippi River basin, each being comprised of smaller watersheds to which the method can also be applied and for which runoff data are available. It is to be anticipated that there will be substantial 'noise' in the relationship between RTT calculations and observed runoff at the small basin scale even in monthly average values, but that this will progressively reduce and then stabilize as the dimensions of the sampled watershed approach and then exceed those of the GCM grid.

Finally, it should be remembered that the Mississippi River basin is in fact a comparatively flat catchment. Therefore, additional evaluation of the performance of the RTT-based interface is required outside the strict geographical limits of the GCIP study area to investigate potential topography-generated discrepancies. Watersheds with marked topography in the western USA are clearly the most obvious target study sites.

5. Concluding Comments

The central objective of this paper is to provide guidance and perspective, but also reassurance and motivation to hydrometeorological modelers who are faced with the task of applying their skills, experience, and process understanding to the task of improving and applying global environmental change predictions.

My central message is that, in approaching this challenge, we should use our physical insight to distinguish between the objectives we need to address and define what model components truly matter in order to fulfill those objectives separately and with minimum complexity. We cannot, and need not, apply small-scale hydrological modeling skills mechanically and universally across the globe, nor do such models need to be absorbed into GCMs in their present complex form and run in real time. If we recognize this, our task becomes tractable and, returning to the theme of this workshop, we can approach 'the trials and tribulations of hydrological models and measurements' in the context of GEC with increased vigor and focus.

6. Acknowledgments

During preparation of this paper, the author was supported under NASA Grant No. 2485-MD/BGE-0011. Using the Reynolds Transport Theorem as an interface between hydrologic and atmospheric model data fields was refined into the form presented in discussion with David Maidment during a visit to The University of Arizona sponsored by the UA Coordinating Committee on Climate Change.

7. References

Andre JC, JP Goutorbe, A Perrier, F Becker, P Bessemoulin, P Bougeault, Y Brunet, W Brutsaert, T Carlson, R Cuenca, J Gash, J Gelpe, P Hildebrand, JP Lagouarde, C Lloyd, L Mahrt, P Mascart, C Mazaudier, J Noilhan, C Ottle, M Payen, T Phulpin, R Stull, J Shuttleworth, T Schmugge, O Taconet, C Tarrieu, RM Thepenier, C Valencogne, D Vidal-Madjar, A Weill (1988) Evaporation over land surfaces: first results from HAPEX-MOBILHY special observing period, *Annales Geophysicae*, 6(5):477-492.

Arain AM (1994) Spatial aggregation of vegetation parameters in a coupled land surface-atmosphere model, Masters Thesis, The University of Arizona, Tucson, Arizona.

Avissar R, F Chen (1993) Development and analysis of prognostic equations for mesoscale kinetic energy and mesoscale (subgrid scale) fluxes for large-scale atmospheric models, *Journal of Atmospheric Science*, 50:3751-3774.

Aris R (1962) Vectors, Tensors, and the Basic Equations of Fluid Mechanics, Dover Publications, New York.

Band LE, EF Wood (1988) Strategies for large-scale, distributed hydrological modeling, *Applied Mathematics and Computation*, 27:23-37

Beven KJ, MJ Kirkby (1979) A physically based variable contributing area model of basin hydrology, *Hydrology Science Bulletin*, 24:43-49.

Beven KJ, EF Wood (1983) Catchment geomorphology and the dynamics of runoff contributing areas, *Journal of Hydrology*, 65:139-158.

Blyth EM, AJ Dolman (1993) Effective resistance to sensible and latent heat flux in heterogeneous terrain, *Quarterly J.R. Meteorological Society*, 19:423-442

Bondedid TR, SA Hanson, PL Taylor (1991) Technical Description of the Reach File, Horizon System Corporation, Herndon, Virginia, June 1991.

Chow VT, DR Maidment, LW Mays (1988) Applied Hydrology. McGraw-Hill, New York.

Dalu GA, RA Pielke (1993) Vertical heat fluxes generated by mesoscale atmospheric flow induced by thermal inhomogeneities in the PLB, *Journal of Atmospheric Science*, 50:919-926.

Dickinson RE, A Henderson-Sellers, PJ Kennedy, MF Wilson (1986) Biosphere-Atmosphere Transfer Scheme (BATS) for the community climate model, NCAR Technical Note NCAR/TN-275+STR, 69 p.

Dulaney RA, Hewitt MJ (1991) RF3-Arc/Info bridge, GIS Technical Memorandum, Environmental Monitoring and Systems Laboratory, US EPA, Las Vegas, Nevada, June 1991.

Famiglietti JS, EF Wood (1994a) Multi-scale modeling of spatially variable water and energy balance processes, *Water Resources Research* (in press).

Famiglietti JS, EF Wood (1994b) Effect of spatial variability and scales on areally averaged evapotranspiration, *Water Resources Research* (in press).

Hall FG, KF Huemmrich, SJ Goetz, PJ Sellers, JE Nickerson (1992) Satellite remote sensing of surface energy balance: success, failures, and unresolved issues in FIFE, *Journal of Geophysical Research*, 97(D17):19061-19089.

Hay LE, MD Branson, GH Leavesley (1992) Simulation of precipitation in the Gunniston River Basin using an orographic precipitation model, *Proceedings of AWRA Symposium on Managing Water Resources During Climate Change*, Reno, Nevada, pp. 651-660, November 1992.

Henderson-Sellers A, RE Dickinson (1993) Intercomparison of land surface parameterizations launched, *EOS*, 73:195-196

Leavesley GH, MD Branson, LE Hay (1992) Using coupled atmospheric and hydrology models to investigate the effects of climate change in mountainous regions, *Proceedings of AWRA Symposium on Managing Water Resources During Climate Change*, Reno, Nevada, pp. 691-700, November 1992.

Liston G (1993) Private communication, NASA/GSFC, Greenbelt, Maryland.

Lozar RC (1992) Global climate management by watershed basin units, Construction Engineering Research Laboratory, US Army Corps of Engineers, Champaign, Illinois, May 1992.

Mason PJ (1988) The formation of aerially averaged roughness lengths. *Quarterly Journal of Royal Meteorological Society*, 114, 399-420.

Moore ID, AK Turner, JP Wilson, SK Jenson, LE Band (1992) GIS and land surface-subsurface process modeling, *Proceedings of First International Conference on Integrating GIS and Environmental Modeling*, Boulder, Colorado, September 1991.

Noilhan J, P Lacarrere (1994) GCM gridscale evaporation from mesoscale modeling, *Journal of Climate* (in press).

Sellers PJ, Y Mintz, YC Sud, A Dalcher (1986) A Simple Biosphere (SiB) for use within general circulation models, *Journal of Atmospheric Science*, 43:505-531.

Sellers PJ, MD Heiser, FG Hall (1992) Relations between surface conductance and spectral vegetation indices at intermediate (100 m^2 to 15 km^2) length scales, *Journal of Geophysical Research*, 97D:19033-19059.

Shuttleworth WJ (1989) Macrohydrology: the new challenge to process hydrology, *Journal of Hydrology*, 100:31-56.

Shuttleworth WJ (1991) The modellion concept, *Reviews of Geophysics*, 29:585-606.

Shuttleworth WJ (1992) Evaporation, In Handbook of Hydrology (D Maidment, editor), McGraw-Hill, New York.

Shuttleworth WJ (1993) The Soil-Vegetation-Atmosphere Interface, In Energy and Water Cycles in the Climate System (E Raschke and D Jacob, editors), Proceedings of the NATO-ASI, Glucksburg, Germany.

WMO (1992) Science Plan for the GEWEX Continental-Scale International Project (GCIP), WCRP-67, WMO/TD No. 461, Available from the International GEWEX Project Office, Washington, D.C.

Wood EF, M Sivapalan, KJ Beven, LE Band (1990) Similarity and scale in catchment storm response, *Reviews of Geophysics*, 28(1):1-18.

Zeng X, RA Pielke (1995) Landscape-induced atmospheric flow and its parameterization in large-scale numerical models, *Journal of Climate* (in press).

Estimating Areal Evaporation Using Multispectral Satellite Observations

Bhaskar J. Choudhury
Hydrological Sciences Branch
NASA/Goddard Space Flight Center
Greenbelt, Maryland 20771, USA

Introduction

A quantitative knowledge of evaporation is required in many studies of hydrologic, meteorologic, and ecologic processes (Brutsaert, 1982). Evaporation (E) directly couples land surface heat and water balance equations:

$$R_n = \lambda E + C + \Delta\Sigma \tag{1a}$$

$$P = E + Q + \Delta S \tag{1b}$$

where, in the heat balance Equation (1a), R_n is the net radiation, C is the sensible heat flux, $\Delta\Sigma$ is the change in heat storage within the canopy and soil, and λ is the latent heat of vaporization. In the water balance Equation (1b), P is precipitation, Q is runoff, and ΔS is the change in water storage (soil moisture and ground water). The storage terms are neglected in long-term studies to obtain:

$$R_n = \lambda E + C \tag{2a}$$

$$P = E + Q \tag{2b}$$

Budyko (1986) introduced the following empirical equation for climatologic annual evaporation, which has been recently evaluated theoretically by Milly (1994):

$$E = P\left[D\tanh D^{-1}\,(1 - \cosh D + \sinh D)\right]^{0.5} \tag{3}$$

where D, called the dryness index, is defined as the ratio of net radiation (R_n) and latent heat equivalent of precipitation (λP), i.e., $D = (R_n/\lambda P)$.

NATO ASI Series, Vol. I 46
Land Surface Processes in Hydrology
Trials and Tribulations of Modeling and Measuring
Edited by S. Sorooshian, H. V. Gupta, and S. C. Rodda

Korzun's (1978) monograph can provide an appreciation for the care and the effort involved in computing the heat and water balance terms, which were arrived at as 800 mm precipitation, 485 mm evaporation, and 315 mm runoff, while 66 W m^{-2} for net radiation and 28 W m^{-2} for sensible heat flux. These estimates provide a useful guide for relative magnitudes of different terms in the water and heat balance equations in the global context. Long-term annual evaporation for different continents given in Table 1 calculated by different authors shows a fair degree of agreement (generally within 20%), although these estimates may not be considered totally independent because of the common nature of some of the basic input data.

Table 1. A Selected Summary of Estimated Annual Evaporation (mm) for Different Continents. There are some uncertainties in the regions included in the evaporation estimates.

	Author(s)				
	Lvovitch	Baumgartner and Reichel (1975)	Korzun (1978)	Budyko (1986)	Henning (1989)
Continent					
Africa	547	582	587	519	502
Asia	433	420	416	402	417
Australia	510	534	511	486	403
Europe	415	375	507	469	362
N. America	383	403	418	402	394
S. America	1065	946	910	854	883

Although climatologic averages of heat and water balance fluxes can be computed without making explicit reference to the structure and functioning of terrestrial plant communities because of long-term equilibrium of the soil-plant-atmosphere system, such a reference appears to be imperative for short-term (e.g., hourly to seasonal) analysis of evaporation or to assess impact of land cover change. The structure of plant communities plays a major role in determining the radiation balance and the exchange coefficients for heat, mass, and momentum between land surface and the atmosphere. Evaporation from plant canopies (transpiration), which is the major component of total evaporation (the other components being soil evaporation and evaporation of intercepted water), is a *consequence* of a plant's need for atmospheric carbon dioxide for its maintenance and growth via photosynthesis (Cowan, 1982; Monteith, 1988). Tanner and Sinclair (1983) showed from basic theoretical consideration that the ratio of the rates of biomass production (net primary productivity) and transpiration for a given plant type (broadly grouped, based upon the mechanism of photosynthesis, such as C_3, C_4, CAM) is almost totally determined by the canopy-to-air vapor pressure deficit, which has been found to hold for small-scale field studies over varied plant canopies. Plants partition the assimilated carbon dioxide between root and shoot in a way to maintain a balance for its competing needs for carbon dioxide, nutrients, and soil water. The ratio of root to shoot dry matter production is generally in the range 0.1-0.3 at harvest for agricultural crops, but this ratio is often significantly higher for most other plant communities, as illustrated in Figure 1 (e.g., Singh and Krishnamurthy, 1981; Wielgolaski et al., 1981; Raich and Nadelhoffer, 1989).

While the heat balance equation applies most directly to the areal parts of the plant communities (performing photosynthesis), the water balance equation applies prominently to the root-zone (soil water to be extracted for transpiration primarily to exchange for atmospheric carbon dioxide; plant nutrient requirements and avoidance of thermal injury also require soil water extraction). Significant carbon allocation to roots (Figure 1) makes amply evident the important role of roots in plant physiological processes and hence evaporation. Estimates of annual biomass accumulation (above and below ground) of varied plant communities at regional and continental scales (Yefimova, 1976; Ajtay et al., 1979) appears to be fairly well correlated with mean annual evaporation (Table 2, $r^2 = 0.96$). The average water use efficiency is about 2 g m^{-2} mm^{-1} (annual biomass accumulation per unit area per unit of evaporation), although very little data are available for plant communities of hot, arid regions.

Dry matter production and evaporation are less strongly correlated for observations at individual sites (Figure 2a), and water use efficiency is more variable, signifying differing dependencies of evaporation and biomass accumulation on site factors (e.g., soil depth and nutrients, exposure). Dry matter production of agricultural crops is also correlated with evaporation (Figure 2b), with variable water use efficiency (c.1-6 g m^{-2} mm^{-1}) depending upon crop, fertilizer use, and irrigation.

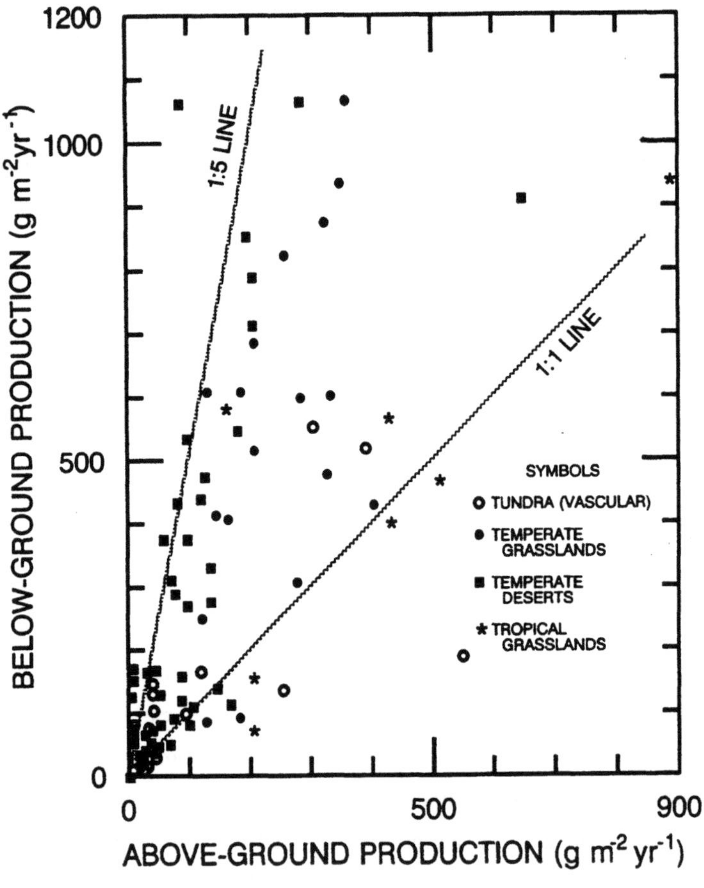

Figure 1. Observed variations of root and shoot dry matter production for different plant communities extracted from varied sources.

Considering the above discussion on the important role of photosynthesis in *determining* a significant part of total evaporation, quantitative data on spatial and temporal changes of vegetation characteristics should be of considerable value for estimating evaporation. Remote sensing from visible to microwave regions can provide radiative characteristics

Table 2. Estimates of annual evaporation (e.g., Korzun, 1978) and dry matter production per unit area (above and below ground) for different continents and plant communities extracted from various sources (e.g., Yefimova, 1976; Ajtay et al., 1979), and water use efficiency, which is the ratio of production and evaporation (g m^{-2} mm^{-1}).

Continent/ Community	Evaporation (mm y^{-1})	Dry Matter Production (g m^{-2} y^{-1})	Water Use Efficiency
Africa	587	1030	1.75
Asia	416	880	2.11
Australia	511	860	1.68
Europe	507	850	1.68
North America (excluding Greenland)	457	820	1.79
South America	910	2090	2.30
Rain Forest	1100	2300	2.10
Closed Boreal Forest	400	850	2.10
Tundra and Open Boreal Forest	180	313	1.73
Temperate Deciduous/ Mixed Forest	650	1300	2.00
Temperate Grasslands	365	800	2.19
Temperate Deserts	165	315	1.91
Global Land Surface (excluding Antarctic)	533	985	1.86

related to some soil and vegetation parameters, together with environmental data, such as solar radiation and air temperature, to aid estimation of evaporation (Choudhury, 1994; Choudhury et al., 1994).

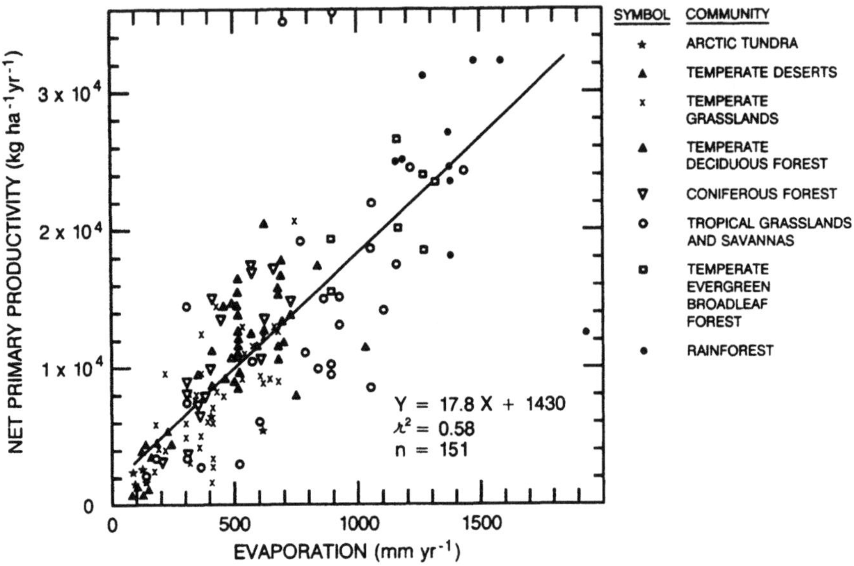

Figure 2a. Variations of dry matter production (above and below ground) and evaporation at individual locations for different plant communities extracted from various publications. The evaporation values are generally based on rainfall and model calculations (e.g., Baumgartner and Reichel, 1975).

The objective of this paper is to provide a short review of the residual and the water use efficiency methods for estimating evaporation using multispectral satellite data. Then, some implications of the observed correlations among multispectral data in the optical region are discussed in the context of estimating evaporation. Significant contribution from microwave observations (cf., Choudhury, 1991a) is not presented here due to space limitation.

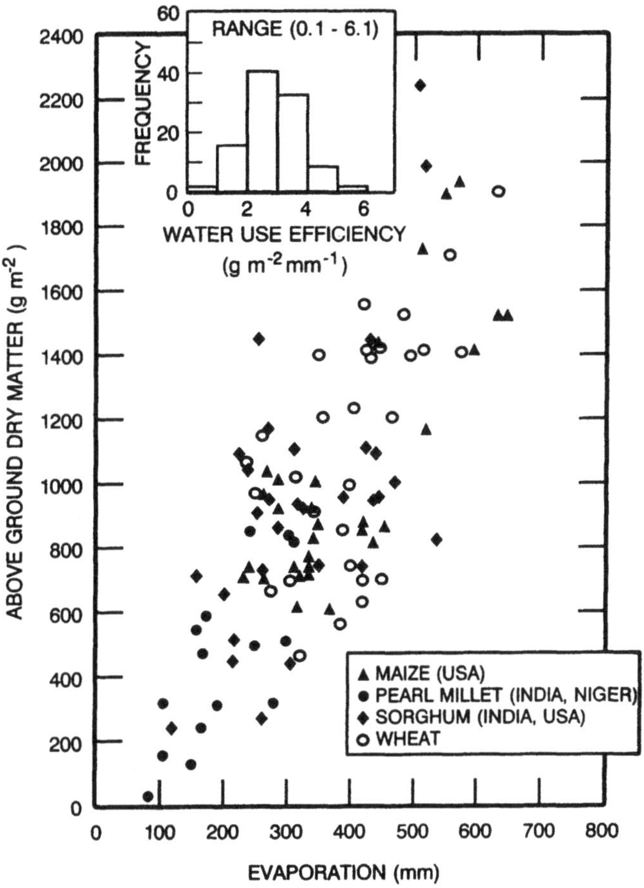

Figure 2b. Selected data for above ground dry matter production, actual evaporation, and water use efficiency of some agricultural crops.

Residual Method for Estimating Evaporation

One of the most widely studied approaches for estimating evaporation at local scales using remotely sensed data is the computation of this flux as the residual component of the energy balance equation (e.g., Jackson et al., 1977). Equation (1a) is rewritten, ignoring canopy heat storage, to calculate λE as:

$$\lambda E = R_n - G - C \tag{4}$$

where G is the soil heat flux. Calculation of the fluxes on the righthand side of Equation (4) using satellite observations is discussed below.

Net Radiation (R_n)

Both physically based and empirical methods have been developed to compute R_n using satellite observations (Choudhury, 1991b). It appears possible to estimate monthly mean incident solar radiation with an accuracy of about 15 W m^{-2} and albedo to about 10%, and global maps of incident solar radiation and continental scale albedos have been produced from satellite observations (Nacke, 1991; Darnell et al., 1992). Figure 3, from Darnell et al. (1992), shows a comparison of mean daily insolation derived from satellite observations with pyranometer data for globally distributed locations. Figure 4, from Nacke (1991), shows albedo over Africa derived from the METEOSAT data. Monthly variations of the net solar radiation globally for a five-year period have been produced recently by Li and Leighton (1993) and checked against the surface observations (Li et al., 1993).

Several attempts have been made to estimate the net longwave flux at the surface from satellite sounding data (Darnell et al., 1992). Estimation of this flux from *only* the top of the atmosphere satellite observations, however, is significantly more difficult and error-prone than absorbed solar radiation because there is virtually *no correlation* between the net longwave radiation at the top of the atmosphere and that at the surface except under restricted conditions (Figure 5). For cloudy atmospheres, the upward infrared radiation at the top of the atmosphere is dependent upon the temperature and humidity at and above the cloud top, while the downward radiation is dependent upon air temperature and humidity below clouds, cloud temperature, and cloud liquid water content. Thus, calculation of net longwave flux requires information about cloud characteristics and temperature and humidity variations within the atmosphere. Observations by sounders (Tiros Operational Vertical Sounder, TOVS, on board NOAA satellites), providing measurements of the variation of atmospheric characteristics at several heights, have given some promising results for longwave flux, with an estimated error of about 20 W m^{-2} (Darnell et al., 1992).

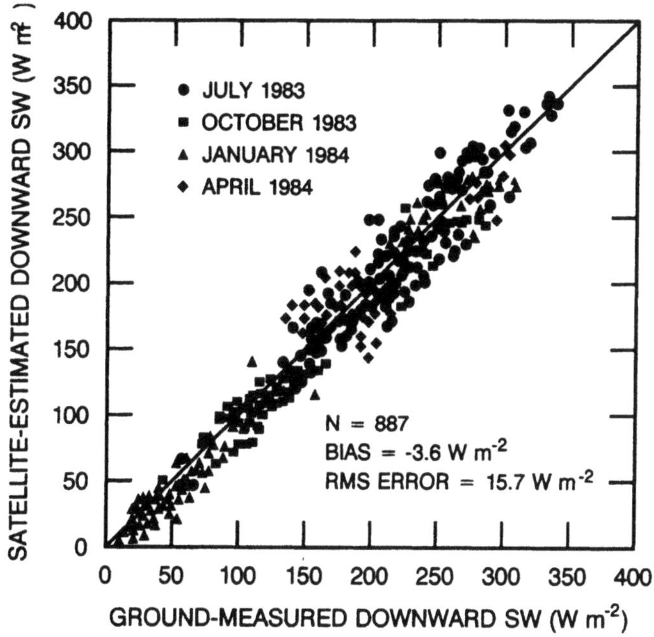

Figure 3. Comparison of mean daily solar radiation derived from satellite observations with ground measurements (Darnell et al., 1992).

An international effort is underway to develop accurate values of the radiative fluxes globally from satellite and other observations (WCRP-69, 1992).

Soil Heat Flux (G)

The soil heat flux is a significant component of net radiation for bare and partially vegetated dry soils on instantaneous basis; the daily average values are rather small. Field observations provide much evidence for a direct relation between R_n and G, the proportionality factor being largely determined by soil surface moisture and vegetation cover. Choudhury et al. (1987) found the following relation for a wheat crop:

$$G = R_n\,[0.4\exp(-0.5L)] \tag{5}$$

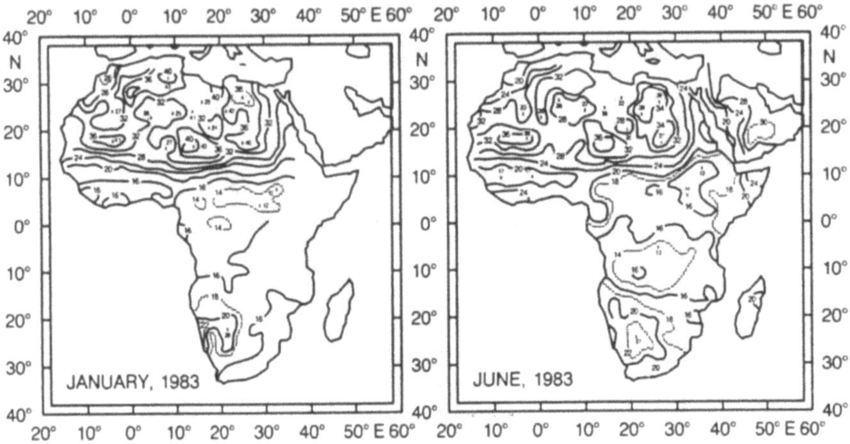

Figure 4. Shortwave (0.3-2.6 mm) albedo for Africa for two months (January and June) of 1983 derived from METEOSAT observations (Nacke, 1991).

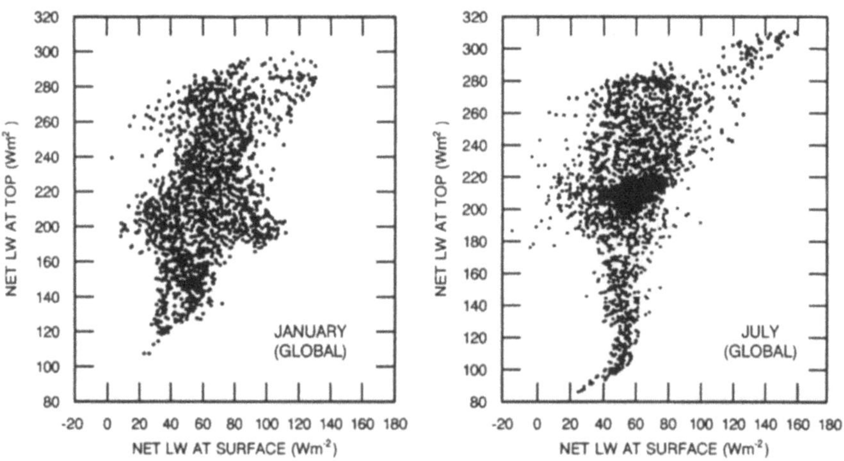

Figure 5. Relation between net longwave flux at the top of the atmosphere and at the surface (January and June) based on model simulations (Harshvardhan et al., 1990).

where L is the canopy leaf area index. The exponential term in Equation (5) provides an approximation for the attenuation of net radiation by the canopy. The dependence of G on soil moisture does not appear in Equation (5); the constant 0.4 appears to correspond to the prevailing fairly dry conditions. The magnitude of G decreases with increasing surface soil moisture and vegetation cover. A minimum value of 0.05 for G/R_n could be used for wet soils or large L.

Multispectral data in the optical region can provide information about radiation attenuation by vegetation, while soil moisture information could be obtained from microwave observations (Choudhury, 1991a). For example, to account for vegetation attenuation, Moran et al. (1989) expressed field observations of G as:

$$G = R_n\ [0.58 \exp(-2.13 ND)] \tag{6a}$$

while Kustas et al. (1993) obtained:

$$G = R_n\ [0.40 - 0.33 ND] \tag{6b}$$

where ND (normalized difference) is derived from visible and near-infrared reflectances (ρ_v and ρ_n, respectively) as:

$$ND = \frac{\rho_n - \rho_v}{\rho_n + \rho_v} \tag{7a}$$

The ND values for bare soils are generally in the range 0.1-0.2, while the range for dense green crops is 0.90-0.92. For ND = 0.15 (i.e., bare soils; L = 0), Equation (6a) gives G/R_n = 0.42, which does not differ significantly from 0.4 in Equation (5). For ND = 0.9 (i.e., dense green vegetation), Equation (6a) gives G/R_n = 0.09. A range of space-borne sensors (e.g., advanced very high resolution radiometer (AVHRR) on board NOAA satellites) can be used to calculate ND or other vegetation indices (VI) like the soil adjusted vegetation index (SAVI) which minimizes the effect of soil reflectance variations on the relation between VI and leaf area index (Huete, 1988; Baret and Guyot, 1991), defined as:

$$\mathrm{SAVI} = \frac{1.5(\rho_n - \rho_v)}{\rho_n + \rho_v + 0.5} \tag{7b}$$

A relation between VI and the ratio G/R_n can be derived theoretically considering the dependencies of net radiation attenuation and VI on L. A semi-empirical model relating R_n to that at the soil surface (R_{ns}) (Choudhury et al., 1987; Choudhury and Monteith, 1988) is:

$$R_{ns} = R_n \exp(-\kappa L) \tag{8a}$$

where the extinction coefficient κ is generally in the range 0.4-0.7. Similarly, one can express VI in terms of L of green canopies (Choudhury, 1987; Baret and Guyot, 1991) as:

$$\mathrm{VI} = \mathrm{VI}_{max} - (\mathrm{VI}_{max} - \mathrm{VI}_{min})\exp(-\kappa' L) \tag{8b}$$

where VI_{max} and VI_{min} are, respectively, the maximum and minimum values of VI achieved for dense green vegetation (c. $L > 4$) and bare soils ($L = 0$), and κ' is the extinction parameter, the value of which depends upon the choice of VI. The value of κ' is generally in the range of about 0.8-1.2 when VI is chosen to be ND and 0.6-0.8 for SAVI (Baret and Guyot, 1991). Now, by eliminating L between Equations (8a) and (8b), one obtains:

$$R_{ns} = R_n \left[\frac{\mathrm{VI}_{max} - \mathrm{VI}}{\mathrm{VI}_{max} - \mathrm{VI}_{min}}\right]^{\delta} \tag{9}$$

where $\delta = \kappa/\kappa'$. Thus, G can be expressed as:

$$G = R_n \left[0.05 + 0.35 \left\{\frac{\mathrm{VI}_{max} - \mathrm{VI}}{\mathrm{VI}_{max} - \mathrm{VI}_{min}}\right\}^{\delta}\right] \tag{10}$$

In writing Equation (10), it has been explicitly recognized that G/R_n approaches a finite minimum value (c. 0.05) for dense vegetation. In most applications, one can assume that $\delta = 1$.

A direct application of Equation (5) to partially vegetated areas (e.g., clumped vegetation) might not be appropriate because the ratio G/R_n depends nonlinearly on L; a fractional area weighted relation might be a better approximation, e.g.:

$$G/R_n = 0.4\left[f_c + (1-f_c)\exp(-0.5L_c)\right] \tag{11}$$

where L_c is the clump leaf area index, and f_c is the fractional exposed soil (the leaf area index, L, on a unit soil area basis is $L = (1 - f_c)\,L_c$). The exponential term in Equation (11) approximates radiation transmission through the clumps.

However, VI could also be an indicator of the fractional cover. The fractional exposed soil (f) observed from nadir can be expressed in terms of L for a Poisson model of canopy leaves (Choudhury, 1987; Choudhury et al., 1994) as:

$$f = \exp(-G_\ell L) \tag{12a}$$

where G_ℓ is the leaf angle distribution function {$G_\ell = 0.5$ when leaf angle distribution is uniform or random; it is a little higher than 0.5 (say, 0.7) when leaf orientation tends towards horizontal, and a little lower than 0.5 (say, 0.4) when leaf orientation tends towards vertical}. By eliminating L between Equations (8b) and (12a), one obtains the following relation between VI and f:

$$VI = VI_{max} - \left(VI_{max} - VI_{min}\right) f^{\kappa''} \tag{12b}$$

where $\kappa'' = (\kappa'/G_\ell)$. When ND is used as the VI, then the value of κ'' would be about 1.5 for highly reflective soils; it increases to about 2 for darker soils, while κ'' would be about 1.2 when SAVI is used for VI. The above models would suggest that the relation between ND and f would be more nonlinear than that between SAVI and f (Choudhury et al., 1994). This nonlinearity and the effect of soil reflectance on the ND vs. f relationship are evident in the field observations of Huete et al. (1985) for cotton (Figure 6). However, Ahlrichs and Bauer (1983) found that, for wheat, a linear relation between ND and f (i.e., $\kappa'' = 1$) can explain 90% of the variance.

Field observations and models relating G to R_n and VI illustrate the feasibility of estimating G from multispectral satellite observations. Variations of G with soil moisture could be assessed using microwave observations.

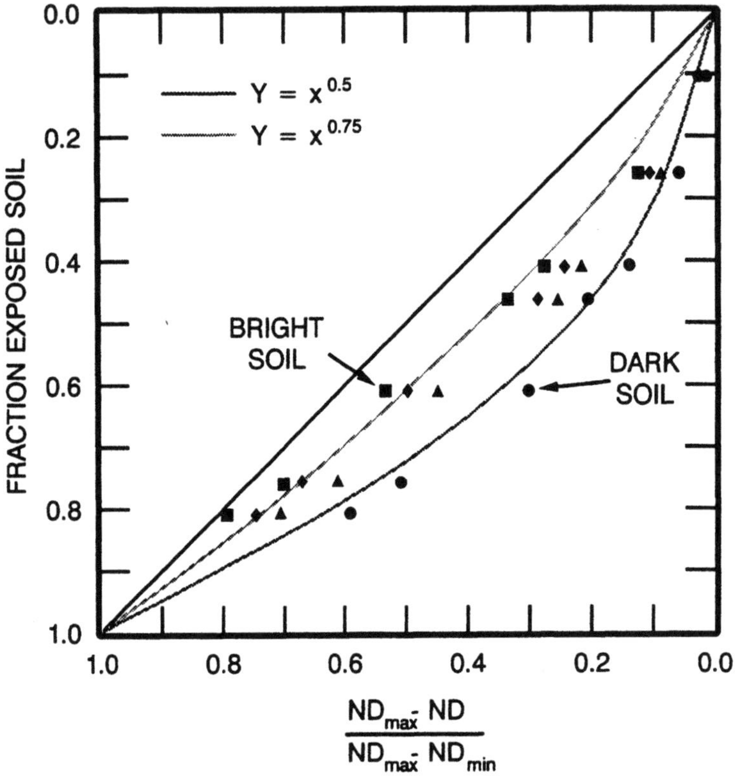

Figure 6. Relation between normalized difference and fractional exposed soil with varied reflectance. The observations are for a cotton crop (Huete et al., 1985) and curves are based on Equation (12b).

Sensible Heat Flux (C)

Surface temperature (T_s) observations by infrared radiometers have been used to calculate C using models of varied complexity or spatial extrapolation of local measurements (cf., Seguin et al., 1989; Vidal and Perrier, 1990; Moran and Jackson,

1991). The simplest of these models used to calculate C is the following aerodynamic relation:

$$C = \rho c_p (T_s - T_a)/r_H \tag{13}$$

where ρ is the density, and c_p is the specific heat of air at constant pressure, r_H is the resistance for heat exchange (which should be distinguished from the resistance for momentum exchange), and T_a is air temperature measured at a reference height (c. generally taken to be 1.5 m above the canopy for analysis of field observations of C). Calculation of the resistance requires additional data for wind speed and roughness heights, and inclusion of stability correction makes the resistance dependent on C, when the above equation needs to be solved iteratively. Infrared temperatures should be corrected for emissivity (i.e., to obtain the kinetic temperature), particularly for observations over sparsely vegetated and bare soil surfaces because of lower values of emissivity and its spectral variation.

Calculations of C using infrared temperature observations have generally given good results in field studies as illustrated in Figure 7 (Kohsiek et al., 1993), although in a few cases, the results are less satisfactory. Kustas (1990) compared estimates of C based on Equation (13) with those calculated from a more complex energy balance method against observations over a furrowed cotton field with partial ground cover, which showed that the root mean square error for λE was less than that calculated from a complex model.

Meteorological variables measured at 1-3 m above the surface are affected by the surface conditions. They represent local interactions over upwind distances of the order of 100 meters. Infrared temperature measurements by Landsat, at a spatial resolution of 120 m, might be commensurate with the spatial scales represented by standard meteorological data at these reference heights. Infrared observations by GOES, METEOSAT, and AVHRR at the spatial scales of kilometers, would be more appropriate for regional studies of the fluxes. Meteorologic data commensurate with these coarse spatial resolution satellite observations can be obtained from measurements in the atmospheric boundary layer, representing spatial scales of the order of 10 km. This spatial scale follows from the fact that the horizontal length scales are an order of magnitude larger than the vertical scales. The surface roughness parameters need to be consistent with these spatial scales (Mahrt and Ek, 1993).

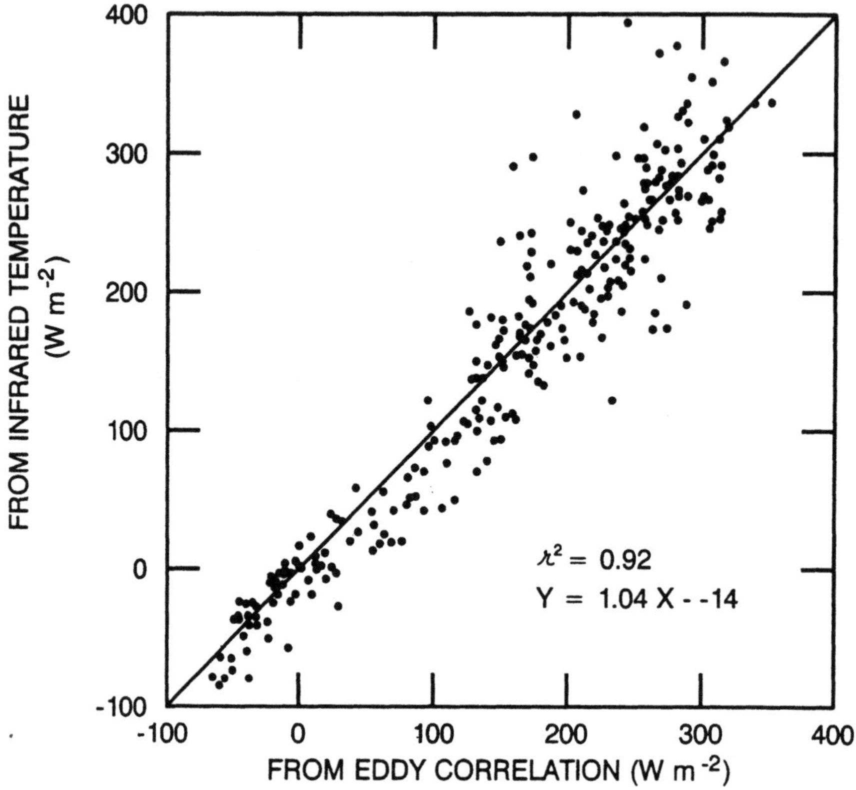

Figure 7. Comparison of the observed (Eddy correlation) and calculated (from infrared temperature) sensible heat flux over a semi-arid area (Kohsiek et al., 1993).

Brutsaert and Sugita (1992a,b) calculated regional (c. 15 km) values of C using infrared temperature observations in the bulk similarity approach to the atmospheric boundary layer. The calculated C values from radiosondes were found to be in fairly good agreement with the average surface observations when soil moisture conditions over the region were such as to allow potential evaporation (quasi-uniform condition). However, the agreement with the average surface flux data was less satisfactory when the soil dried to become spatially nonuniform, and evaporation became less than the potential rate (Figure 8). It was concluded that, under such nonuniform soil moisture conditions, it is difficult to aggregate local measurement of the fluxes to form a regional value. We need to develop a better understanding of the relation between local and regional fluxes under nonuniform conditions. Menenti and Choudhury (1993) proposed using coarse resolution infrared temperature data together with boundary layer measurements to quantify the ratio of actual and potential evaporation.

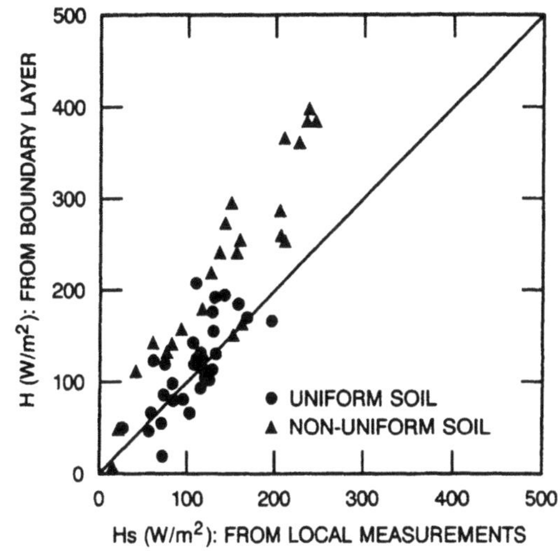

Figure 8. Comparison of sensible heat flux derived from boundary layer observations with infrared temperature and surface observations of the flux at isolated locations over 15 x 15 km area for varying soil moisture conditions (Brutsaert and Sugita, 1992b). These two estimates of the sensible heat flux agree better for uniform soil moisture cases.

Diak and Whipple (1993), Tarpley (1994), among others, used diurnal variations of infrared temperature observed from geostationary satellites within boundary layer models to estimate C, while Taconet et al. (1986), among others, developed methods involving only one time of day observations of surface temperature from polar-orbiting satellites. The method proposed by Brunet et al. (1991) used early-morning upper air sounding data as the initial condition to integrate a boundary layer model and calculate surface temperature from Penman-Monteith type surface energy balance equations. The surface resistance term in the energy balance equation was adjusted to match early-afternoon infrared temperature observations. Diak and Whipple (1993) also used radiosonde data to initialize a boundary layer model, but used the observed temporal variation of surface temperature to compute sensible heat flux. Temporal changes in the boundary layer height was shown to be directly related to surface sensible heating for a wide range of surface parameters and atmospheric conditions. Remote sensing of the boundary layer height using liders can be used very effectively to estimate regional values of C (Gal-Chen et al., 1992). Tarpley (1994) combined diurnal variation infrared temperature with vegetation index (VI) within a boundary layer model to compute monthly evaporation over Kansas, the VI being used to estimate the surface roughness parameter for computing the aerodynamic resistance.

Thus, infrared temperature observations have been used in models of varied complexity to estimate C. Aggregation of local flux measurements to obtain regional values may not always be accurate. Use of coarse resolution infrared temperature data with boundary layer models appears to be a promising approach for estimating regional sensible heat fluxes. Estimation of the resistance (r_H) from remotely sensed data remains a difficult problem. Correlations of T_s and VI discussed below could provide an approach for estimating regionally representative values of r_H.

Water Use Efficiency Method

This method of estimating evaporation using remotely sensed data has been proposed by Wiegand and Richardson (1990). The concept of water use efficiency (i.e., the ratio of biomass accumulation and transpiration or total evaporation) was discussed briefly in the Introduction. Modeling and field observations (cf., Tanner and Sinclair, 1983; Gregory et al., 1992) show that the ratio of biomass accumulation (Y; g m^{-2}) and transpiration (T; mm) during a period (e.g., month) is given by the following equation:

$$Y/T = K/D \tag{14}$$

where D is canopy-to-air vapor pressure deficit (approximately the air vapor pressure deficit), and K is a constant which depends primarily upon the mechanism of photosynthesis (C_3, C_4, CAM, etc.) used in biomass production. The value of K is about 3 for C_3

photosynthesis crops (e.g., wheat and barley) and about 6 for C_4 crops (e.g., maize and sorghum) when D is expressed in units of kPa.

Modeling and field observations (e.g., Russell et al., 1989; Muchow and Sinclair, 1994) also showed that biomass accumulation during a period is linearly related to absorbed or intercepted photosynthetically active radiation (APAR; MJ m^{-2}) corresponding to that period:

$$Y = \epsilon APAR \tag{15}$$

where ε (g MJ^{-1}), called the radiation use efficiency, is about 2 for C_3 cereals and 3 for C_4 crops. The efficiency depends to a degree upon the nature of incident radiation (e.g., ratio of direct and diffuse component or cloudiness), and it decreases with stresses of soil water, nutrients, and temperature.

Expressing APAR as the product of incident photosynthetically active radiation and the fraction of this radiation absorbed by a canopy, one finds that both of these quantities can be obtained from satellite observations (Pinker and Laszlo, 1992; Pinter, 1993; Ruimy et al., 1994). Thus, an approach for estimating transpiration using remotely sensed data is obtained by combining Equations (14) and (15) as:

$$T = \epsilon K^{-1} D \ APAR \tag{16}$$

Estimation of D from remotely sensed data remains a difficult problem. However, an analogous equation can be derived which replaces D by potential evaporation, which is more amenable to estimation using satellite data (see below). Very little has been done to estimate evaporation using this method. Estimation of biomass accumulation using satellite data, which is a required component of this method, is an active area of research (Ruimy et al., 1994).

Relations Among Multispectral Data

The above discussion pertained to the use of multispectral satellite data for estimating evaporation using the residual and the water use efficiency methods. Multispectral data may also provide other approaches and input parameters for estimating evaporation.

Spectral observations from satellites, when corrected for atmospheric effects, directly provide the radiative characteristics of the surface from which biological parameters have to be derived. Estimation of biophysical parameters from the spectral data is difficult because, in general, any spectral measurement is not determined by a unique surface parameter (several surface parameters with varied sensitivity determine spectral observations). Correlations among multispectral data might be useful for developing a hypothesis regarding relative sensitivities of spectral data to varied surface parameters, which would have to be confirmed by field observations.

A number of studies have found fairly high linear correlation (with negative slopes) between infrared temperature (T_s) and normalized difference vegetation index (ND; Equation (7a)) (e.g., Choudhury, 1989; Nemani et al., 1993; Carlson et al., 1994). A high correlation between these measurements appears to have fair generality considering different sensors, times of the year, and locations involved in these studies. Smith and Choudhury (1990a) also found high linear correlation (positive slope) between visible reflectance and T_s (Figure 9), which was observed previously in field observations by Wendler and Eaton (1983) over partially vegetated areas (Figure 10).

An understanding of the correlation seen in Figure 9 could be based on the hypothesis that fractional vegetation cover is a significant determinant of both the reflectance and the temperature. If the pixel values of T_s and visible reflectance (ρ_s) are assumed to be primarily determined by the fractional vegetation cover ($f' = 1 - f$; with f defined in Equation (11) or (12a)) as:

$$T_s = f' T_v + (1-f') T_b \tag{17a}$$

$$\rho_s = f' \rho_v + (1-f') \rho_b \tag{17b}$$

where the subscripts v and b refer to vegetated and bare fractions; eliminating f' between Equations (17a) and (17b) gives a linear relation between T_s and ρ_s with positive slope (a theoretical regression model):

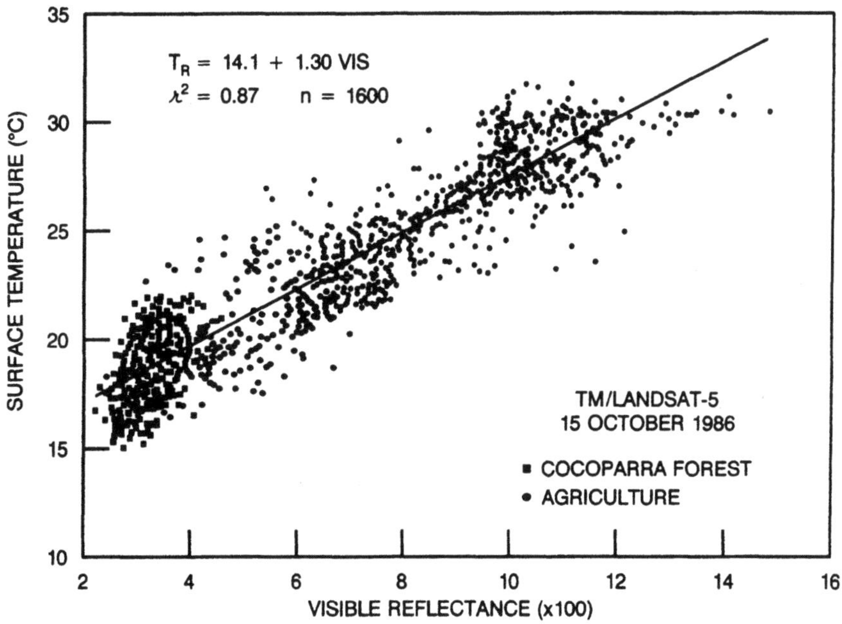

Figure 9. Observed relation between infrared surface temperature and visible reflectance over an area in Australia (Smith and Choudhury, 1990b).

$$T_s = a + b\rho_s \tag{18a}$$

where:

$$a = (T_v \rho_b - T_b \rho_v)/(\rho_b - \rho_v) \tag{18b}$$

$$b = (T_b - T_v)/(\rho_b - \rho_v) \tag{18c}$$

While Equation (18a) provides an overall regression model for average slope and intercept, pixel-to-pixel variations of soil and canopy temperatures and reflectances will introduce scatter about this regression line. The above equations would also suggest diurnal and seasonal variations of the slope and intercept resulting from variations of the component temperatures (T_v and T_b) due to solar heating and air temperature variations. Observations of T_s and ρ_s during midday for dry soil conditions are likely to give a higher correlation than during early morning or wet soil conditions because of larger difference between T_b and T_v. Seasonal and diurnal aspects of Equation (18a) for a limited data set have been discussed by Smith and Choudhury (1990a).

Now, by eliminating fractional cover appearing in Equations (12b) and (17a), one can obtain the following relation between VI and T_s:

$$VI = VI_{max} - (VI_{max} - VI_{min}) \left[\frac{T_s - T_v}{T_b - T_v} \right]^{k''} \tag{19}$$

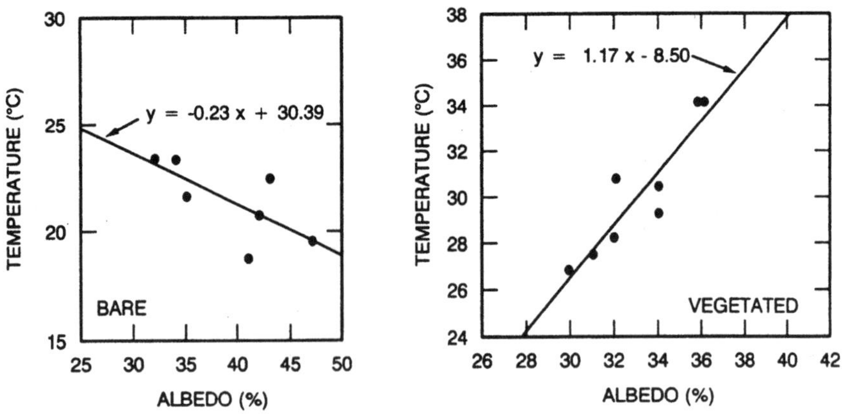

Figure 10. Relations between surface temperature and albedo observed in the field (Wendler and Eaton, 1983). Note that the slope is negative for bare soils, while it is positive for partially vegetated soils, as would be expected from surface heat balance analysis.

Considering the likely range of values for κ'' when VI is either ND or SAVI, an almost linear relation between T_s and VI with negative slope would not be inconsistent with the above model, although nonlinearity in the relation would be expected more for ND than SAVI.

The surface energy balance equation has also been used to study the relationship between T_s and VI, which provides some further understanding of this relationship (Smith and Choudhury, 1990b). A possible direct relation between evaporation and VI is discussed below, followed by a reassessment of the correlation between T_s and VI. Theoretical basis for relating evaporation to VI has been discussed by Choudhury et al. (1994).

Observations and modeling show that the ratio of unstressed transpiration (T) and potential evaporation (E_p) is related by an exponential function of the leaf area index (e.g., Choudhury and Monteith, 1988; Al-Kaisi et al., 1989; Choudhury et al., 1994) as (Figure 11):

$$\frac{T}{E_p} = 1 - \exp(-\xi L) \tag{20}$$

where ξ is in the likely range 0.5-0.7. One can combine Equations (8b) and (20) by eliminating L to express the ratio (T/E_p) as a function of VI as (Figure 12):

$$\frac{T}{E_p} = 1 - \left[\frac{VI_{max} - VI}{VI_{max} - VI_{min}}\right]^{\eta} \tag{21}$$

where $\eta = (\xi / \kappa')$, which could be 0.6-0.8 when VI is ND and 0.8-1.0 when VI is SAVI, suggesting that a relation between T/E_p and SAVI would be closer to being a linear relation than a relation between T/E_p and ND (Choudhury et al., 1994), although deviation from linearity is not expected to be large.

Alternately, one can combine Equations (12a) and (20) by eliminating L to express the ratio (T/E_p) as a function of fractional cover (f) as:

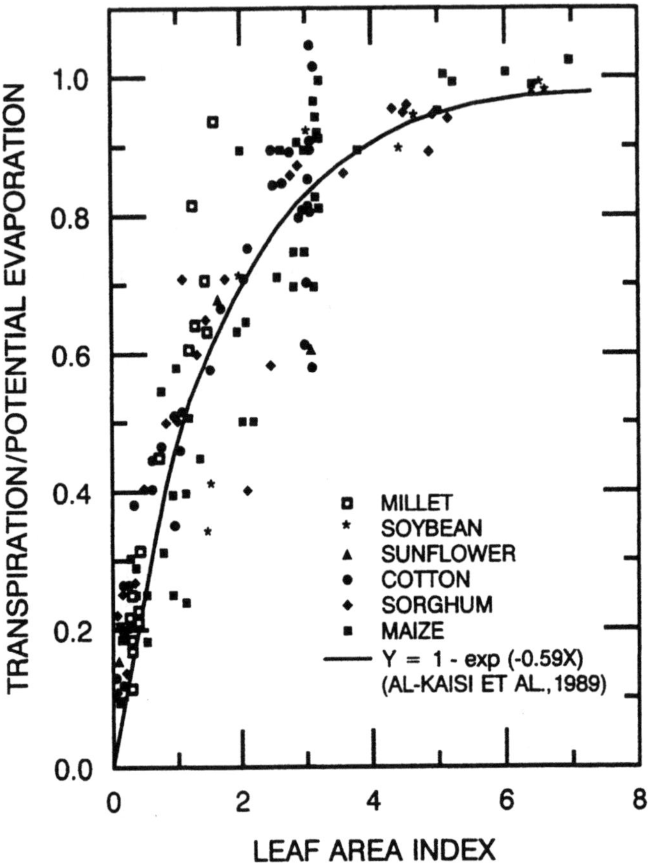

Figure 11. Variation of the ratio of transpiration and potential evaporation with leaf area index as observed in fields studied on several crops (for further discussion see Choudhury et al., 1994).

$$\frac{T}{E_p} = 1 - f^{\alpha'} \tag{22}$$

where $\alpha' = (\xi/G_\ell)$, which is likely to be in the range 0.8-1.3. Field observations over an alfalfa crop (Heilman et al., 1982) showed a nearly linear relation, i.e., $\alpha' = 1$.

The ratio (T/E_p) will be close to the ratio of actual and potential evaporation (E and E_p, respectively) when soil evaporation is small. If the ratio E/E_p is approximated by a linear function of VI (i.e., $\eta = 1$ in Equation (21)) as:

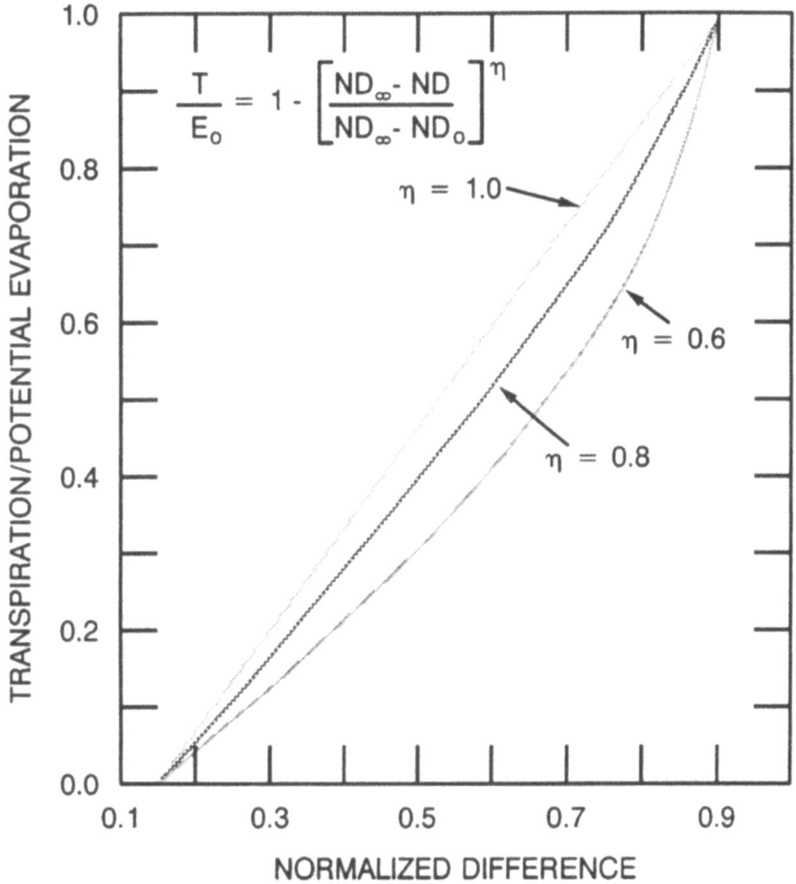

Figure 12. Theoretically expected variation of the ratio of transpiration and potential evaporation with normalized difference as determined by semi-empirical Equation (21).

$$\frac{E}{E_p} = \psi \left[\frac{VI - VI_{min}}{VI_{max} - VI_{min}}\right] \tag{23}$$

where ψ is canopy stress factor (varying between 0 and 1); then from Equations (4), (13), and (23) one obtains:

$$T_s = T_a + \left[R_n - G - \lambda E_p \psi \left\{\frac{VI - VI_{min}}{VI_{max} - VI_{min}}\right\}\right] \frac{r_H}{\rho c_p} \tag{24}$$

For a linear approximation for G in terms of VI similar to Equation (12) or (6b) as:

$$G = 0.4\ R_n \left[\frac{VI_{max} - VI}{VI_{max} - VI_{min}}\right] \tag{25}$$

in Equation (24) and rearranging the terms, one can obtain the following linear relationship between T_s and VI with negative slopes, as has been generally observed for T_s vs. ND (Figure 13):

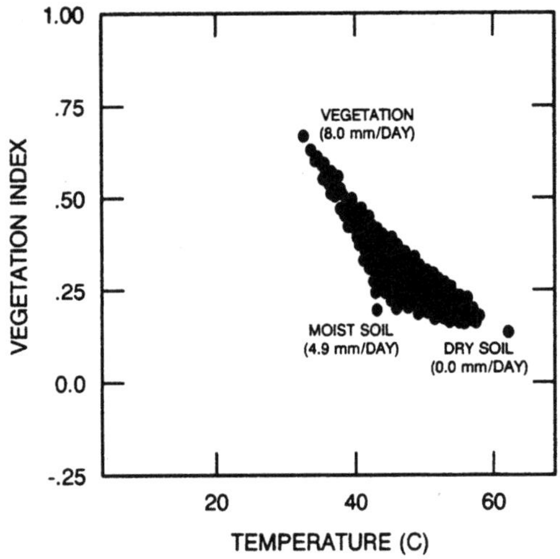

Figure 13. Relation between normalized difference vegetation index and surface temperature derived from AVHRR observations over the southern United States, together with an assessment of daily evaporation associated with this relation (Price, 1990).

$$T_s = A - B\ VI \tag{26a}$$

where:

$$A = T_a + \left[R_n + \frac{\lambda E_p \psi VI_{min} - 0.4 R_n\ VI_{max}}{VI_{max} - VI_{min}}\right] \frac{r_H}{\rho c_p} \tag{26b}$$

$$B = \left[\frac{\lambda E_p \psi - 0.4 R_n}{VI_{max} - VI_{min}} \right] \frac{r_H}{\rho c_p} \tag{26c}$$

The limiting values of T_s would be those observed when $VI = VI_{min}$ (i.e., $E = 0$) and $VI = VI_{max}$ (i.e., $E = E_p$ in the absence of stress). Equation (26) is an explicit theoretical model for pixel-wise linear regression between T_s and VI. The above formulation could be extended for wet soil surface, when soil evaporation is not negligible.

The above equations suggest that the slope of T_s vs. VI depends upon the magnitude of crop stress (ψ), environmental conditions (E_p and R_n) and the efficiency of heat exchange between the surface and the atmosphere (r_H). The parameter values in these equations (e.g., r_H) need to be representative of the spatial resolutions of T_s and VI. A possible application of the above theoretical analysis is that, when crop stress is assumed to be negligible ($\psi = 1$), remotely sensed data providing R_n and E_p can be combined with the estimated slope of T_s vs. VI relation to calculate r_H. Alternately, this slope could provide an indication of crop stress when other variables in Equation (26) are known. Satellite-observed infrared temperatures should be corrected for emissivity, and the reflectances should be corrected for atmospheric effects for an accurate interpretation of T_s vs. VI relationship.

In practical applications, one may use the following Priestley-Taylor equation to calculate E_p:

$$\lambda E_p = \alpha \left[\frac{\Delta (R_n - G)}{\Delta + \gamma} \right] \tag{27}$$

where α is a proportionality factor (generally taken to be 1.26 for daily evaporation; a collection of α values from varied observations can be found in Choudhury et al., 1994); Δ is the slope of saturation vapor pressure evaluated at air temperature; and γ is the psychrometer constant. The soil heat flux (G) could be approximated as 0.05 R_n for instantaneous measurements. The applicability of the Priestley-Taylor equation for estimating regional evaporation has been discussed by several investigators (cf., McNaughton and Spriggs, 1989), and remotely sensed data (air temperature and net radiation) can be used to estimate E_p (Choudhury, 1991b; Choudhury et al., 1994).

It is pertinent to note that the slope of T_s vs. ND for nonvegetated areas can be positive, because higher soil reflectance will decrease absorption of solar radiation and hence will lead to a lower surface temperature compared to that for a darker soil. In this case (i.e., bare soils), the range of ND values would be considerably smaller (c. 0.05-0.20) than found under partial vegetation cover (0.1-0.8). The contrasting slopes of surface temperature vs. albedo variations for vegetated and bare soils observed in the field are shown in Figure 10 (Wendler and Eaton, 1983). Thus, in going from desert to vegetated areas, the slope of T_s vs. ND can change from positive (for essentially nonvegetated areas) to negative (for partially vegetated areas) as observed in the satellite data for a transect from the Sahara Desert to rainforest (Figure 14). Locations of the intersection of positive and negative slopes could provide an objective demarcation of a boundary separating radiation-controlled and evaporation-controlled regions of surface temperatures (Becker and Seguin, 1985). In interpreting the correlation seen in Figure 14, one should note that there are gradients of soil reflectance, air temperature, and λE_p associated with the desert to rainforest transect affecting both ND and T_s, apart from fractional cover. A clear discontinuity of slope appearing in Figure 14, however, does not appear when the variation of these data along the transect is expressed in terms of albedo (or average reflectance) and surface temperature as in Wendler and Eaton (1983), partly because a decrease in visible reflectance with increasing vegetation cover can get compensated by an increase in near infrared reflectance (the average reflectance could be less variable over a range of vegetation cover). It should be noted that the discontinuity of slope appearing in Figure 14 is partly due to ignoring the effect of spectral variation of emissivity in computing the surface temperatures. Very little quantitative data are currently available for spectral variation of emissivity to correct infrared temperature data to obtain the kinetic temperature needed for accurate interpretation.

Summary and Conclusions

Structure and functioning of plant communities brings about an intimate relation between heat, water, and carbon exchange at the land surface. Although heat and water balance have been studied at regional and global scales without making explicit reference to carbon (and nutrient) relations of plant communities, such a reference allows a better appreciation for the biological control on heat and water balance and a more accurate representation of the processes, particularly at short time scales and to quantify the effect of land use change. There is evidence that stomata functions in a way to optimize carbon gain in relation to water loss. A close relation appears to exist between dry matter accumulation by plant communities and evaporation (transpiration). A better quantification of evaporation could be obtained by separating it into its component fluxes (transpiration, soil evaporation, and evaporation of intercepted water), because of differing controlling factors and spatial and temporal scales involved with these fluxes.

Multispectral satellite data can provide spatially and temporally representative data for some of the surface parameters and fluxes for estimating evaporation. The residual and water use efficiency methods for computing evaporation using multispectral satellite data were reviewed. The net solar radiation can be computed with high accuracy from the top of the atmosphere satellite observations. Calculation of longwave flux requires data from sounders, and the uncertainty is about 20 W m^{-2}. Estimates of net radiation, combined with surface soil moisture from microwave observations and vegetation attenuation from vegetation indices, could provide an estimate of the soil heat flux. Infrared temperature observations, combined with boundary layer model and data, could provide regional values of the sensible heat flux. As compared to radiation balance, much less analysis is currently available for soil and sensible heat fluxes at regional scales using coarse resolution satellite data. We also need spatially representative data for heat exchange parameters, which might be possible through the observed correlation between surface temperature and vegetation indices. Very little work has been done to estimate evaporation using the water use efficiency method; thus, its domain of applicability and accuracy are not well established. Estimation of biomass production using satellite data, which is a required component of this method, is an active area of research.

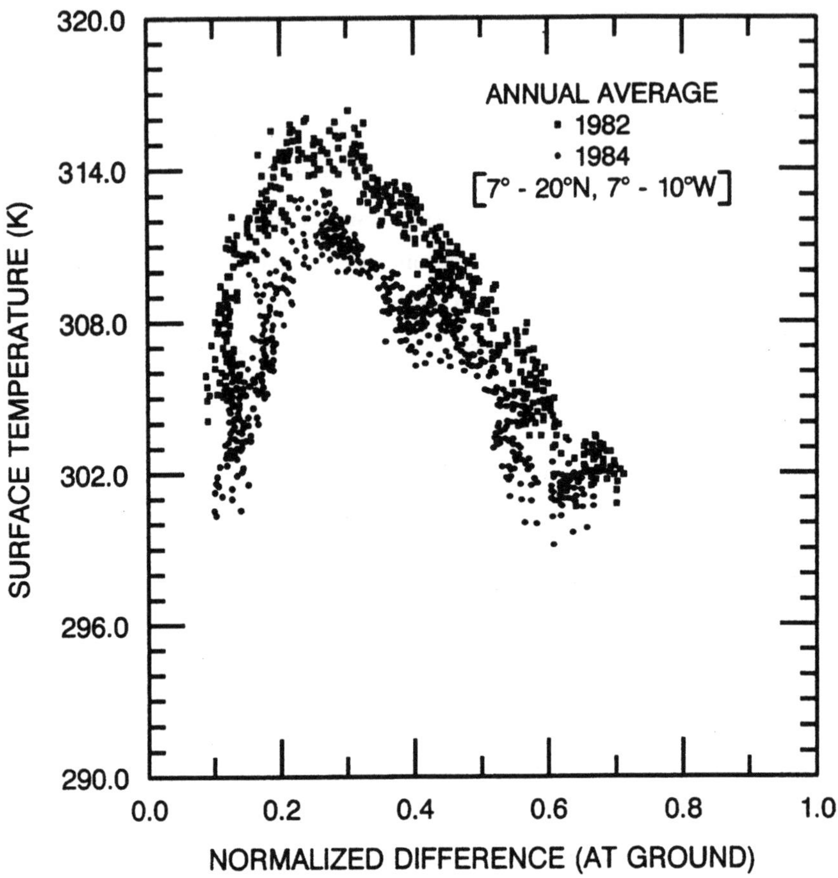

Figure 14. Relation between surface temperature and normalized difference vegetation index (ND) from the AVHRR-GAC data for 1982 and 1984 along a transect in northern Africa from the Sahara Desert to rainforest passing through drought affected Sahel and Sudan zones. A split window method has been used to compute surface temperature without considering spectral variation of emissivity. The ND values are based on spectral reflectances using post-launch sensor calibration and corrected for atmospheric effects. The observed interannual variation is due primarily to the satellite orbit degradation (i.e., time of observation).

Multispectral satellite data may also provide other approaches and some input parameters for more complex models for evaporation. Fairly strong correlations have been observed between spectral reflectance, vegetation index, and surface temperature, and models have been developed to interpret these correlations. Field validation of various hypothesis associated with these correlations is needed for their use to infer surface parameters and fluxes.

Acknowledgements

A substantial part of this paper has appeared in a special issue of *Remote Sensing of the Environment* (Choudhury, 1994), published by Elsevier Science, Inc. The work was supported by the Radiation, Dynamics, and Hydrology Branch of the NASA Headquarters and EOS interdisciplinary research.

References

Ahlrichs JS, ME Bauer (1983) Relation of agronomic and multispectral reflectance characteristics of spring wheat canopies, *Agronomy Journal*, 75:987-993.

Ajtay GL, P Ketner, P Duigneand (1979) Terrestrial primary production and phytomass, pp. 129-181, In, The Global Carbon Cycle, B Bolin, ET Degans S Kempe, P Ketner (Eds.), SCOPE Report 13, John Wiley, New York.

Al-Kaisi M, LJ Brun, WEnz JW (1989) Transpiration and evapotranspiration from maize as related to leaf area index, *Agricultural and Forest Meteorology*, 48:111-116.

Baret F, G Guyot (1991) Potentials and limits of vegetation indices for LAI and APAR assessment, *Remote Sensing of the Environment*, 35:161-173.

Baumgartner A, E Reichel (1975) The World Water Balance, Elsevier, New York.

Becker F, B Seguin (1985) Determination of surface parameters and fluxes for climate studies from space observations: methods, results, and problems, *Advances in Space Research*, 5:299-317.

Becker F, BJ Choudhury (1988) Relative sensitivity of normalized difference vegetation index (NDVI) and microwave polarization difference index (MPDI) for vegetation and desertification monitoring, *Remote Sensing of the Environment*, 24:297-311.

Brunet Y, M Nunuz, J-P Lagouarde (1991) A simple method for estimating regional evapotranspiration from infrared surface temperature data, *ISPRS Journal of Photogram Remote Sensing*, 46:311-327.

Brutsaert W (1982) Evaporation into the Atmosphere, Reidel, Boston, Massachusetts.

Brutsaert W, M Sugita (1992a) Regional surface fluxes from satellite-derived surface temperatures (AVHRR) and radiosonde profiles, *Boundary Layer Meteorology*, 58:355-366.

Brutsaert W, M Sugita (1992b) Regional surface fluxes under nonuniform soil moisture conditions during drying, *Water Resources Research*, 28:1669-1674.

Budyko MI (1986) The Evolution of the Biosphere, Elsevier, New York.

Carlson TN, RR Gillies, EM Perry (1994) A method to make use of thermal infrared temperature and NDVI measurements to infer surface soil water content and fractional vegetation cover, *Remote Sensing Reviews*, 9:161-173.

Choudhury BJ (1987) Relationships between vegetation indices, radiation absorption, and net photosynthesis evaluated by a sensitivity analysis, *Remote Sensing of the Environment*, 22:209-233.

Choudhury BJ (1989) Estimating evaporation and carbon assimilation using infrared temperature data: vistas in modeling, pp. 628-690, In, G Asrar (Ed.) Theory and Applications of Optical Remote Sensing, John Wiley, New York.

Choudhury BJ (1991a) Passive microwave remote sensing contribution to hydrological variables, *Surveys of Geophysics*, 12:63-84.

Choudhury BJ (1991b) Multispectral satellite data in the context of land surface heat balance, *Review of Geophysics*, 29:217-236.

Choudhury BJ (1992) Multispectral satellite observations for arid land studies, *ISPRS Journal of Photogram Remote Sensing*, 47:101-126.

Choudhury BJ (1994) Synergism of multispectral satellite observations for estimating regional land surface evaporation, *Remote Sensing of the Environment*, 49:264-274.

Choudhury BJ, JL Monteith (1988) A four-layer model for the heat budget of homogeneous land surfaces, *Quart. JR Meteor. Soc.*, 114:373-398.

Choudhury BJ, SB Idso, RJ Reginato (1987) Analysis of an empirical model for soil heat flux under a growing wheat crop for estimating evaporation by an infrared-temperature based energy balance equation, *Agricultural and Forest Meteorology*, 39:283-297.

Choudhury BJ, NU Ahmed, SB Idso, RJ Reginato, CST Daughtry (1994) Relations between evaporation coefficients and vegetation indices studied by model simulations, *Remote Sensing of the Environment* (forthcoming).

Cowan IR (1982) Regulation of water use in relation to carbon gain in higher plants, pp. 589-613, In, OL Lange, PS Nobel, CB Osmond, H Ziegler (Eds.), Physiological Plant Ecology, Vol. 12B, Springer-Verlag, New York.

Darnell WL, WF Staylor, SK Gupta, NA Ritchey, AC Wilber (1992) Seasonal variation of surface radiation budget derived from International Satellite Cloud Climatology Project C1 data, *Journal of Geophysical Research*, 97:15741-15760.

Diak GR, MS Whipple (1993) Improvements to models and methods for evaluating the land-surface energy balance and 'effective' roughness using radiosonde reports and satellite-measured 'skin' temperature data, *Agricultural and Forest Meteorology*, 63:189-218.

Gal-Chen T, M Xu, WL Eberhard (1992) Estimation of atmospheric boundary layer fluxes and other turbulence parameters from Doppler lidar data, *Journal of Geophysical Research*, 97:18409-18423.

Gregory PJ, D Tennant, RK Belford (1993) Root and shoot growth, water, and light use efficiency of barley and wheat crops grown on a shallow duplex soil in a Mediterranean-type environment, *Aust. Journal of Agricultural Research*, 43:555-573.

Harshvardhan, DA Randall, DA Dazlich (1990) Relationship between the longwave cloud radiative forcing at the surface and the top of the atmosphere, *Journal of Climate*, 3:1435-1443.

Henning D (1989) Atlas of the Surface Heat Balance of the Continents, Gebruder Borntraeger, Berlin, Germany.

Heilman JL, WE Heilman, DG Moore (1982) Evaluating the crop coefficient using spectral reflectances, *Agronomy Journal*, 74:967-971.

Huete AR, RD Jackson, DF Post (1985) Spectral response of a plant canopy with difference soil background, *Remote Sensing of the Environment*, 17:37-53.

Jackson RD, RJ Reginato, SB Idso (1977) Wheat canopy temperatures: a practical tool for evaluating water requirements, *Water Resources Research*, 13:651-656.

Kohsiek W, HAR DeBruin, H The, B van den Hurk (1993) Estimation of the sensible heat flux of semi-arid areas using surface radiative temperature measurements, *Boundary Layer Meteorology*, 63:213-230.

Korzun VI (1978) World Water Balance and Water Resources of the Earth, UNESCO, Paris, France.

Kustas WP (1990) Estimates of evapotranspiration with a one- and two-layer model of heat transfer over partial canopy cover, *Journal of Applied Meteorology*, 29:704-715.

Kustas WP, CST Daughtry, PJ van Oevelen (1993) Analytical treatment of the relationships between soil heat flux/net radiation ratio and vegetation indices, *Remote Sensing of the Environment*, 46:319-330.

Li Z, HG Leighton (1993) Global climatologies of solar radiation budgets at the surface and in the atmosphere for 5 years of ERBE data, *Journal of Geophysical Research*, 98:4919-4930.

Li Z, HG Leighton, RD Cess (1993) Surface net solar radiation estimated from satellite measurements: comparison with tower observations, *Journal of Climate*, 6:1764-1772.

Lvovitch MI (1973) The global water balance, *US/IHD Bulletin*, 23:28-42.

Mahrt L, M Ek (1993) Spatial variability of turbulent fluxes and roughness lengths in HAPEX-MOBILHY, *Boundary Layer Meteorology*, 65:381-400.

McNaughton KG, TW Spriggs (1989) An evaluation of the Priestley and Taylor equation and the complementary relationship using the results from a mixed-layer model of the convective boundary layer, IAHS Publication No. 177:89-104.

Menenti M, BJ Choudhury (1993) Parameterization of land surface evaporation by means of location dependent potential evaporation and surface temperature range, IAHS Publication No. 212:561-568.

Milly PCD (1994) Climate, soil water storage, and the average annual water balance, *Water Resources Research*, 30:2143-2156.

Monteith JL (1988) Does transpiration limit the growth of vegetation or vice versa? *Journal of Hydrology*, 100:57-68.

Moran MS, RD Jackson (1991) Assessing the spatial distribution of evapotranspiration using remotely sensed inputs, *Journal of Environmental Quality*, 20:725-737.

Muchow RC, TR Sinclair (1994) Nitrogen response of leaf photosynthesis and canopy radiation use efficiency in field-grown maize and sorghum, *Crop Science*, 34:721-727.

Nacke G (1991) Surface albedos derived from METEOSAT imagery with application to Africa, *Journal of Geophysical Research*, 96:18581-18601.

Nemani R, L Pierce, S Running, S Goward (1993) Developing satellite-derived estimates of surface moisture status, *Journal of Climate*, 32:548-557.

Pinker RT, I Laszlo (1992) Global distribution of photosynthetically active radiation as observed from satellites, *Journal of Climate*, 5:56-65.

Pinter PJ Jr (1993) Solar angle independence in the relationship between absorbed PAR and remotely sensed data for alfalfa, *Remote Sensing of the Environment*, 46:19-25.

Price JC (1990) Using spatial context in satellite data to infer regional scale evapotranspiration, *IEEE Transactions on Geosciences and Remote Sensing*, 28:940-948.

Raich JW, KJ Nadelhoffer (1989) Below ground carbon allocation in forest ecosystems: global trends, *Ecology*, 70:1346-1354.

Ruimy A, B Saugier, G Dedieu (1994) Methodology for the estimation of terrestrial net primary production from remotely sensed data, *Journal of Geophysical Research*, 99:5263-5283.

Russell G, PG Jarvis, JL Monteith (1989) Absorption of radiation by canopies and stand growth, pp. 21-39, In, G Russell, B Marshall, and PG Jarvis (Eds.) Plant Canopies: Growth, Form, and Function, Cambridge University Press, New York, New York.

Seguin B, E Assad, JP Freteaud, J Imbernon, Y Kerr, J-P Lagouarde (1989) Use of meteorological satellites for water balance monitoring in Sahelian regions, *International Journal of Remote Sensing*, 10:1101-1117.

Singh JS, L Krishnamurthy (1981) Analysis of structure and function of tropical grassland vegetation of India, *Indian Review of Life Sciences*, 1:225-270.

Smith RCG, BJ Choudhury (1990a) On the correlation of indices of vegetation and surface temperature over southeastern Australia, *International Journal of Remote Sensing*, 11:2113-2120.

Smith RCG, BJ Choudhury (1990b) Relationship of multispectral satellite data to land surface evaporation from the Australian continent, *International Journal of Remote Sensing*, 11:2069-2088.

Taconet O, R Bernard, D Vidal-Madjar (1986) Evapotranspiration over an agricultural region using a surface flux temperature model based on NOAA-AVHRR data, *Journal of Climate and Applied Meteorology*, 25:284-307.

Tanner CB, TR Sinclair (1983) Efficient water use in crop production: research or re-search?, pp. 1-28, In, HM Taylor, WR Jordan, TR Sinclair (Eds.) Limitations to Efficient Water Use in Crop Production, American Society of Agronomy, Madison, Wisconsin.

Tarpley JD (1994) Monthly evapotranspiration from satellite and conventional meteorological observations, *Journal of Climate*, 7:704-713.

Vidal A, A Perrier (1990) Irrigation monitoring by following the water balance from NOAA-AVHRR thermal infrared data, *IEEE Transactions on Geosciences and Remote Sensing*, 28:949-954.

WCRP-69 (1992) Radiation and Climate: Report of Fourth Session of the WCRP Working Group on Radiative Fluxes, WMO/TD-No. 471, World Meteorological Organization, Geneva, Switzerland.

Wendler G, F Eaton (1983) On the desertification of the Sahel, *Climate Change*, 5:365-380.

Wiegand CL, AJ Richardson (1990) Use of spectral vegetation indices to infer leaf area, evapotranspiration and yield: results, *Agronomy Journal*, 82:630-636.

Wielgolaski FE, LC Bliss, J Svoboda, G Doyle (1981) Primary production of tundra, pp. 187-225, In, LC Bliss, OW Heal, and JJ Moore (Eds.) Tundra Ecosystems, Cambridge University Press, New York, New York.

Yefimova NA (1976) The maps of productivity of vegetation cover on the continents, *Botan Z*, 61:1641-1657.

The Challenges We Face: Panel Discussion on Evapotranspiration

Edited by Hoshin Gupta and Soroosh Sorooshian

1. Introduction

This panel discussion, moderated by *Bernard Seguin* and *Susan Moran,* provided interesting perspectives on the trials and tribulations (as well as the dreams and realities!) of the measurement and modeling of evapotranspiration (ET). The following is a summary of the major points of discussion, based primarily on notes prepared by participants *Christa Peters, Helene Unland,* and *Jim Washburne*. Their assistance in documenting the discussions is greatly appreciated.

The moderators focused the discussion on (a) the evaluation of evapotranspiration, and (b) the assimilation of ET. Most comments centered around evaluation, including measurement, modeling, and validation. The assimilation of ET data into hydrologic and atmospheric models was discussed only briefly, although this is an area which needs much work. The general consensus, crystallizing out of a number of insightful questions, was that a synergistic approach combining both conventional and remotely sensed measurements with modeling offers the best opportunity for enhancing our understanding of global environmental change and land-surface processes.

2. What is the best approach to measure evapotranspiration?

Evapotranspiration may be measured by either conventional means (i.e., eddy correlation or Bowen ratio instruments) or by the use of remote sensing to infer surface characteristics from which ET may be calculated. Although the participants did not discount the possibility of obtaining useful estimates of ET using remote sensing, it was pointed out that ground-based measurements by conventional methods (including potential ET estimates obtained using

NATO ASI Series, Vol. I 46
Land Surface Processes in Hydrology
Trials and Tribulations of Modeling and Measuring
Edited by S. Sorooshian, H. V. Gupta, and S. C. Rodda

evaporation pans) provide the ground truth data necessary for validation of remote sensing techniques, as well as the long-term historical records needed for investigating global change.

deBruin, Choudhury, Starosolszky and several others felt, therefore, that it is important to maintain and improve ground-based ET measurement systems. *Goodison* suggested that although meteorological offices may not be willing to spend more money on sustaining and expanding existing ET networks, these costs could be borne by current environmental change initiatives. *Rodda* emphasized that both continuity and accuracy (i.e., quality control) are essential in determining the cost effectiveness of any data set.

The need to better define what hydrologists, meteorologists, and climate modelers *want* in terms of spatial and temporal resolution of data--as well as what is *feasible*--was noted by *deBruin* and *Seguin*. For instance, spatial and temporal coverage is severely limited in developing countries. Even in the developed nations, the data are biased toward certain ecological and/or meteorological conditions. *Choudhury* remarked that while we are probably able to get reasonable remotely sensed inferences of total ET, information on the fractional cover at the resolution of the sensor is necessary to distinguish between evaporation and transpiration. Further, we are not fully utilizing the data (such as SAR) that do exist, and we may need to fundamentally change the way we conceptualize and model hydrologic and atmospheric processes. With this in mind, *Engman* suggested that a multi-sensor, multi-resolution approach, combining remote sensing and modeling, may offer the best opportunity for understanding and modeling land-surface processes and environmental change.

3. The limitations of current models

In hydrologic models, knowledge of ET is essential as a boundary condition for determining subsurface moisture content, while in atmospheric models, ET is an important part of complex mass, momentum, and heat exchange boundary conditions. The panel felt that a synergistic approach could lead to improved model performance, although the degree of potential improvement is not clear. A point of discussion that came up repeatedly was that all types of models are limited by the availability of data at the scale of the model. *Moran* pointed out that, while this is true, a more fundamental question is that of the *importance* of various land-surface processes in these models. *Dickinson* remarked that some global scale modelers have little experience with the detailed physics of land processes and, therefore, a great deal more interaction is needed between hydrologists, ecologists, and the GCM community if significant progress is to occur in terms of improving sensitivity and land-process representation. In addition, many processes are currently simplified and misrepresented in GCMs because the global data sets required for parameterization of the land-surface models are typically not in useable form.

4. Challenges for the hydrological community

The moderators *Bernard Seguin* and *Susan Moran* summarized the major conclusions of this panel discussion as follows:

(a) Remote sensing methods for the measurement or inference of ET appears promising, but they must be supplemented by data conventional ground-based networks.

(b) Hydrologists, meteorologists, and climate modelers must define and justify their specific data needs, in terms of cost, resolution, and value of information.

(c) A synergism between ET modeling and measuring in a data assimilation framework is the most promising avenue for improving both process understanding and model performance.

Part V - Runoff

Streamflow Measurement for the 21st Century

R.W. Herschy
CNS Scientific & Engineering Services
5 Alderney Court
Montague Street
Reading, Berkshire RG1 4JW
UNITED KINGDOM

Introduction

Streamflow Measurement

More water is used in the world today than ever before, but the quantity available for use remains the same. This quantity is not known with any accuracy because of inadequate monitoring or, in some cases, no monitoring at all.

Water is precious. Figure 1 shows that only 3% of all water is fresh water but less than 1% is available for use. As many as 12 countries in Africa are currently suffering a severe drought; in five of these countries, the drought is bordering on the catastrophic. In Israel, approximately 95% of the available water resources are utilized; consequently, the national development is now restricted. The present utilization of water in India is already nearly 80% of that available but, by the year 2025, it is estimated from rainfall and streamflow monitoring that the available water will no longer be enough to sustain the population. By the end of the century, global water use is expected to reach twice the level of water use in 1980. The problem is exacerbated by much of the 1% available fresh water being polluted, causing some five million deaths per year through waterborne diseases. Indeed, many millions of people still do not have proper water for drinking and sanitation. These few examples explain a need for worldwide streamflow monitoring and the necessity to ensure the methods available are suitable to meet this need.

Unfortunately, however, there are not enough streamflow stations in the world to assess the water resources, the density over the land mass being on the average about 1,000

NATO ASI Series, Vol. I 46
Land Surface Processes in Hydrology
Trials and Tribulations of Modeling and Measuring
Edited by S. Sorooshian, H. V. Gupta, and S. C. Rodda

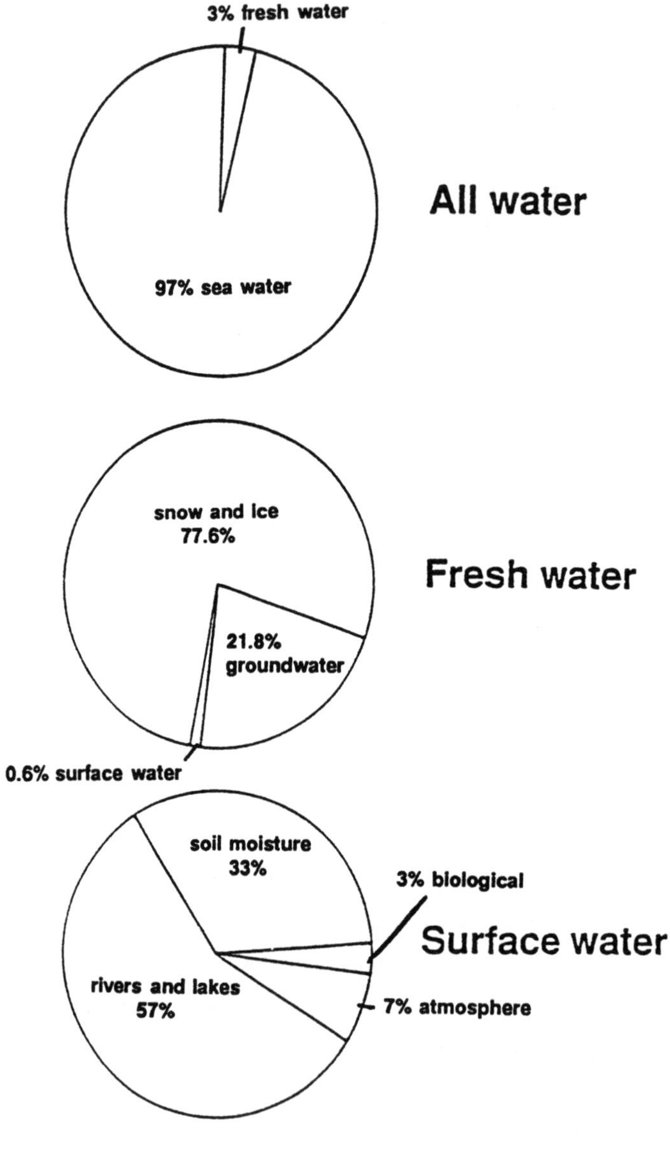

Figure 1. Global water resources.

square miles per streamflow gauge. The network is particularly deficient in India, China, Africa, and South America, in countries where the need is most critical.

Streamflow is the combined result of all climatological and geographical factors that operate in a drainage basin. It is the only component of the hydrological cycle in which the water is confined in well-defined channels that permit the accurate measurement of the quantities involved.

Good water management is founded on reliable streamflow information, but the final reliability of the information depends upon the initial field measurements. Indeed, the successful processing, analysis, and interpretation of the streamflow data depend largely on the quality of these measurements.

Streamflow in Water Management

There are many different uses of streamflow data within the broad context of water management, such as water supply, pollution control, flood control, irrigation, abstractions and discharges, reservoir compensation water, sewage works flows, and dry weather flows. The importance placed on any one or more of these purposes may vary from country to country. In India and China, for example, emphasis might be placed on irrigation and flood control at any one time, whereas in the United Kingdom, abstractions for water supply may be given priority during a particular period and low river flows during another period. Hence, the priorities may change over short or long periods of time, but the necessity for accurate streamflow measurement remains.

Floods and droughts are among the most frequent and costly of natural disasters, both in terms of human hardship and economic loss. Floods strike suddenly, some with little or no warning and raise havoc and destruction along the course of streams and rivers. As much as 90% of damage related to international natural disasters, excluding droughts, is caused by floods. In the United States, the 10-year (1979-88) average annual estimated flood damage is some $2.4 billion, and the long-term (1925-88) annual average for lives lost is 95 (U.S. Geological Survey, 1991). However, the flood damage caused by the 1993 Mississippi flood has been estimated to be at least $10 billion; in the 1993 Asian flood (Bangladesh-India), some 3000 people lost their lives. Droughts on the other hand, affect human activities gradually as rainfall deficits accumulate over a period of months or years. The cumulative effects of these deficits reduce streamflow and reservoir storage at a time when water demand increases. In developing countries, such rainfall deficits and their cumulative effect drastically reduce food supplies, causing widespread famine. However, even in developed countries, droughts can have long-term effects on crops, operating costs, industrial production, and the environment.

Research continues to better understand the physical processes that lead to floods and droughts, and data from streamflow stations worldwide are being made available to support the UN International Decade for Natural Disaster Reduction, which encourages all nations to reduce the loss of life and property damage and to minimize social and economic disruption from natural hazards during the period 1990-2000. As nations' populations and economies expand, demand for available water resources continues to grow, and of increasing concern is the effect of long-term droughts and human activities on the environment. As more and more water is impounded or diverted from rivers for human use, streamflow may be altered significantly. Because the amount of freshwater is fixed and because demand increases commensurate with population and economic growth, global water resources planning will assume even greater importance, and problems will arise as growth continues to push the limits of supply to a point where, because of the vulnerability of water supply to drought, demand regions may eventually exceed supply. Water resources planning for the future, however, is aggravated by uncertainty about the effect of potential global warming on available freshwater.

Applications of Streamflow Information

Current-use data have an immediate high return value because the data are invariably required for operation and control. Current-use streamflow stations are operated for as long as the need remains.

Furthermore, designers of water control and water-related facilities, however, increasingly use the statistical characteristics of streamflow rather than flow over specific historic periods. The probability that the historical sequence of flow history at a given site will occur again is remote. Indeed, when a hydrologist makes just one measurement of discharge, it is probable that the exact conditions under which the discharge occurred may rarely happen again.

It is often desirable to consider the future, not in terms of specific events but in terms of probability of occurrence over a span of years. For example, many highway bridges are designed on the basis of the flood that will be exceeded on the average only once in 50 years. Storage reservoirs are designed on the basis of the probability of failure of a particular capacity to sustain a given draft rate. The water available for irrigation, dilution of waste, or other purposes may be stated in terms of the mean flow, or probability of flow magnitudes, for periods of a year, season, month, week, or day. In addition, there is a trend towards flow simulation based on statistical characteristics such as the mean and standard deviation. To define statistical characteristics adequately, however, a record of at least 30 years is desirable for reliable results.

Therefore, streamflow data are analyzed for a multitude of applications, including probability distributions, hydrological and water resources models, streamflow routing, reservoir routing, low flow analysis, flow duration curves, low flow frequencies, statistical analysis of droughts, recession curve analysis, flood flow frequencies, statistical analysis of floods, peak flows, flood warning, hydrological forecasting, time series analysis, flood forecasting, drought forecasting, storm surges in rivers, reservoir design, wastewater treatment works design, water demand estimates, water power generation, pollution control, fish and wildlife conservation, irrigation, navigation, sediment transport, river training, etc.

Summary of Methods

The methods of streamflow measurement may be summarized as follows, and guidelines for the selection of methods are given in the Appendix.

Velocity Area Method

The discharge is derived from the sum of the products of stream velocity, depth of flow, and distance between observation points (verticals), the stream velocity being normally obtained by current meter. A unique relation may often be established between discharge and stage (water level). Once established, this stage-discharge relation is used to derive discharge values from recordings of stage values.

Weirs and Flumes

The relation between stage (or head) and discharge over a weir or through a flume is established from laboratory or field calibration. The discharge is subsequently derived from this rating equation.

The Ultrasonic Method

The velocity is measured by transmitting an ultrasonic pulse diagonally across the channel in both directions simultaneously. The difference in time transits is a measure of the velocity. This velocity is required to be multiplied by the flow area to derive discharge.

Electromagnetic Method

The discharge is estimated by measuring the electromotive force (emf) produced by a moving conductor (the flowing water) through a magnetic field produced by a coil placed either below or above the open channel. The emf is proportional to the average velocity; the stage measurement and length of conductor (width of the channel) are related to area of flow.

Dilution Method

A tracer liquid is injected into the channel, and the water is sampled at a point further downstream where turbulence has mixed the tracer uniformly throughout the cross-section. The change in concentration between the solution injected and the water at the sampling station is a measure of the discharge. The dilution method is arguably the only method which does not use the velocity-area principle.

Moving Boat Method

A current meter is suspended from a boat which traverses the channel normal to the streamflow. The component of the velocity in the direction of the stream is computed from the resultant velocity and the angle of this resultant. The discharge is the sum of the products of the stream velocity, depth, and distance between observation points (verticals).

Slope-Area Method

The discharge is derived from measurements of the slope of the water surface and the area of flow over a fairly straight reach, assuming a roughness coefficient for the channel boundaries. The slope-area method is normally used to measure floods or high discharges.

Stage-Fall-Discharge Method

In a stable open channel affected by backwater, a relation is established between stage, fall (slope), and discharge, the latter being determined by current meter.

Float Gauging

The water velocity is measured by recording the time taken for a float to travel a known distance along the channel. Observations are made using floats at different positions across the channel, and discharge is derived from the sum of the products of velocity and area of flow. Generally, this method is used only when the velocity is either too fast or too slow to use a current meter.

Sewers and Culverts

New methods are now available to measure the flow in sewers and culverts by means of a sensor installed in the invert of the pipe which measures the depth of flow and velocity and combines these two components by the velocity area method. The depth component of the sensor is normally a pressure transducer, and the velocity element may operate on the electromagnetic principle, the Doppler or acoustic-Doppler principle.

Velocity Times Area Equals Discharge

With the exception of dilution techniques, the principle of all streamflow measurement is based upon velocity x area, so that:

$$Q = v \times A \quad (1)$$

where:

Q = the discharge (m^3/s or ft^3/s);

v = the average velocity (m/s or ft/s); and

A = the cross-sectional area of flow (m^2 or ft^2).

Velocity is found directly (e.g., by current meter), or indirectly (e.g., using either the Chezy or Manning equations). Area is determined directly (depth and width of the cross-section), or indirectly (e.g., stage vs. cross-section area relation). Various electronic instruments are now available to determine velocity and area. The following methods of flow measurement demonstrate the application of the velocity area principle.

Velocity-Area by Current Meter (Herschy, 1985)

Velocity is determined by rotating element (or electromagnetic, solid state), current meter; area is determined by wading, cableway, boat, or moving boat (Figure 2), and discharge, Q, is derived from Equation (1).

Weirs and Flumes (Ackers et al., 1978)

It can be shown that the general equation of discharge:

$$Q = (2/3)^{3/2}\, C_D\, C_V \sqrt{g}\, b\, h^{3/2} \quad (2)$$

is established from critical depth theory, from which the area of flow (b2/3H) over the structure is determined, and velocity is found from Figure 3.

$$\bar{v} = \sqrt{2gH} \quad (3)$$

Now:

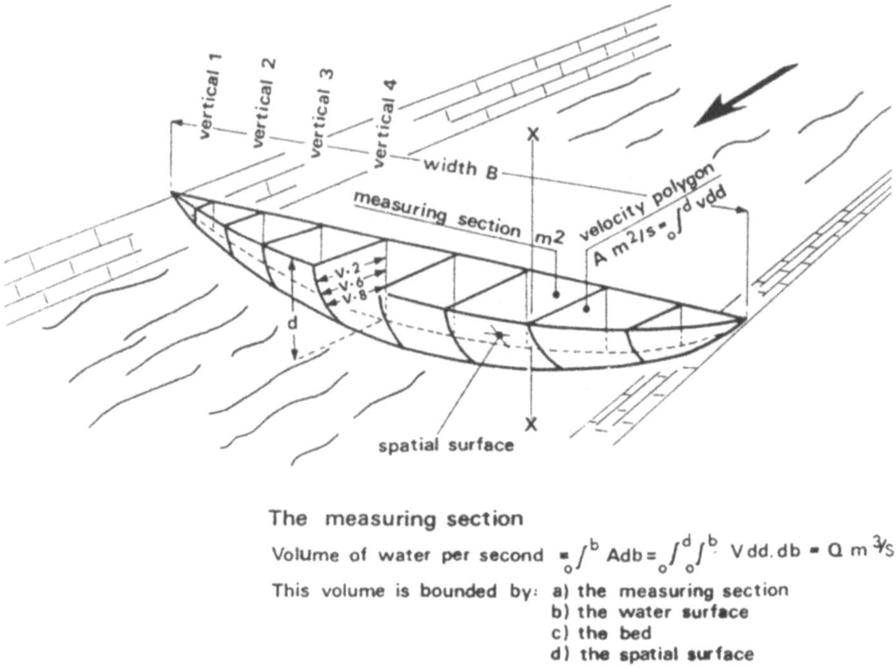

Figure 2. The measuring section. The volume of water is bounded by the measuring section, the water surface, the bed, and the spatial surface, as shown schematically. At any section *XX*, the area of the velocity polygon is the integral *VDD* (with limits from 0 to *d*) and equal to *A* m^2/s. The volume of water passing per second is then found from the integral of *Adb* (with limits from 0 to *b*) and equal to the integral of *vdd db* (with limits from 0 to *d* and 0 to *b*) which is equal to the total flow *Q* in m^3/s.

$$A = b2/3H \tag{4}$$

and:

$$\bar{v} = \sqrt{2gH/3} \tag{5}$$

so that:

$$Q = \bar{v} \times A \tag{6}$$

$$= (2/3)^{3/2} \sqrt{g} b H^{3/2}$$

where:

b = the length of the weir crest (or flume throat) (m);

H = the total head (m);

h = the gauged head (m);

C_d = the coefficient of discharge;

C_v = the coefficient of velocity; and

g = 9.81 m/s^2.

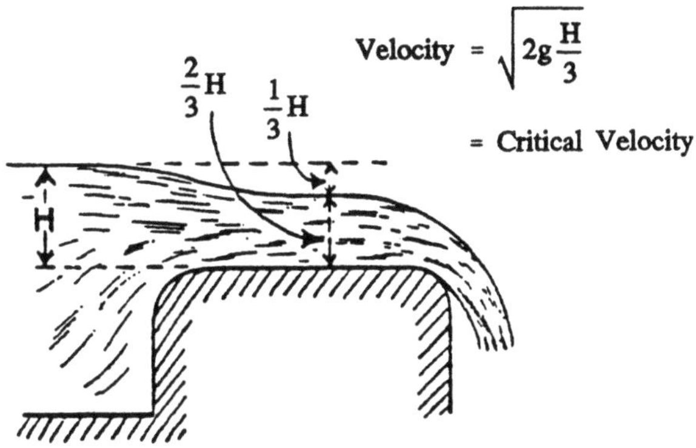

Figure 3. Schematic illustration of the principle and theory of flow over weirs.

When the coefficients of discharge and velocity and the gauged head are inserted, Equation (6) takes the form of Equation (2).

The Ultrasonic Method (Herschy, 1985)

Velocity is found by transmitting ultrasonic pulses across the channel in both directions simultaneously (Figure 4). Area is determined from a stage-cross-sectional area relationship, stage being measured by an ultrasonic water level gauge. For one ultrasonic flight path, the resolved component of the mean velocity of the water along the path can be computed as follows:

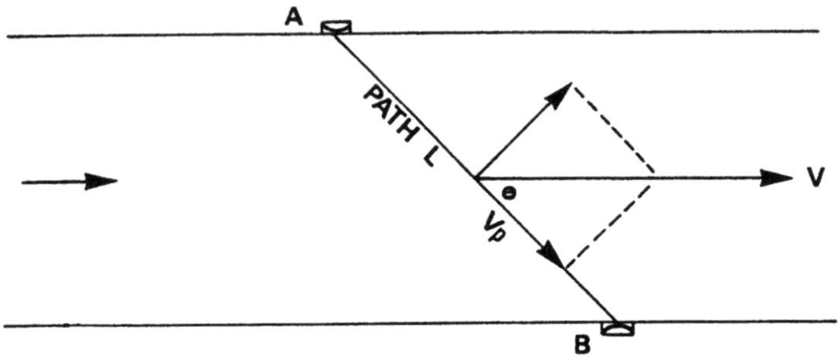

Figure 4. Ultrasonic method; velocity components.

$$t_1 = \frac{L}{(C + V_p)} \qquad t_2 = \frac{L}{(C - V_p)} \tag{7}$$

where:

L = the length of the path (m or ft);

C = the velocity of sound in water (m/s or ft/s);

V_p = the resolved mean of velocity (m/s or ft/s); and

t_1 and t_2 = the flight path times (s).

The equations combine to give:

$$V_p = \frac{L}{2}\left[\frac{1}{t_1} - \frac{1}{t_2}\right] \tag{8}$$

The average line velocity of the water VI is given by:

$$V = \frac{V_p}{\cos\theta} \tag{9}$$

where θ is the angle between the direction of flow and the flight path. The velocity is measured in a similar manner at several levels between the surface and the river bed, and the flow is computed in various "slices" of the river by multiplying the velocity by the width and the thickness of the "slice". The total discharge is found by summing these separate measurements. Depth is measured with a similar single transducer mounted under the water with its beam pointing vertically towards the service. The ultrasonic pulse transit time to the surface and its reflection to the same transducer give a measure of the river depth. For this depth measurement, compensation for variation in the speed of sound in water (due to either temperature or density changes with dissolved salts) is made by measuring the speed of sound at each level using the horizontal paths. The compensation allows for different speeds of sound at different depths to give a precise measurement.

The Electromagnetic Method (Herschy, 1985)

Figure 5 gives a schematic view of the system, and it can be shown that:

$$E \quad \alpha \quad \bar{v}bI \tag{10}$$

where:

- E = emf (volts);
- $\bar{v}$ = average velocity (m/s);
- b = channel width (m); and
- I = coil current (amps).

From Equations (1) and (10):

$$Q \quad \alpha \quad \frac{E}{bI} \times A \tag{11}$$

$$\therefore \quad Q \quad \alpha \quad \frac{E}{bI} \times bh \tag{12}$$

or:

$$Q = K\frac{Eh}{I} \tag{13}$$

where:

- h = the depth of flow, and;
- K = a dimensionless coefficient.

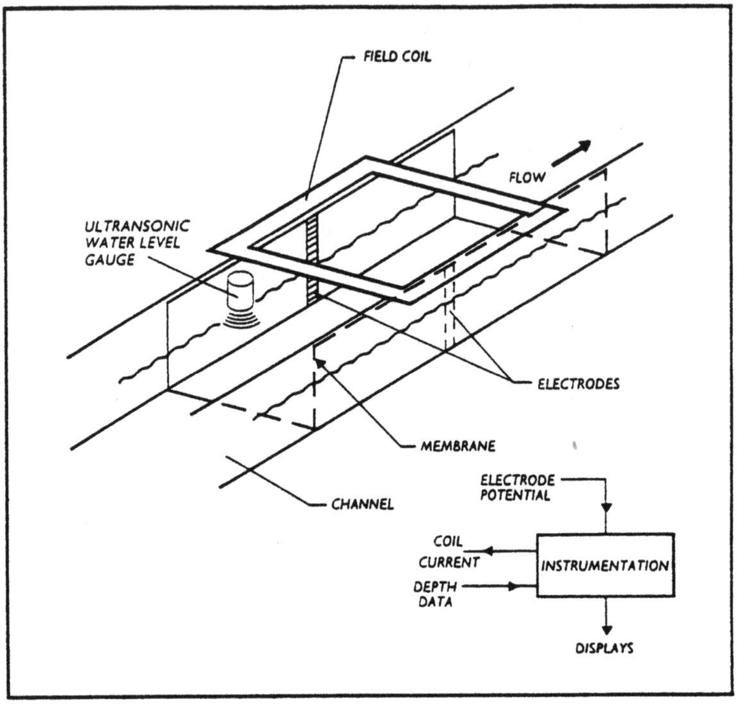

Figure 5. Schematic view of an electromagnetic gauge with coil bridged over the channel (courtesy of Sarasota Automation).

Orifices (Bos, 1976)

From Equation (1):

$$Q = C_1 \, A\bar{v} \tag{14}$$

$$= C_1 \, A\sqrt{2gh} \tag{15}$$

where:

A = the cross-sectional area of the orifice;

h = the head above the center of the orifice, and;

C_1 = the coefficient of discharge.

For a submerged orifice:

$$Q = C_2 A \sqrt{2g(h_1 - h_2)} \tag{16}$$

where:

$(h_1 - h_2)$ = the head differential across the orifice; and
C_2 = the coefficient of discharge.

Sluice Gates (Bos, 1976)

From Equation (1):

$$Q = C_3 A \sqrt{2g(h_1 - h_2)} \tag{17}$$

where:

h_1 = the upstream depth of flow;
h_2 = the downstream depth of flow (at the *vena contracta*); and
C_3 = the coefficient of discharge.

For submerged flow, $(h_1 - h_2)$ is the head differential across the gate.

Closed Conduit Flow (under pressure) (ISO 5167 1980)

The flow through a constriction area A_1, velocity $\bar{v}_1$, pressure head h_1 in a closed circuit under pressure (e.g., in a Venturi meter) results in a lowering of pressure at the constriction area A_2, velocity v_2, and pressure head h_2. The drop in piezometric head, $(h_1 - h_2)$, measured by a differential manometer or pressure gauges, is a function of flow rate so that from Equation (1):

$$Q = A_1 \bar{v}_1 = A_2 \bar{v}_2 \tag{18}$$

and:

$$\bar{v}_1^2 = (A_2/A_1)^2 \, \bar{v}_2^2 \tag{19}$$

Now:

$$h_1 + \frac{\bar{v}_1^2}{2g} = h_2 + \frac{\bar{v}_2^2}{2g} \tag{20}$$

and:

$$\frac{\bar{v}_2^2}{2g} - \frac{\bar{v}_1^2}{2g} = h_1 - h_2 \tag{21}$$

From Equation (19):

$$\frac{\bar{v}_2}{2g} \left\{1 - (A_2/A_1)^2\right\} = h_1 - h_2 \tag{22}$$

and:

$$Q = A_2 \, \bar{v}_2 \tag{23}$$

$$= \frac{C A_2 \sqrt{2g(h_1 - h_2)}}{\sqrt{1 - (D_2/D_1)^4}} \tag{24}$$

where:

D_1 = the diameter of the pipe; and

D_2 = the diameter of the contraction.

Standards: The Way Forward? (Herschy, 1993)

Standards are rules that specify how something should be done, just as laws are rules that specify how individuals should behave in society. Civilization would never have evolved without acceptance of laws and standards. As the world's population increased, so did the need for more specific laws. As technology became more complex, the need for more detailed and effective standards increased accordingly. Standardization in the field of new technology presents innumerable difficulties because of the rapid advances taking place in this field. Consequently, a standard written today may be out of date tomorrow.

Water in a stream in a specific locality knows no jurisdictional boundary, local, national, or international. That same water may eventually move to any other part of the Earth through the hydrologic cycle. Therefore, water data are needed from all parts of the Earth to enable hydrologists to discover the quantity and quality of the Earth's water resources on a comprehensive and continuous basis. Standards for streamflow measurement are particularly important in the case of international rivers which cross or form frontiers (e.g., the Danube, the Indus, the Ganges, etc).

Hydrologists cannot validly compile water data that have been gathered by nonstandardized methods. In many cases, the data on streamflow or quality of water are not even compatible. This makes it impossible to compare them in an effort to discern patterns or trends or to draw conclusions. Consequently, many individuals and organizations who are concerned with hydrometry and water quality have recognized the need for standardization. A system of developing standards that is founded on voluntary consensus works best. No one element of society, whether industrial or governmental, is competent to keep pace with a rapidly changing technology. A standard written by representatives from all affected segments is more likely to be unbiased, authoritative, and, above all, utilized. In the process of formulating such a standard, people with diverse interests come together, resolve their differences, and make mutual concessions that result in the greatest good for the greatest number. The standards produced by such a process protect everyone's interests. Some, however, are apprehensive of standardization, taking the viewpoint that it may lead to uniformity, inhibit innovation, and generally retard progress. These fears are unfounded. Standards are not meant to be restrictive, but to establish a measure of uniformity so that data derived will be compatible and comparable. The hydrologist may choose the method that best suits a particular situation from the many available. The door should be, and is, left open for new developments in technology. The fact that there is a standard method in no way deters hydrologists from searching for an even better method.

Cost Effectiveness: A Recent Study (DOE, UK, 1990)

In a study carried out for the United Kingdom Department of the Environment of the UK Streamflow Network, while it was established that individual gauging stations had benefit-cost ratios greater than unity, it is interesting to note that the total benefit derived from having a network of gauging stations was greater than the sum of the benefits of the individual station. Streamflow gauging revenue costs in the study included the cost of collection, processing, and publication of the data. They also included telemetry and flood warning costs where streamflow stations were involved. To the basic revenue cost was added 150% to allow for overheads, which included management, superannuation, office accommodation, etc. The main quantified benefits were found to come from the following:

(a) water authority operations, including the supply of potable water;
(b) the risk of climatic change or other factors causing a significant change of river flow (assuming that the streamflow network had been abandoned);
(c) irrigation;
(d) flood alleviation; and
(e) flood warning.

Only small benefits could be quantified from flood warning and flood alleviation and none at all from consents to discharge effluents. It was considered, however, that the latter had considerable potential benefit in the future, but this was not evaluated.

The annual benefits were found to be in the range of US $16.5-$90 million, depending on how these were quantified, the best estimate being US $31.5 million! The annual cost of operating the streamflow network was US $13.5 million. The benefit-cost ratio was, therefore, in the range of 1.2-7, with a best estimate of 2.3. It was concluded that, even at the lowest level of benefit-cost ratio, the UK streamflow network represented a sound economic investment.

Sensitivity Analysis: A Forgotten Indicator! (March 1993)

The relationship between rate of change of stage and rate of change of discharge at a controlled streamflow station is governed by the physical characteristics of the channel at, or near, the gauging site. The sensitivity of the control and, therefore, of the stage-discharge relation is the response of an increase in discharge to an increase in stage. Where a small increase in discharge produces a relatively large increase in stage, the relation is said to be sensitive; where a relatively large increase in discharge produces a small increase in stage, the relation is said to be nonsensitive. In a broad streamflow program where a wide variety of stations are operated, it is likely that variations in sensitivity between the two extremes will be found. For this purpose, the definition of sensitivity may be taken as the percentage error in discharge arising from a 10 millimeter systematic error in stage at the corresponding 95% exceedance flow.

Table 1 shows a randomly selected group of stations from the UK where the percentage sensitivity has been calculated at the 95 percentile flow. Hence, the sensitivity values will be better than the adopted values for 95% of the time (347 days in a year on the average). Table 1 also shows (in the last column) the required accuracy of stage measurement in millimeters to restrict errors in the 95 percentile flow to 5%.

Remote Sensing: Can It Help? (NRA, 1993)

Streamflow cannot be measured directly by remote sensing methods, but there are a number of ways in which remote sensing data can be used to estimate runoff and gain insight into the hydrograph characteristics. One such method has long been used in air photo interpretation by fluvial morphologists. This is the quantitative analysis of drainage basin and channel networks, which involves stream ordering and the measurement of a number of basic and stream-related parameters. These measurements fall into three groups: areal extent and shape measurements, which determine the size of the input in terms of precipitation; shape and overall slope measurements, which determine the rapidity of runoff; and drainage density, which determines the efficiency

Table 1. The Sensitivity of UK Gauging Stations.

(Note: The "sensitivity error" referred to in this table relates to the percentage change in flow associated with a 10 mm change of water level at a stage corresponding to the 95% exceedance flow. The "required accuracy of stage measurement" is that necessary to restrict errors in computed flows (from this source), at the 95% exceedance, to less than 5%.

Gauging Station Number	River	Station Type	Catchment Area km²	Mean Flow m^3s^{-1}	95%ile Flow m^3s^{-1}	Sensitivity error %	Required accuracy of stage measurement
004001	Conon	VA	961.8	45.62	8.59	5.5	9
007001	Findhorn	VA	415.6	13.22	2.05	13.9	4
012001	Dee	VA	1370.0	36.40	8.40	5.2	10
015006	Tay	VA	4587.1	158.10	42.84	1.9	26
021009	Tweed	VA	4390.0	76.71	14.02	4.5	11
023001	Tyne	VA	2175.6	43.87	5.44	6.5	8
024005	Browney	CB	178.5	1.73	0.34	13.7	4
024009	Wear	FV	1008.3	14.78	3.29	7.8	6
025019	Leven	FV	14.8	0.20	0.06	25.0	2
027035	Aire	VA	282.3	6.04	0.52	15.9	3
027041	Derwent	C	1586.0	17.53	4.92	5.5	9
027051	Crimple	FV	8.1	0.11	0.01	54.0	1
027055	Rye	C	131.7	2.36	0.55	22.1	2
028003	Tame	VA	408.0	5.84	2.70	3.3	15
028012	Trent	VA	1129.0	12.52	5.04	3.6	14
028025	Sence	C	169.4	1.51	0.25	22.4	2
028044	Poulter	C	65.0	0.33	0.17	21.2	2
031006	Gwash	C	150.0	0.86	0.29	23.3	2
033012	Kym	CB	137.5	0.63	0.02	65.0	< 1
036006	Stour	FL	578.0	2.83	0.50	7.9	6
038007	Canons Brook	FL	21.4	0.20	0.05	32.0	2
039016	Kennet	C	1033.4	9.65	3.98	6.4	8
039019	Lambourn	C	234.1	1.72	0.79	13.3	4
039020	Coln	C	106.7	1.34	0.38	21.3	2
043005	Avon	C	323.7	3.43	1.15	8.9	6
048005	Kenwyn	CC	19.1	0.38	0.05	15.6	3
049004	Gannel	C	41.0	0.69	0.10	38.2	1
052010	Brue	C	135.2	1.89	0.26	21.5	2
053017	Boyd	FV	48.0	0.57	0.05	27.5	2
054004	Sowe	C	262.0	2.94	1.03	8.6	6
054012	Tern	FV	852.0	7.09	2.41	4.2	12
054019	Avon	C	347.0	2.50	0.48	15.0	3
056001	Usk	VA	911.7	27.67	4.34	5.1	10
065005	Erch	C	18.1	0.60	0.09	45.0	1
090003	Nevis	VA	76.8	6.28	0.57	8.8	6

Station types

VA	Velocity-area	C	Crump Weir
CB	Compound broad-crested weir	CC	Compound Crump Weir
FV	Flat Vee weir	F	Flume

with which the runoff is removed. Thus, the various characteristics of the hydrograph can be estimated: peakedness, lag time, and recession limb, as well as mean annual flow and maximum peak. The resolution capabilities of air photographs often mean that channel characteristics can be estimated to provide an estimate of bank-full flow and information about over-bank flow. Channel patterns in terms of meanders, braided systems, etc. can be recognized, measured, and used to refer to developed relationships between channel morphometry and runoff frequency. Although satellite resolution does not permit such detailed measurement in most rivers, transformations and enlargements enable channel networks to be identified by vegetation differences and enable quite detailed network analysis to be accomplished, which can be of great value in regions of scarce hydrologic data.

Conventional measurement in hydrology is a point-time phenomenon, whereas hydrology itself is a time-space phenomenon. It is towards the rectification of the shortcomings of the former that the main thrust of remote sensing applications is directed by supplying the essential, but missing, space element. There are a number of significant achievements in satellite remote sensing, including operational systems for flood mapping, delta development and, more recently, areal rainfall and snow mapping of area, depth, and water equivalent (Figures 6, 7, 8, and 9).

International Rivers (Herschy, 1993)

There are many instances where international water rights can lead to tension and subsequent conflict. The most serious situation presently causing concern is the international rivers of the Middle East. The water issues in this region could either be a catalyst for peace and development or remain a focal point of tension. The great rivers Euphrates and the Tigris feed Turkey, Iran, Iraq, and Syria; however, with the construction of the Ataturk Dam, Turkey now controls the water of the Euphrates (Figure 10). There is an agreement to share the waters of the Tigris and Euphrates, and already a regional water summit of interested parties has been held in Turkey. At the Middle East Peace Conference, water rights are always included in the talks. Options

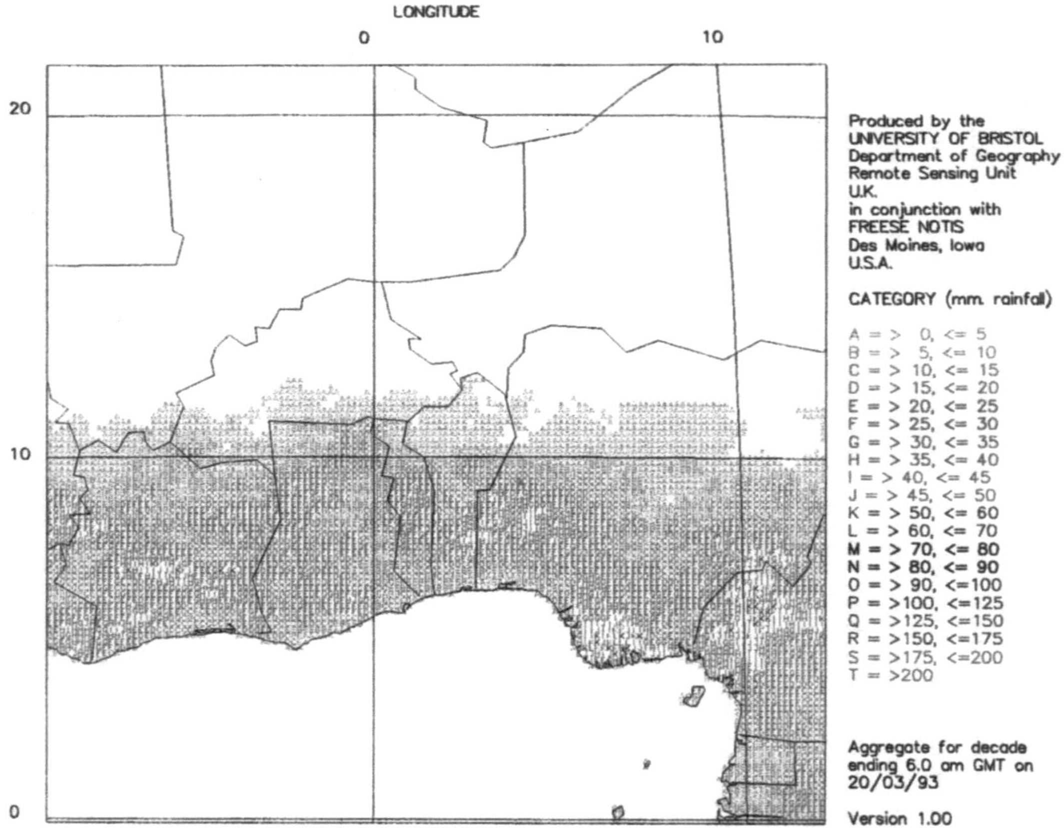

Figure 6. Total rainfall for 10 days over West Africa, determined by remote sensing from *Meteosat*. (Figure reproduced courtesy of the University of Bristol).

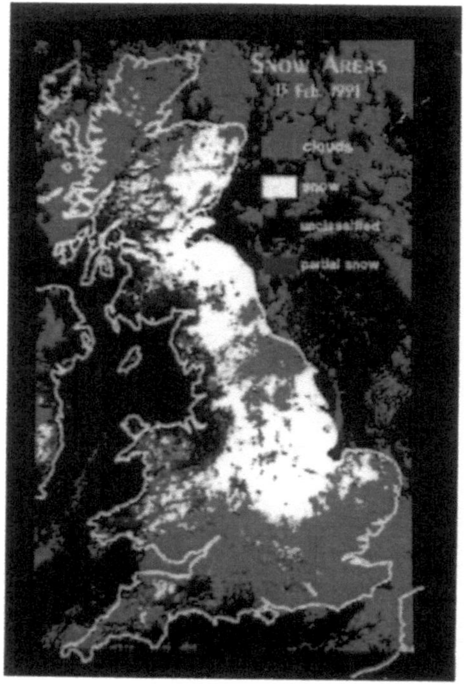

Figure 7. Lying snow area over the United Kingdom.

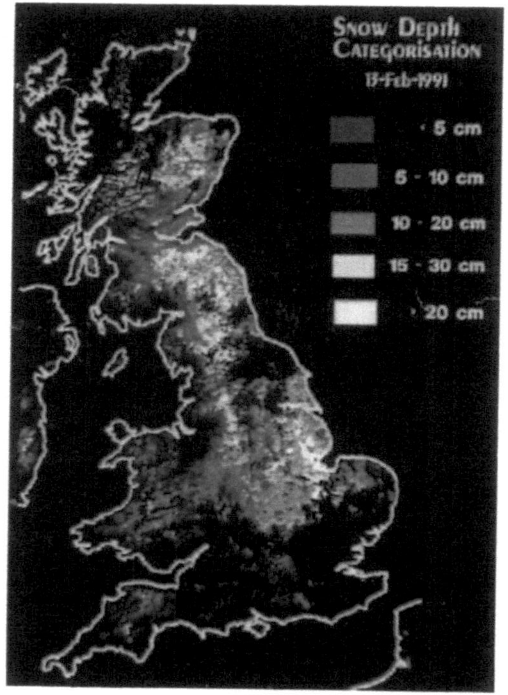

Figure 8. Snow depth survey of the United Kingdom.

(Both figures were obtained by remote sensing and are reproduced courtesy of the University of Bristol).

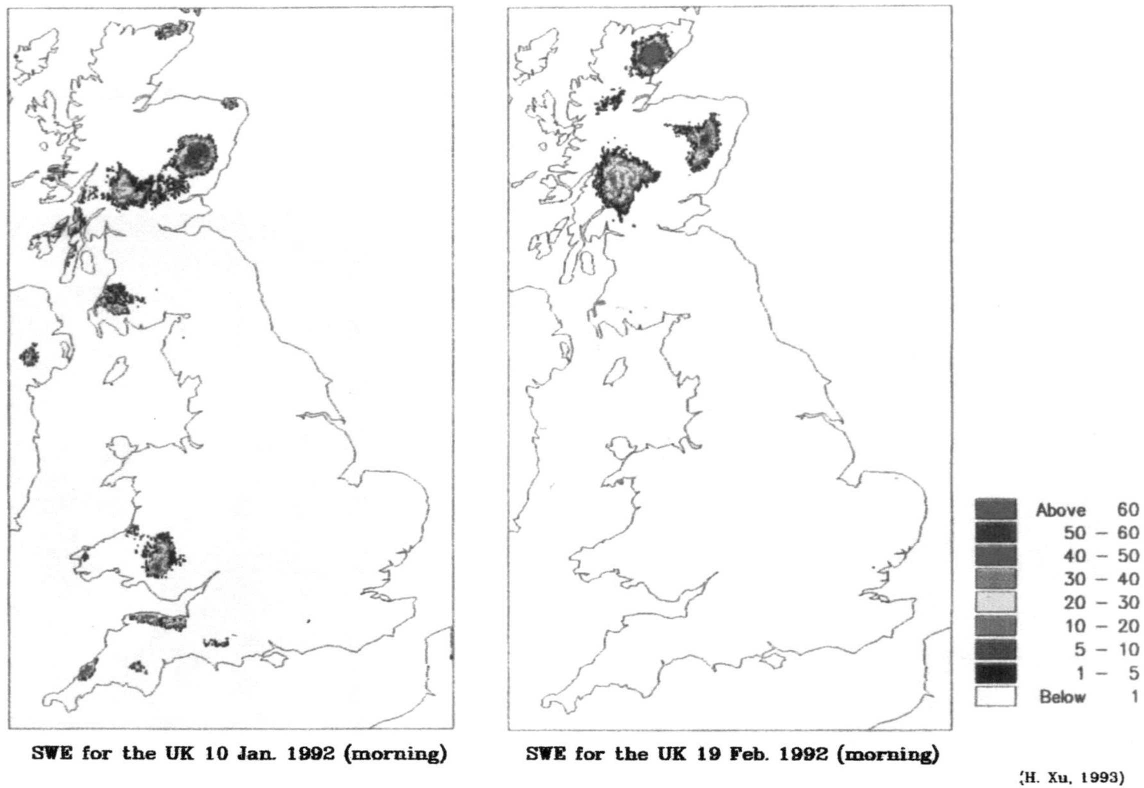

Figure 9. Snow-water equivalent of the United Kingdom by remote sensing. (Figure reproduced courtesy of the University of Bristol).

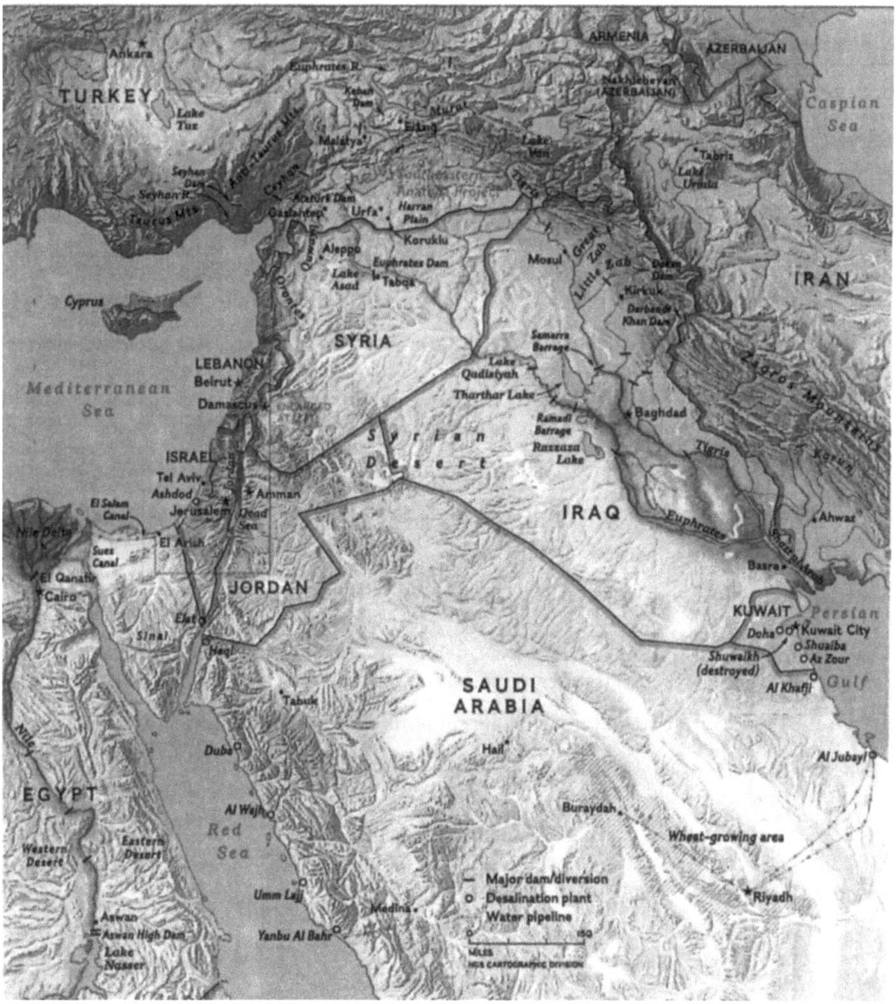

Figure 10. The River Euphrates showing the dams in Anatolia and the proposed dams/diversion on the Tigris and Euphrates in Iraq. (Figure courtesy of *National Geographic*).

debated include a "peace pipeline" from the Turkish mountains to the Gulf which, it is believed, would overcome the present problems. Some 60% of Israel's water comes from the occupied areas, the three main sources of the country's supply coming from Lake Kinneret (Sea of Galilee) which is fed by the River Jordan, a coastal aquifer which extends south to the occupied Gaza Strip, an aquifer under the occupied West Bank, and a pumped supply of 400 Mm^3 per year from Lebanon's Litani River. Israel's water demand of over 2000 Mm^3 per year, serving a population of 6.5 million, now exceeds the sustainable yield of the available water resources. The 120 mile-long Shatt al Arab waterway, into which the Tigris and Euphrates empty at Al Qurnah, has been the subject of a longstanding frontier dispute between Iran and Iraq. This dispute dates back some 400 years or more and has been featured in over 25 wars. A 1937 treaty established the frontier at low water mark on the Iranian side of the waterway, but the line did not satisfy Iran, and it was not until 1975 that the thalweg was agreed upon. This arrangement, however, lasted only four years--until the start of the war between the two countries. After the war in 1988, the frontier returned to the thalweg of the waterway.

It is clear that an adequate number of sound streamflow stations will be needed in the Middle East to ensure that water rights agreements, or treaties still to be formulated, are preserved by the participating nations.

Advanced Technology

The most important technological advances taking streamflow measurement into the 21st Century are:

- The ultrasonic total flow streamflow system.
- The electromagnetic total flow streamflow system.
- Solid state recorders, particularly those combined with shaft encoders recording stage or head.
- The broad-band acoustic Doppler current profiler (ADCP) (Figure 11).
- The electromagnetic (point velocity current meter).

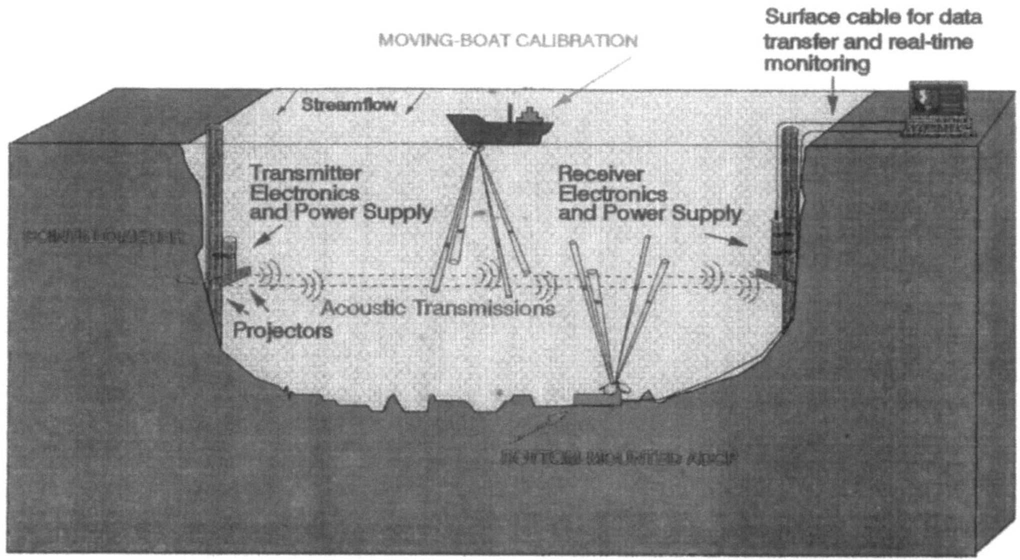

Figure 11. The broad-band acoustic Doppler current profile (BB-ADCP) and the scintillometer. (Figure reproduced courtesy of USGS).

- The ultrasonic "look-down" transducer to measure head in small open channels (Figure 12) (Current design still needs improvement).
- Flow in sewers and culverts by combined velocity/depth module (Figure 13).
- Lap top computers.

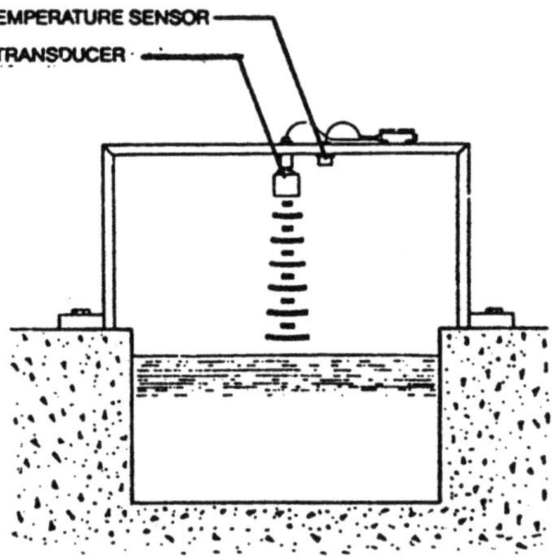

Figure 12. Schematic diagram of the ultrasonic "look-down" transducer with temperature compensation for measuring head in open channels.

Each of the above systems are now in operation, some have been for several years, others are at an earlier stage of development. In addition, weather radar and satellites will have indirect, but important, input to streamflow measurement.

Research Requirements

It is important that, notwithstanding the new technological advances, research into existing and new methods of streamflow should continue into the next century. There are still (and probably always will be) problems to be solved, methods to be improved, and advantage to be taken from new and improved electronics technology. There is also still testing (and improvement) to be carried out on the electromagnetic total flow system and the electromagnetic point velocity meter. Continued testing is also needed on the ADCP. More examination and testing is required on the ultrasonic look-down systems, especially with respect to temperature effects, wind effects, and damping.

New research is needed on the calibration of compound measuring weirs without divide walls. Present international standards restrict the use of compound weirs to those having divide piers. Compound weirs without divide piers suffer from complex three-dimensional flow, due to the interaction between the flows in adjacent sections. Orifices and sluice gates have found increased use in measuring compensation water, river abstractions, or discharges, but new work is required into shape standardization and related coefficients of discharge.

It is hoped that further research will be carried by the U.S. Geological Survey out on the scintillometer. The present ultrasonic total flow system directs pulses of ultrasound diagonally across the river in order to achieve a component velocity in the direction of the current. The scintillometer employs a set of transmit and receive transducers along two parallel acoustic paths perpendicular to the current. The principle of the method is that the downstream path will see the same fluctuations as the upstream path, but delayed in time according to the velocity perpendicular to the paths. This time delay, along with the spacing of the parallel paths, allows the computation of the current velocity perpendicular to the acoustic beam (Figure 11).

Another system which demands further research is the SENSANET, so called because it consists of a number of submerged vertical "strings", flexible or rigid, anchored to the river bed along the cross-section (Figure 14). Sensors attached to the "strings" measure velocity and depth. In principle, the system design allows each sensor to maintain automatically its position in the vertical, irrespective of the changing water levels.

In addition to the above possibilities for further research, attention should be given to improving existing equipment such as suspension equipment, cableways, rotating element current meters' performance. and their rating in tanks. Although there is, as yet, no dedicated satellite for water resources (Hydrosat ?), research will continue into possible operational systems, using existing and proposed satellites to assess and exploit the world's water resources.

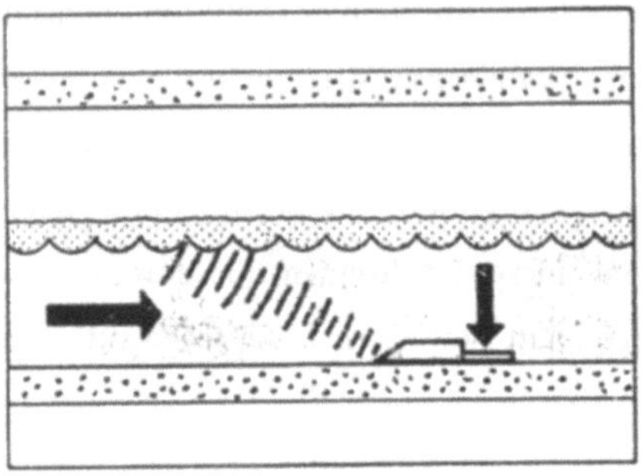

Figure 13. Flow measurement in sewers and open channels using dual sensors to measure depth of flow ("look-up") and velocity (courtesy of Detectronic, Ltd.)

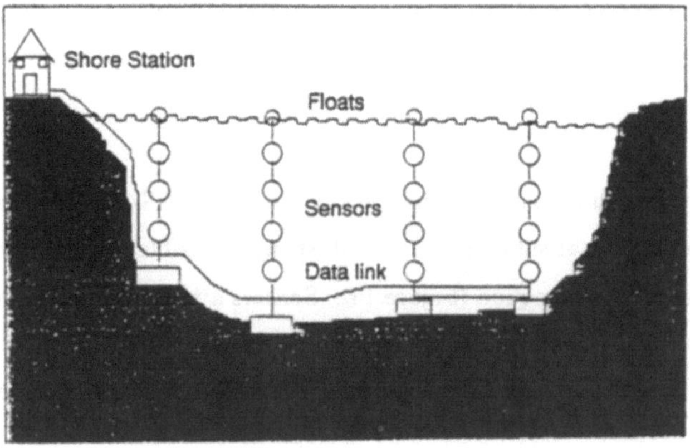

Figure 14. Schematic illustration of a fully automated system for flow gauging capabilities (*SENSANET D*). (Figure reproduced courtesy of Aqua Data Services, Ltd.).

References

Ackers, P, WR White, JA Perkins, AJM Harrison (1978) Weirs and Flumes for Flow Measurement, John Wiley and Sons.

Bos, MG (1976) "Discharge measurement structures", Delft Hydraulics Laboratory, No. 161 Wageningen, The Netherlands.

Cobb, ED (1993) "Acoustic Doppler Current Profiler", Vol. 4, No. 6, Butterworth-Heinemann, Oxford, United Kingdom.

Department of the Environment, United Kingdom (1990) "The benefit-cost of hydrometric data: river flow gauging", Report by CNS Scientific and Engineering Services, Reading; Published by Foundation for Water Research, United Kingdom.

Herschy, RW (1985) Streamflow Measurement, Elsevier Applied Science Publishers, London and New York.

Herschy, RW (1993) "National and International Standards in Streamflow Measurement", Vol. 4, No. 1, Butterworth, Heinemann, Oxford, United Kingdom.

Herschy, RW (1993) Disputed Frontiers, The Book Guild, Lewes, United Kingdom.

ISO 6157 (1980) "Measurement of fluid flow by means of orifice plates, nozzles, and venturi tubes inserted in circular cross-section conduits running full", Geneva, Switzerland.

Marsh, TJ (1993) "Aspects of river flow measurement and hydrometric data interpretation in the UK with particular reference to low flows", Vol. 4, No. 1, Butterworth-Heinemann, Oxford, United Kingdom.

National Rivers Authority (NRA) (UK) (1993) "Remote Sensing of Snow by Satellite", Report of research for the NRA, University of Bristol, United Kingdom.

Reynolds, P (1991) "Liquid Gold", New Civil Engineer, Thomas Telford, London 14 November.

United States Geological Survey (1991), National Water Summary 1988-89, Hydrological Events and Floods and Droughts, Water Supply Paper 2375, Washington DC.

APPENDIX

Table 1. Guidelines for the Selection of an Appropriate Streamflow Method (courtesy of the British Standards Institution)

Method			Criteria						Uncertainty	
No.	Description	Relevant International Standard	Width	Depth	Velocity	Sediment load	Approach condition	Time factor	Minimum percentage	Comment
1	Velocity-area, by wading	ISO 748	L, M, S	S	S, M		b, c, d	J, K	± 3	A, B
2	Velocity-area, from bridge	ISO 748	M, L	M, L	M, L		b, c, d	K	± 3	A, B, C, D
3	Velocity-area, cableway	ISO 748	M, L	M, L	M, L		b, c, d	K	± 4	A, B, C
4	Velocity-area, static boat	ISO 748	M, L	M, L	M, L		b, c, d	K	± 4	A, B, C, E
5	Velocity-area, moving boat	ISO 4369	L	M, L	M, L		b, c, d	K	± 6	A, B, E
6	Velocity-area, floats	ISO 748	M, L	M, L	M, L, S		b, c, d	K	± 10	F
7	Slope-area	ISO 1070	M, L	M, L	M, L		b, c, d	K, N	± 10	Q
8	Ultrasonic	ISO 6416	M, L	M, L	M, L, S	R	b, c, d	G, J, H	± 5	
9	Electromagnetic	ISO/TR 9213	M, S	S, M	S, M		b, d	G, H, J	± 5	T
10	Dilution, chemical, continuous injection	ISO 555/1	S, M	S, M	S, M		c, g, k	K, N	± 3	
11	Dilution, chemical, sudden injection	ISO 555/2	S, M	S, M	S, M		c, g, k	K	± 3	
12	Dilution, radioactive, tracer, sudden injection	ISO 555/3	S, M	S, M	S, M		c, g, k	K	± 3	
13	Dilution, radioactive tracer, continuous injection	ISO 555/3	S, M	S, M	S, M		c, g, k	K, N	± 3	
14	Cubature	ISO 2425						K	± 10	H
15	Thin-plate weirs, sharp crest, V-notch	ISO 1438/1	S	S	M, S	I	a, b, e, j	J, G	± 3	
16	Thin-plate weirs, sharp crest, rectangular, suppressed	ISO 1438/1	S	S	M, S	I	a, b, e, f, j	J, G	± 1	
17	Thin-plate weirs, sharp crest, rectangular	ISO 1438/1	S	S	M, S	I	a, b, e, f, j	J, G	± 1	
18	Weirs, broad-crested with sharp upstream edge	ISO 3846	M, S	S	M, S	I	a, b, e, h, j	J, G	± 5	
19	Weirs, broad-crested with rounded upstream edge	ISO 4374	M, S	S	M, S	I	a, b, e, h, j	J, G	± 5	
20	Weirs, triangular profile	ISO 4360	M, S	S	M, S	I	a, b, e, j	J, G	± 5	
21	Weirs, triangular profile, flat-V	ISO 4377	M, S	S	M, S	I	a, b, e, j	J, G	± 5	
22	Weirs, V-shaped, broad-crested	ISO 8333	M, S	S	M, S	I	a, b, i	J, G	± 5	
23	Flumes, rectangular	ISO 4359	M, S	S	M, S	I	a, b	J, G	± 5	
24	Flumes, trapezoidal	ISO 4359	M, S	S	M, S	I	a, b	J, G	± 5	
25	Flumes, U-shaped	ISO 4359	M, S	S	M, S	I	a, b, i	J, G	± 5	
26	Free overfalls, rectangular channels (end-depth method)	ISO 3847	M, S	M, S	M, S		a, b	J, G	± 10	
27	Free overfalls, non-rectangular channels (end-depth method)	ISO 4371	M, S	M, S	M, S		a, b	J, G	± 10	

Table 2. Explanation of the Symbols Used in Table 1.

Symbol	Definition
a	Flow should be subcritical
b	Flow should have no cross-currents
c	Channel should be relatively free from vegetation
d	Channel should be fairly straight and uniform in cross-section
e	Channel should be fairly straight and symmetrical in cross-section for about 10 channel widths upstream
f	Channel should have vertical walls and a level floor for a distance upstream of not less than 10 times the width of the nappe at maximum head
g	Flow in the channel should be turbulent (even including a hydraulic jump) to ensure mixing
h	Channel should be rectangular to a distance upstream of at least twice the maximum head
i	Channel should be nearly U-shaped
j	Velocity distribution should be fairly uniform
k	Channel should be free from recess in the banks and depressions in the bed
A	For velocity-area method, with velocity observed at 0.6 times the depth, or with two-point method, the minimum uncertainy may be up to 5%
B	For velocity-area method, with velocity observed at surface, the minimum uncertainty may be up to 10%
C	Corrections may be required because of distance or air- and wet-line effects
D	Major error can be caused by pier effects
E	Major error can be due to drift, obstruction of boat, and heaving action
F	This method is recommended for use only when the effect of the wind is small and when no other will serve. Such conditions are likely to be so variable that no representative accuracies can be quoted; however, the accuracy of this method usually is lower than conventional methods using current meters and higher than the slope-area method.
G	Method suitable for more frequent discharge measurements
H	Method suitable for tidal waterways
I	Heavy sediment concentration not permissible
J	Quick method (less than 1 h)
K	Slow method (1 to 6 h)
L	Large width (more than 50 m) or high velocity (more than 3 m/s) or large depth (more than 5 m)
M	Medium width (between 5 and 50 m) or medium velocity (between 1 and 3 m/s) or medium depth (between 1 and 5 m)
N	Very slow method (more than 6 h)
Q	Approximate method used when velocity-area method is not feasible and slope can be determined with sufficient accuracy
R	Suspended material concentration should continue to be low in order to avoid too large a loss of acoustic signal; for the same reason, the flow should be free from bubbles.
S	Narrow width (less than 5 m) or shallow depth (less than 1 m) or low velocity (less than 1 m/s)
T	May be used in rivers with weed growth and moving bed material

Model and Data Requirements for Simulation of Runoff and Land Surface Processes

Jens Christian Refsgaard
Chief Hydrologist
Danish Hydraulic Institute (DHI)
Agern Alle 5
DK-2970 Horsholm
DENMARK

1. Introduction

Hydrological models are usually applied under stationary climatological conditions and for purposes other than predicting the effects of climate change. Hence, important questions with regard to application of hydrological models within the global environmental change context are: (a) whether such models are applicable for this purpose; (b) which special test schemes are necessary in order to validate hydrological models for this purpose; (c) which special requirements this type of application puts to hydrological models; and (d) what are the data requirements for application of hydrological models in this context.

Traditionally, hydrological simulation models are classified in two main groups, namely the lumped, conceptual type and the distributed, physically based type. In Section 2, a brief description of three different hydrological models are given with respect to model structure and data requirements. Two of the models are typical representatives of the above two main groups, while the third one represents a mixed approach.

In Section 3, a test procedure recommended by Klemes (1985) is described briefly. This procedure is considered suitable for testing the applicability of hydrological models for predicting the effects of climate change.

In order to test the capability of different types of hydrological models to predict the hydrological regime under changed climatological conditions, a comprehensive and rigorous testing scheme has been carried out using three hydrological models on data from catchments in Zimbabwe. The data from Zimbabwe were chosen so that data sets

NATO ASI Series, Vol. I 46
Land Surface Processes in Hydrology
Trials and Tribulations of Modeling and Measuring
Edited by S. Sorooshian, H. V. Gupta, and S. C. Rodda

both from very wet and very dry periods were available. The models were then subject to calibration in the wet period and subsequent validation in the dry period in the same catchment. The capability of the models to simulate runoff without prior calibration on data from the same catchment was also investigated. Some key results are presented in Section 4.

Section 5 presents views on how remote sensing data can most optimally be coupled with hydrological models and provide mutual benefits. Finally, in Section 6, a discussion is given on the requirements to hydrological models for coupling to Global Circulation Models.

2. Brief Description of Three Different Types of Hydrological Models

The following three models have been used for the Zimbabwe study:

- NAM: a lumped, conceptual rainfall-runoff model
- WATBAL: a semi-distributed hydrological model
- MIKE SHE: a fully distributed, physically based model

The NAM and MIKE SHE models can be characterized as very typical of their respective classes, while the WATBAL falls in between these two classes. All three models are being used on a routine basis at the Danish Hydraulic Institute (DHI) in connection with consultancy and research projects.

2.1 NAM

NAM is a traditional hydrological model of the deterministic, lumped, conceptual type operating by continuously accounting for the moisture contents in four mutually interrelated storages. The NAM model was originally developed at the Technical University of Denmark (Nielsen and Hansen, 1973) and has been modified and applied extensively by DHI in a large number of engineering projects covering all climatic regimes of the world. Furthermore, the NAM has been transferred to more than 100 other organizations worldwide as part of DHI's MIKE 11 generalized river modeling package.

The structure of the model is illustrated in Figure 1. The input data consist of:

- Meteorological time series of precipitation on a daily or finer basis, potential evaporation on at least a monthly basis, and, if the snow routine is activated, temperature.
- A set of model parameters defining the hydrological characteristics of the catchment.
- Groundwater abstractions (optional).

In its present version, the NAM has a total of 17 parameters; however, in most cases, only about 10 of these are adjusted during calibration.

2.2. *WATBAL*

WATBAL was developed in the early 1980s by DHI in an attempt to provide a physically based, distributed hydrological model capable of using readily available data on land surface properties (topography, vegetation, soil), yet simple enough to allow large-scale applications within reasonable computational requirements. In the following, the model is introduced briefly, while more detailed information may be obtained in Knudsen et al. (1986).

WATBAL has been designed to account for the spatial and temporal variations of soil moisture. Based on distributed information on meteorological conditions, topography, vegetation, and soil types, the catchment area is divided into a number of hydrological response units as illustrated in Figure 2, with each unit being characterized by a different composition of the above features.

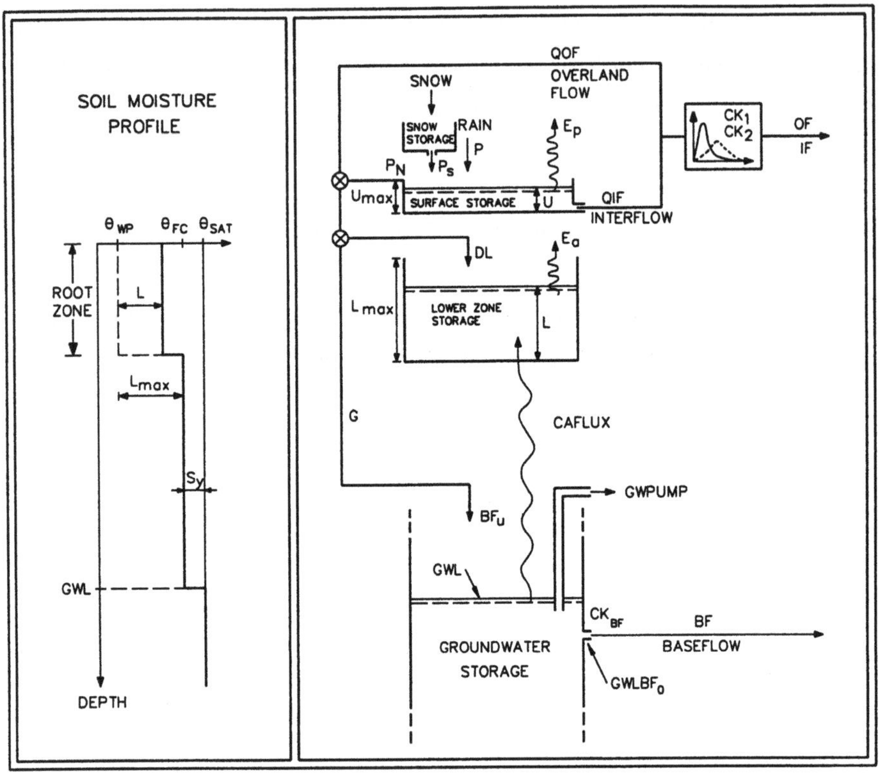

Figure 1. Structure of the NAM model.

These units are used to provide the spatial representation of soil moisture, while temporal variations within each unit are accounted for by means of empirical relations for the processes affecting soil moisture, using physical parameters particular to each unit.

For a description of subsurface flows, a simple lumped, conceptual approach is applied, using a cascade of linear reservoirs to account for the interflow and baseflow components (Figure 3).

In summary, WATBAL provides a physically based, distributed description of the processes affecting soil moisture (interception, infiltration, evapotranspiration, and percolation), while a conceptual approach is used to represent subsurface flows. In this way, WATBAL may be viewed as a compromise between limitations on data availability, the complexity of hydrological response at catchment scale, and the advantages of model simplicity.

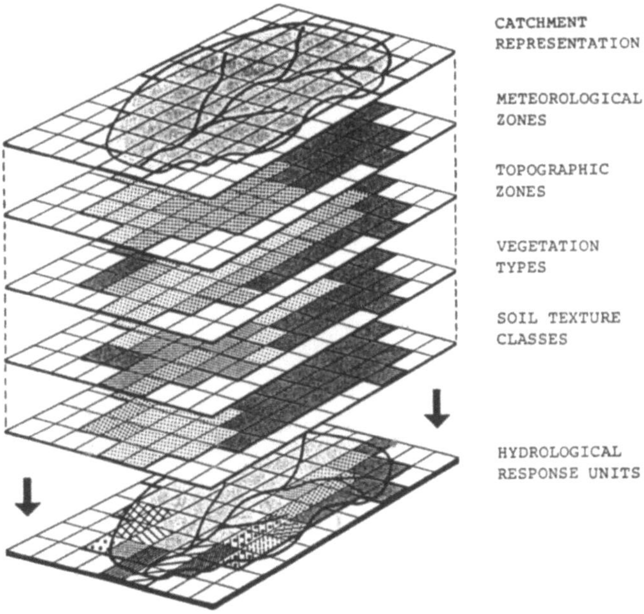

Figure 2. WATBAL representation of catchment characteristics and definition of hydrological response units.

The exogenous data required to operate the model comprise time series of rainfall on at least a daily basis and monthly evaporation data, while necessary model parameters are given in Table 1.

2.3 MIKE SHE

The European Hydrological System (SHE) was developed in a joint effort by the Institute of Hydrology (UK), SOGREAH (France), and the Danish Hydraulic Institute (DHI). It is a deterministic, fully distributed, and physically based modeling system for describing the major flow processes of the entire land phase of the hydrological cycle. A description of the SHE is given in Abbott et al. (1986a,b). Since 1987, the SHE has been further developed independently by the three respective organizations which now are University of Newcastle (UK), Laboratoire d'Hydraulique de France, and DHI. DHI's version of the SHE, known as the MIKE SHE, represents significant new developments with respect to user interface, computational efficiency, and process descriptions.

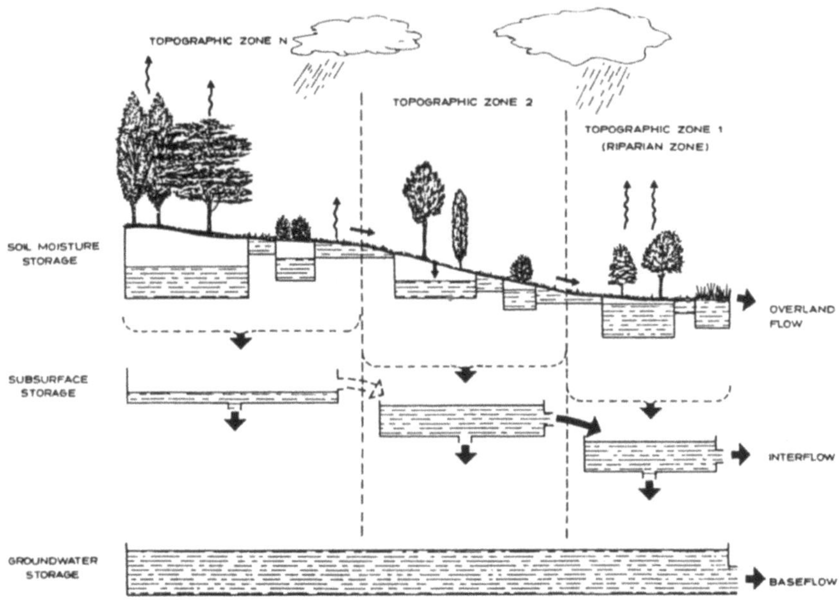

Figure 3. Principal structure of WATBAL.

MIKE SHE solves the partial differential equations for the processes of overland and channel flow and unsaturated and saturated subsurface flow. The model is completed by a description of the processes of snow melt, interception, and evapotranspiration. The flow equations are solved numerically using finite difference methods.

In the horizontal plane, the catchment is discretized in a network of grid squares. The river system is assumed to run along the boundaries of these. Within each square, the soil profile is described in a number of nodes which, above the groundwater table, may become partly saturated. Lateral subsurface flow is only considered in the saturated part of the profile. Figure 4 illustrates the structure of the MIKE SHE. A description of the methodology and some experiences of model application are presented in Refsgaard et al. (1992) and Jain et al. (1992).

Table 1. Model Parameters Required by WATBAL

TOPOGRAPHY	Within each topographic zone: (depends on selected mode) o length of flow plane o slope o manning number o depression storage
VEGETATION	For each type of vegetation: o leaf area index (time varying) o root depth (time varying)
SOIL TYPES	For each texture class: o wilting point o field capacity o total porosity o saturated conductivity o average suction
SUB-SURFACE REGIME	For each topographic zone: o threshold value for interflow generation o two time constants (interflow/percolation outlets) Groundwater storage: o groundwater area relative to catchment area o time constant of base flow outlet

The spatial and temporal variations in the catchment characteristics and meteorological input are provided in a series of two-dimensional matrices of grid square codes. A number of attributes describing either parametric data or input data is attached to each code. A list of the most important data and parameters is shown in Table 2.

The distributed description in the MIKE SHE allows the user to include and test against spatially varying data. MIKE SHE is a multioutput model which, besides discharge in any river link, also produces information about water table elevations, soil moisture contents, infiltration rates, evapotranspiration, etc. in each grid square.

MIKE SHE is usually categorized as a physically based model. Strictly speaking, the characterization is only correct if the model is applied on an appropriate scale. A number of scale problems arise when the MIKE SHE is set up on a regional scale. In addition, if there is a considerable uncertainty attached to the basic information, and if the spatially and temporal varying variables (such as groundwater table elevations) cannot be validated against observations, MIKE SHE is, on that particular setup, not physically based but rather a detailed conceptual model. In this case, the calibration procedure is usually to adjust the parameters with the largest uncertainties attached, within a reasonable range.

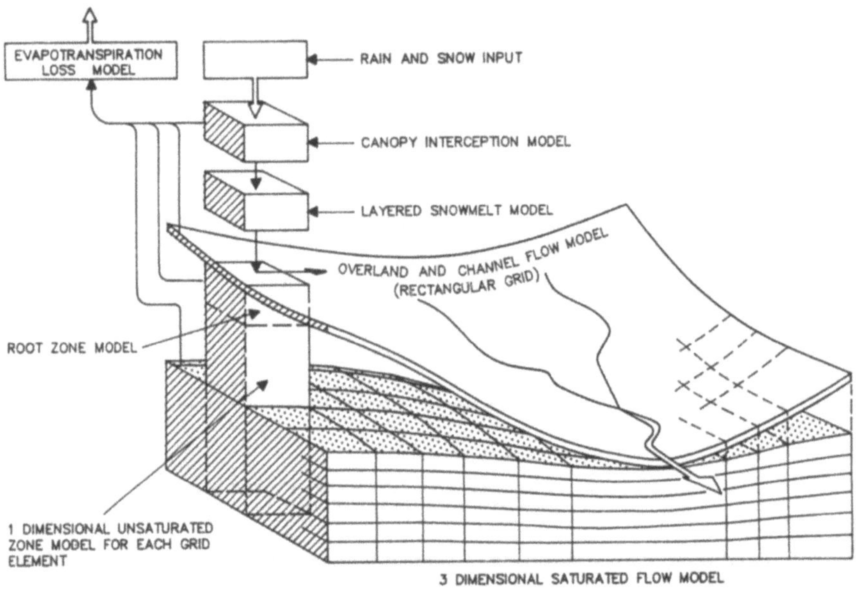

Figure 4. Schematic presentation of the MIKE SHE.

An illustration of the outputs simulated by a distributed model like the MIKE SHE is given in Figure 5. In this figure, selected results from a two-year simulation of the 422 km^2 Karup catchment are shown. To the left, the spatial distribution of actual evapotranspiration, vegetation type, and depth to groundwater table on July 6, 1976 is shown. The right column shows the temporal variation of rainfall input, potential and actual evapotranspiration, leaf area index, and soil moisture content in the upper 1.5 m of one of the 500 m x 500 m grids.

Table 2. Important Data and Parameters Required by the MIKE SHE

GENERAL	For each grid square or river link: o topography o impermeable bed o channel geometry o meteorological station code o vegetation type code o soil profile code
VEGETATION	For each vegetation type: o leaf area index (time varying) o rooting depth (time varying)
SOIL (unsaturated zone)	For each soil type: o retention curve o hydraulic conductivity function
SOIL (saturated zone)	Saturated hydraulic conductivity in horizontal and vertical directions: o storage coefficients o drainage depth o drainage time constant
OTHER	o roughness coefficients in each grid square (surface) and for river links o groundwater abstraction data

3. Testing Schemes for Validation of Hydrological Models

For a long time, the testing of hydrological models through validation on independent data has been emphasized by the World Meteorological Organization (WMO). In their pioneering studies (WMO, 1975, 1986), several hydrological simulation models were tested on the same data from different catchments. The actual testing, however, only included the standard split-sample test comprising an initial calibration of the model and subsequent validation based on data from an independent period.

WMO and UNESCO have included a project for "Testing the transferability of hydrological simulation models" (Project D.5) as part of the World Climate Programme--Water, WMO (1985). The study by DHI (1993) may be viewed as a contribution to the WCP-Water/D.5-project.

The hierarchial testing scheme of Klemes (1985) appears to be suitable for testing the capability of a model to predict the hydrological effects of climate change. Klemes (1985) distinguished between simulations conducted at the same station (catchment) used for calibration and simulations conducted for ungauged catchments. He also distinguished between cases where climate and land use are stationary and cases where they are not.

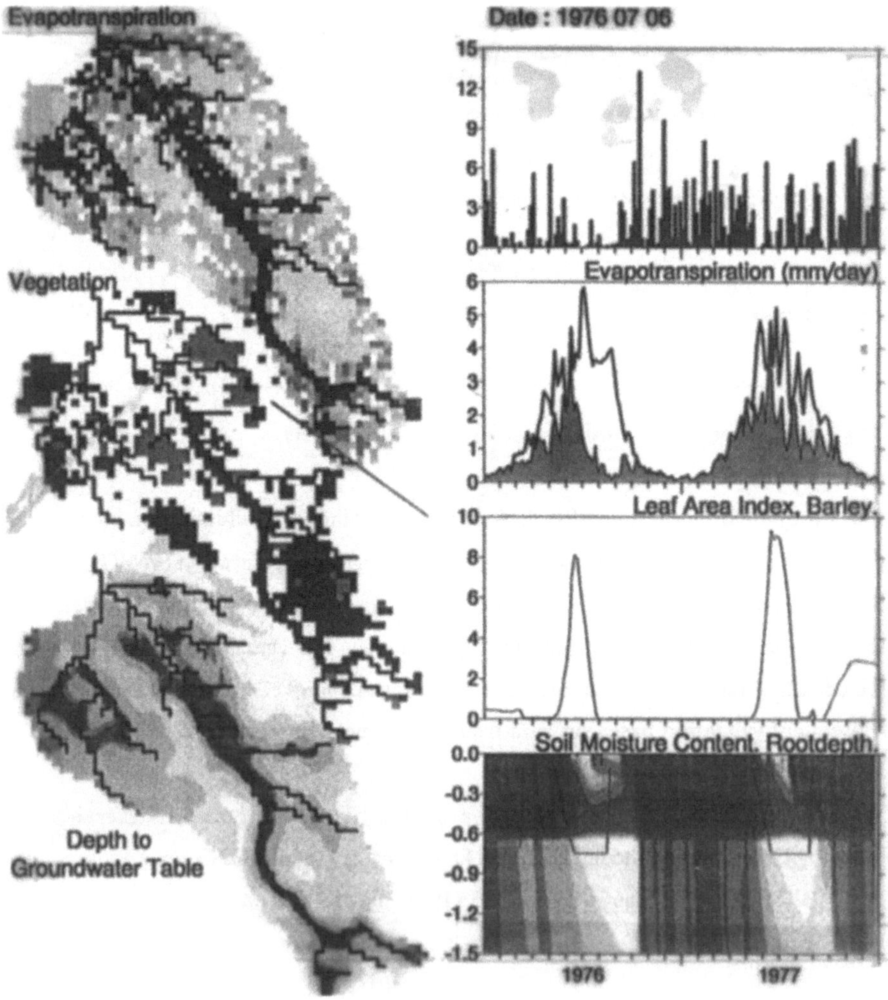

Figure 5. Illustration of selected output generated by MIKE SHE for the 422 km² Karup catchment in Denmark.

This combines to the definition of four basic categories of typical modeling tests:

(a) Split-sample test: Calibration of model based on 3-5 years of data and validation on another period of similar length.

(b) Differential split-sample test: Calibration of model based on data before catchment change occurred, adjustment of model parameters to characterize the change, and validation on the subsequent period.

(c) Proxy-basin test: No direct calibration allowed, but advantage of information from other gauged catchments may be taken. Hence, validation will comprise identification of a gauged catchment deemed to be of similar nature as the validation catchment, initial calibration, transfer of model including adjustment of parameters to reflect actual conditions within validation catchment, and validation.

(d) Proxy-basin differential split-sample test: Again, no direct calibration is allowed, but information from other catchments may be used. Thus, validation will comprise initial calibration on other relevant catchment, transfer of model to validation catchment, selection of two parameter sets to represent the period before and after the change, and subsequent validations on both periods.

4. Test of Hydrological Models for Runoff Simulations Under Varying Climate Conditions in Zimbabwe

The following is based on results from a research project conducted at Danish Hydraulic Institute (DHI, 1993).

4.1 Selected Catchment

The three catchments which have been selected for the model tests in Zimbabwe are Ngezi-South, Lundi, and Ngezi-North. The locations of the catchments are shown in Figure 6.

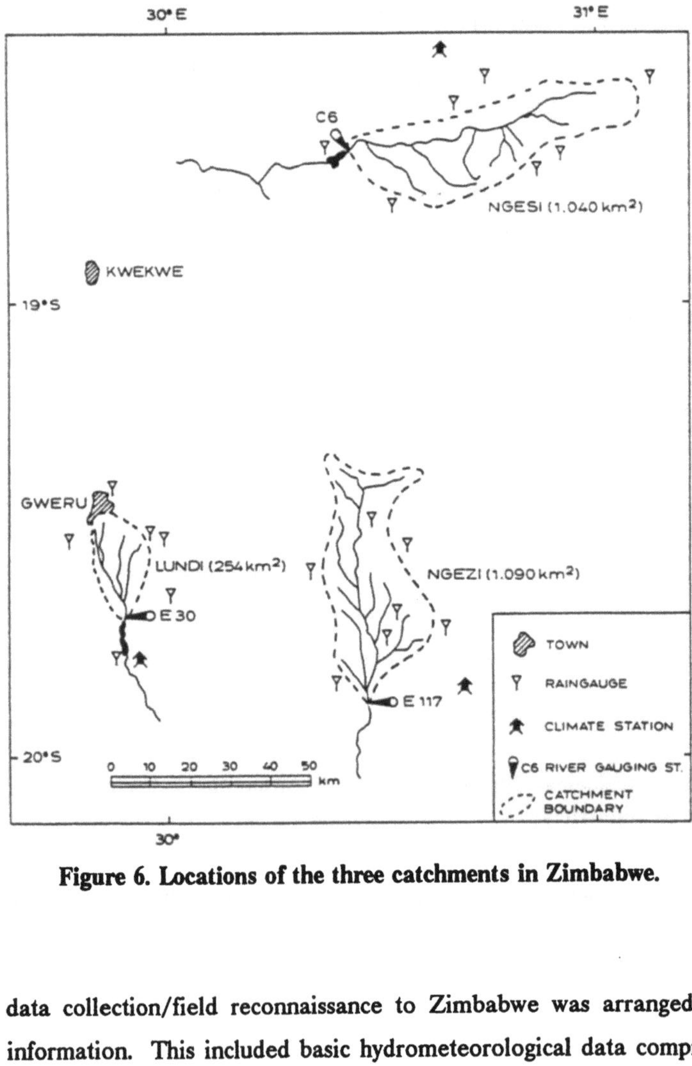

Figure 6. Locations of the three catchments in Zimbabwe.

A brief data collection/field reconnaissance to Zimbabwe was arranged to obtain relevant information. This included basic hydrometeorological data comprising daily series of rainfall and runoff and monthly series of pan evaporation. Detailed information on land use was obtained through subcontracting the University of Zimbabwe to prepare land use maps based upon 1:25,000 aerial photographs, while information on soil and vegetation characteristics was collected from various institutions and relevant literature. Furthermore, 1:50,000 topographical maps were collected and digitized.

Finally, available data on vegetation characteristics, soil characteristics, hydrogeology, and water rights were obtained. A more detailed description is given in DHI (1993).

Some of the key features for the three catchments are presented in Tables 3 and 4. It is noticed from the rainfall and runoff figures in Table 3 that there are very large variations between some of the periods. From Table 4, it appears that there are significant differences in the vegetation and soil characteristics from catchment to catchment.

Table 3. Catchment Areas, Modeling Periods, Rainfall, and Runoff Values for the Three Zimbabwean Test Catchments

Catchment	Area (km2)	Periods	Average rainfall (mm/year)	Average runoff (mm/year)
Ngezi-South	1090	1971/72 - 74/75	896	179
		1975/76 - 78/79	873	128
Lundi	254	1971/72 - 75/76	915	171
		1981/82 - 83/84	497	8
Ngezi-North	1040	1977/78 - 78/79 + 81/82 - 83/84	607	47

4.2 Testing Scheme

The testing scheme for the Zimbabwe study carried out in DHI (1993) is illustrated in Table 5. The testing of the involved models has been undertaken in parallel and in the following sequence:

A: Split-sample test based on data from Ngezi-South comprising an initial calibration of the models and a subsequent validation using data for an independent period.

C: Proxy-basin test. Transfer of models to the Lundi catchment, adjustment of parameters to reflect the prevailing catchment characteristics, and validation without any calibration.

(C): Same as above, but adjusted by allowing model calibration based on one year of runoff data.

B: Differential split-sample test. Model calibration based on data from an initial calibration period and validation based on data from a subsequent period. The differential nature of this test is justified by the fact that the later independent period includes three successive years (1981/82-1983/84) with a markedly lower rainfall than otherwise and, hence, represents a nonstationary climate scenario.

D: Proxy-basin differential split-sample test. Transfer of models to the Ngezi-North catchment, adjustment of parameters to represent the catchment characteristics, and validation by runoff simulation over a nonstationary climate period.

(D): Same as above, while allowing models to be calibrated using a short-term (one year) record.

In this paper, only the results from tests A, C, and B are presented.

4.3 Performance Criteria

For measuring the performance of the models for each test, a standard set of criteria has been defined. The criteria have been designed with the sole purpose of measuring how closely the simulated series of daily flows agree with the measured series. Due to the generalized nature of the defined model validations, it has been necessary to introduce several criteria for measuring the performance with regard to water balance, low flows, and peak flows.

The standard set of performance criteria comprises a combination of graphical plots and numerical measures. The graphical diagrams used include joint plots of the simulated and observed hydrographs, scatter diagram of monthly runoffs, flow duration curves, and scatter diagram of annual maximum discharges. To support the graphical presentations, various numerical measures are computed, including the overall water balance, the Nash-Sutcliffe coefficient (R2), and an index (EI) measuring the agreement between the simulated and observed flow duration curves. Furthermore, additional measures for each hydrological year are computed.

Table 4. Land Use Vegetation and Soil Characteristics Estimated from Available Information and a Brief Field Visit

	Ngezi-South	Lundi	Ngezi-North
Land use/vegetation (area %)			
Dense/closed woody vegetation	7	13	10
Open woody vegetation	36	25	35
Sparse woody vegetation	14	19	14
Grassland	11	39	16
Cropland	29	3	19
Abandoned cropland	2	0	6
Rock outcrops	1	0	0
Soil depth range (m)	0 - 2.5 m	0 - 1 m	0.5 - 6 m
Saturated hydraulic conductivity in root zone soil (mm/hour)	Range: 1 - 250 Average: 80	Range: 1 - 70 Average: 60	Range: 2-100 Average: 50
Available water content in root zone soil (vol %)	Range: 10 - 14 Average: 12	Range: 10 - 12 Average: 11	Range: 9 - 29 Average: 17

The coefficient R2, introduced by Nash and Sutcliffe (1970), is computed on the basis of the sequence of observed and simulated monthly flows over the entire testing period (perfect agreement for R2 = 1):

$$R2 = 1 - \sum_{m=1}^{M} \left(Q_m^o - Q_m^s\right)^2 / \sum_{m=1}^{M} \left(Q_m^o - \bar{Q}^o\right)^2$$

where:

M = total number of months;

Q_m^s = simulated monthly flows;

Q_m^o = observed monthly flows; and

$\bar{Q}^o$ = average observed monthly flows over the entire period.

Table 5. Model Validations, Zimbabwe

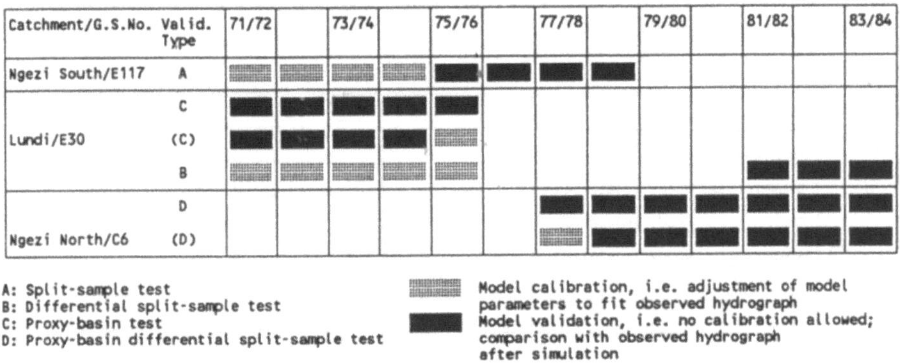

The flow duration curve error index, EI, provides a numerical measure of the difference between the flow duration curves of simulated and observed daily flows (perfect agreement for EI = 1):

$$EI = 1 - \int \left[f_o(q) - f_s(q) \right] dq \,/ \int f_o(q)\, dq$$

where:

$f_o(q)$ = flow duration curve based on observed daily flows; and

$f_s(q)$ = flow duration curve based on simulated daily flows.

Thus, the above criteria measure the extent to which the models are able to provide an accurate representation of the overall water balance, whether major discrepancies for individual years occur, the overall accuracy of the simulated series of monthly flows, and its capability to represent the overall pattern of the daily flows, that is, the frequency of occurrence of low, medium, and high flows. In spite of its incompleteness, the above criteria provide a reasonable summary of the overall model performance.

4.4 Results of Model Validations

(a) Split-sample test: This test is based on data from Ngezi-South and comprises an initial calibration of the models and a subsequent validation using data for an independent period. The main results are summarized in Table 6.

As indicated by this table, the performance of all three models is generally very similar. All models are able to provide a close fit to the recorded flows for the calibration period while, for the independent validation period, the performance is somewhat reduced as expected. The reduction is, however, limited, and all models are able to maintain a very good representation of the overall water balance, the interannual and seasonal variations, as well as the general flow pattern.

(c) Proxy-basin test (ungauged catchment): This test comprises a transfer of models to the Lundi catchment, adjustment of parameters to reflect the prevailing catchment characteristics, and validation without any calibration.

The proxy-basin test was arranged to test the capability of the different models to represent runoff from an ungauged catchment area; hence, no calibration was allowed prior to the simulation. For all models, three alternative runoff simulations were prepared, reflecting an expected low, central, and high estimate, respectively.

All models have used the experience from the Ngezi-South calibrations in combination with the available information on the particular catchment characteristics for Lundi. While the NAM model has used this information in a purely subjective manner to revise model parameters, both the WATBAL and MIKE SHE models have used this information directly for the model setup. The estimates prepared by the latter two models have, however, also been influenced by the individual modelers subjective interpretation of the available information on soil and vegetation characteristics.

Table 6. Ngezi-South. Summary of Split-Sample Test Results. All flows are in mm/year.

CALIBRATION			QSIM - QOBS		
Year	Rain	QOBS	NAM	WATBAL	MIKE SHE
1971/72	890	131	-19	-11	-11
72/73	317	2	1	- 1	7
73/74	1,290	349	14	- 5	-10
74/75	1,087	236	5	8	2
Mean:	896	179	0	- 2	- 3
Mean | QSIM - QOBS| :			10	6	8
R2			.97	.96	.97
EI			.88	.95	.91

VALIDATION			QSIM - QOBS		
Year	Rain	QOBS	NAM	WATBAL	MIKE SHE
1975/76	879	90	3	24	5
76/77	872	116	-28	2	7
77/78	1,131	245	26	37	67
78/79	609	59	-13	2	- 9
Mean:	873	128	- 3	16	18
Mean | QSIM - QOBS| :			18	16	22
R2			.89	.86	.84
EI			.74	.86	.80

As an example of the model performance, the hydrograph simulated by the MIKE SHE is shown in Figure 7 for two of the years, together with the flow duration curves and the scatter diagram of monthly discharges for the entire five year validation period. A summary of the main results of the proxy-basin tests is given in Table 7.

In general, all models provide an excellent representation of the general flow pattern yet with some discrepancies for the small and/or larger peaks. As seen in Table 7, the best

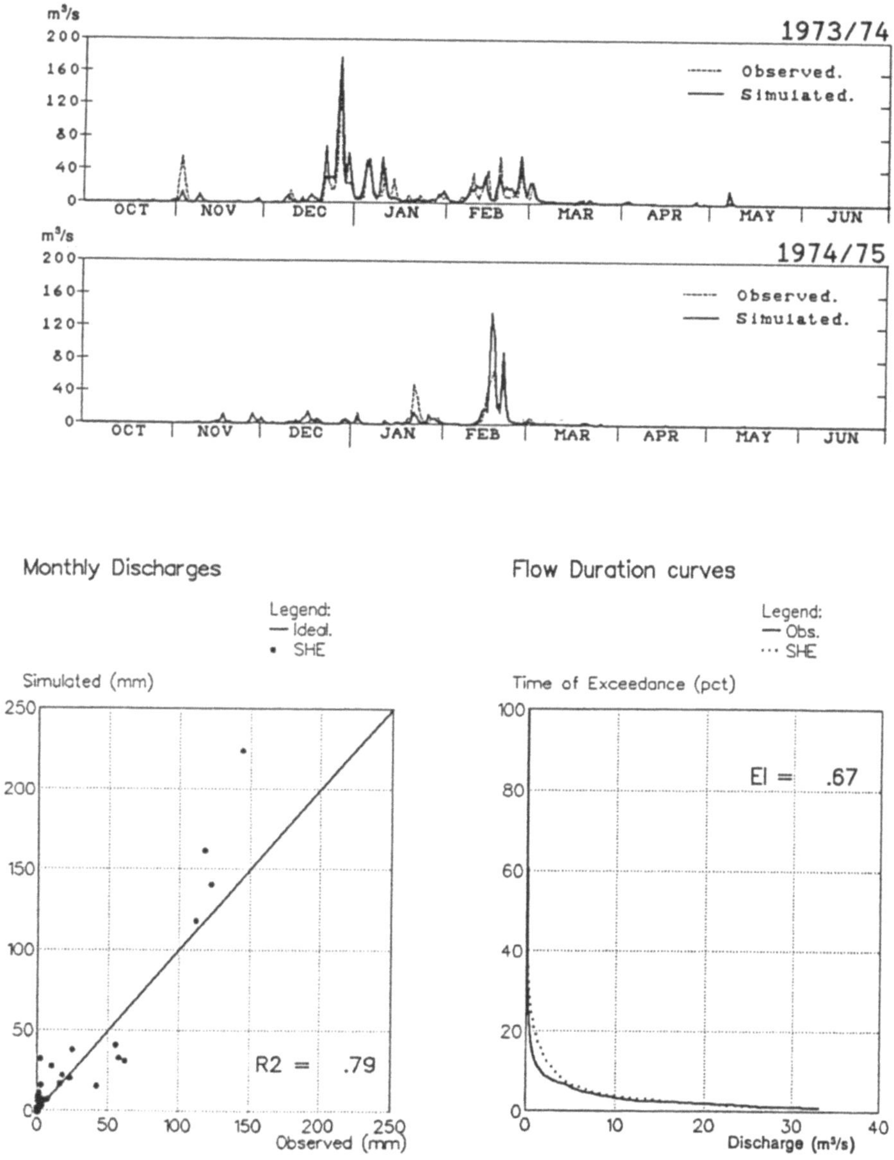

Figure 7. Lundi MIKE SHE (central estimate) proxy-basin test hydrographs from two of the years, together with flow duration curves and scatter diagram of monthly discharges for the entire period.

runs of the individual models provide a good representation of the overall water balance, while maintaining the significant interannual variability to a satisfactory degree and an overall good simulation of flows within individual months.

Table 7. Lundi. Summary of Proxy-Basin Test Results. All flows are in mm/year.

PROXY-BASIN TEST			QSIM - QOBS								
			NAM			WATBAL			MIKE SHE		
Year	Rain	QOBS	Low	Ctr.	High	Low	Ctr.	High	Low	Ctr.	High
1971/72	920	89	5	63	122	1	3	32	- 7	27	29
72/73	371	2	2	5	10	- 1	- 1	6	10	21	21
73/74	1,384	460	-44	72	151	- 78	3	34	43	98	120
74/75	1,046	217	-36	44	116	-55	-17	14	8	45	61
75/76	857	89	-23	23	77	-14	-11	11	-20	13	22
Mean:	915	171	-19	42	95	-29	- 5	19	7	41	51
Mean | QSIM - QOBS| :			22	41	95	30	7	19	18	41	51
R2			.89	.87	.57	.85	.91	.90	.86	.79	.71
EI			.66	.72	.41	.59	.76	.75	.78	.67	.63

The overall performance of the central estimates by the NAM and MIKE SHE models is somewhat reduced compared to validation runs for the Ngezi-South catchment as expected when no calibration is possible. The estimates would, however, still be very valuable for all practical purposes. For the WATBAL model, the central estimate is even better than obtained for the validation period for Ngezi-South, providing for a very accurate representation of observed runoff record.

<u>(b) Differential split-sample test (nonstationary climate)</u>: This test consists of model calibrations based on data from four wet years (1971-76) and validation on data from three very dry years (1981-84). The purpose of this test is to assess the capability of the models to do simulations during nonstationary time periods.

As an example of the model performance, the hydrograph simulated by the MIKE SHE is shown in Figure 8 for two of the years, together with the flow duration curves and scatter diagram of monthly discharges for the entire five-year validation period. A summary of the main results of the differential split-sample tests is given in Table 8.

Table 8. Lundi. Summary of Differential Split-Sample Test Results. All flows are in mm/year.

CALIBRATION			QSIM - QOBS		
Year	Rain	QOBS	NAM	WATBAL	MIKE SHE
1971/72	920	89	23	14	-16
72/73	371	2	- 1	1	- 1
73/74	1,384	460	- 3	14	28
74/75	1,046	217	- 3	- 5	3
75/76	857	89	-17	- 1	-29
Mean:	915	171	0	5	- 3
Mean | QSIM - QOBS| :			9	7	15
R2			.92	.92	.86
EI			.82	.80	.82

VALIDATION			QSIM - QOBS		
Year	Rain	QOBS	NAM	WATBAL	MIKE SHE
1981/82	416	10	- 2	4	- 6
82/83	528	7	- 3	9	0
83/84	547	7	2	8	1
Mean:	497	8	- 1	7	- 2
Mean | QSIM - QOBS| :			2	7	2
R2			.66	-1.65	.13
EI			.55	- .02	.56

As evident from the plots of the associated hydrographs, the MIKE SHE is capable of reproducing the intermittent pattern of the measured hydrograph during the very dry period comprising a small number of very small peaks and intermediate periods with nil

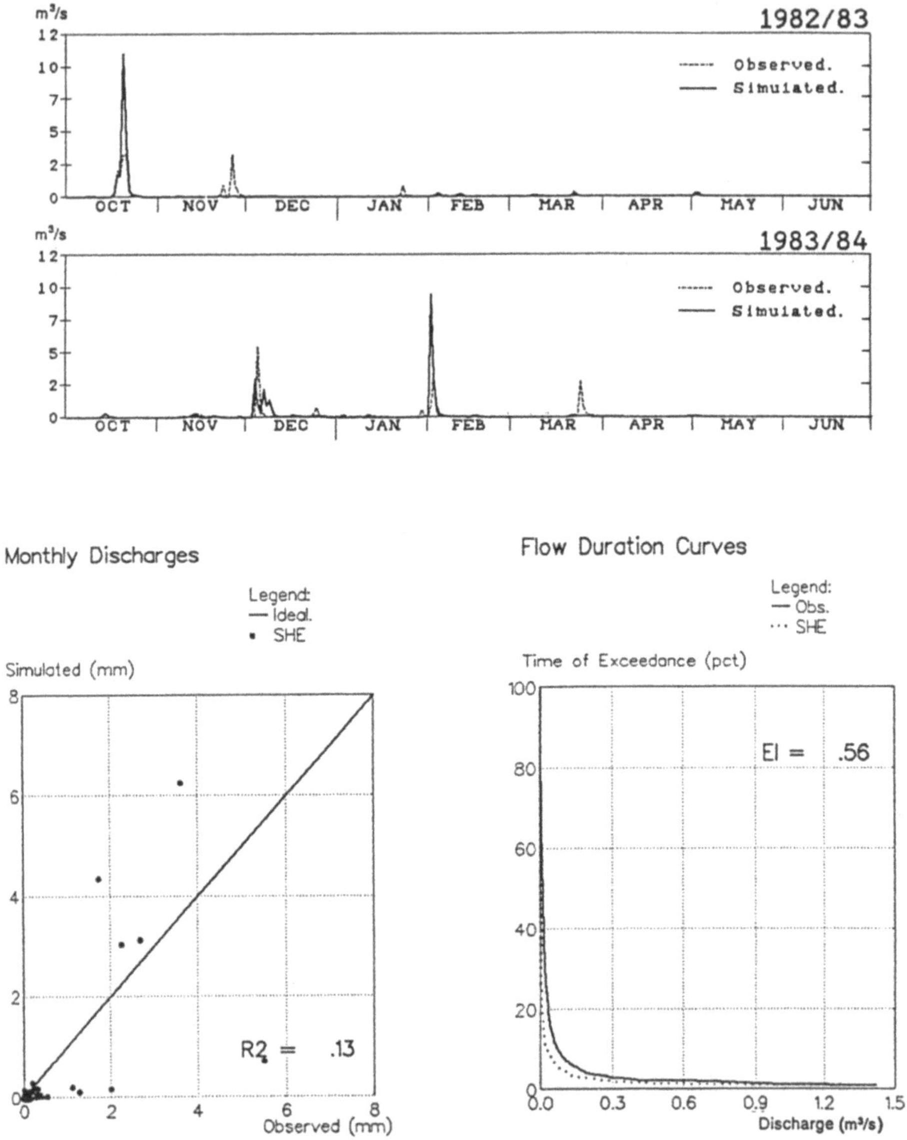

Figure 8. Lundi MIKE SHE differential split-sample test hydrographs from two of the validation years together with flow duration curves and scatter diagram of monthly discharges for the entire validation period.

or insignificant flows only. The same conclusion applies in general to the other two models (DHI, 1993).

The WATBAL model, however, grossly overestimates the peaks in the relative sense, causing the simulated average runoff to be about twice of that measured (15 mm compared to 8). For the NAM and MIKE SHE models, a much better performance is obtained, providing a quite accurate representation of the measured record in general. The related statistics are poorer than in the other testing schemes, but it should be noted that even small deviations causes poor statistics when mean flows are as low as in this case.

4.5 Conclusions on the Model Validations on the Zimbabwe Catchments

The following conclusions are based primarily on the results shown above; however, some of the other results of the original research project (DHI, 1993) are also referred to.

In view of the difficult tasks given to the models involving simulation for ungauged catchments and nonstationary time periods, the overall performance of the models is considered quite impressive. The overall water balance agrees within $\pm$25% in all cases but one, and good results are achieved without balancing out excessive positive and negative deviations within individual years. In most cases, the models score an R2 value at about 0.8 or greater and an EI-index generally above 0.7.

In addition to the above general comments, the following is noted with regard to the specific types of validations undertaken.

For the split-sample test, the NAM, WATBAL, and MIKE SHE models generally exhibit similar performance. All models are able to provide a close fit to the recorded flows for the calibration period, without severely reducing the performance during the independent validation period. Hence, this test suggests that, if an adequate runoff period for a few (3-5) years exists, any of the models could be used as a reliable tool for filling in gaps in such records or used to extend runoff series based on long-term rainfall series. Considering the data requirements and efforts involved in the setup of the different models, however, a simple model of the NAM type should generally be selected for such tasks.

For the proxy-basin tests, designed for validating the capability of the models to represent flow series of ungauged catchments, it had been expected that the physically based models would produce better results than the simple type of models. The results, however, do not provide unambiguous support for this hypothesis.

All three models generated good results, with the WATBAL providing slightly more accurate results than the others. Therefore, for the Zimbabwean conditions, the additional capabilities of the MIKE SHE as compared to the WATBAL, namely the distributed, physically based features relating to subsurface flow, proved to be of little value in simulating the water balance.

For the proxy-basin tests, it is noticed that the uncertainty range represented by the low-high estimates is significantly larger for the NAM than for the WATBAL and MIKE SHE cases. This probably reflects the fact that parameter estimation for ungauged catchments is generally more uncertain for the NAM, the parameters of which are semi-empirical coefficients without direct links to catchment characteristics.

A general experience of the modified proxy-basin tests is that allowing for model calibration based on only one year of runoff data improves the overall performance of all models. The improvement appears to be particularly significant for the NAM model, which also showed the largest uncertainties in the cases where no calibration was possible.

For the differential split-sample tests, all models have been able to simulate flows of the right order of magnitude and correct pattern. Thus, all models have proved to be able to simulate the runoff pattern in periods with much reduced rainfall and runoff as compared to the calibration period.

On the basis of these results, there appears no immediate justification for using an advanced type of model to represent flows following a significant change of rainfall, providing a number of years as available for calibration purposes. It is tempting to extend this finding to suggest that the simple type of model could be used to assess the impact of climate change on water resources. However, it should be recognized, that the above results cannot fully justify such a hypothesis, because a long-term climate change would probably bring about changes in vegetation and their evaporation. This type of unstationarity has not been adequately tested.

In summary, the results of the comprehensive validations suggest that, given a few years of runoff measurements, a lumped model of the NAM type would be a suitable tool from the point of view of technical and economical feasibility. This applies for catchments with homogeneous climatic input as well as cases where significant variations in the exogenous input are encountered.

For ungauged catchments, however, where accurate simulations are critical for water resources decisions, a distributed model is expected to give better results than a lumped model, if appropriate information on catchment characteristics can be obtained.

5. Optimal Interaction Between Hydrological Models and Remote Sensing

A considerable amount of spatial information on several hydrological variables can now be provided by use of remote sensing techniques. Such data represent new sources of information with potentially large scopes for practical application in hydrology. However, with the exception of snowcover data, remote sensing has generally not been used operationally for hydrological purposes until now. One of the reasons for this gap between the potential and the realized utilization appears to be the uncertainties in interpretation of such data.

On the other hand, it is evident that distributed hydrological models badly need spatial data for calibration and validation purposes. This is particularly true in cases where the aim of the hydrological modeling goes beyond simple predictions of catchment runoff.

Thus, it is realized that both remote sensing data and hydrological model predictions of the same variables, e.g., soil moisture or actual evapotranspiration at a certain grid, will contain uncertainty. Therefore, it is suggested that an optimal way of combining the information from those two sources would be to feed in the remote sensing data to the hydrological model by use of some updating (data assimilation) technique such as Kalman filtering. In this way, the most likely value of the hydrological variables can be found by weighting of the remote sensing and the model estimates by explicitly taking the respective uncertainties in the two methods into account.

It appears evident that such interaction between remote sensing data and distributed hydrological models will prove to be of immense importance for the further developments and applications of both methods.

6. Discussion of Requirements to Hydrological Models for Coupling to Global Circulation Models for Climate Change Modeling

In order to ensure and verify accurate predictive capabilities for nonstationary climatic conditions, a hydrological model needs to undergo a more advanced test procedure than the standard split-sample test. In this paper, the hierarchial test scheme, originally proposed by Klemes (1985), has been successfully used.

The present type of hydrological models are, in general, suitable for runoff simulations under nonstationary conditions. This does not mean that all hydrological models are suitable, nor that research and improvements are not required; but rather that it will be possible to identify models which are suitable, either directly or after minor modifications. Whether the same hydrological models are adequate also for simulation of other hydrological variables than runoff under nonstationary conditions has not been addressed in the present paper.

Based on the results presented in this paper as well as the experience of the author, the runoff data requirements for calibration of hydrological models may be summarized as follows:

- In general, 3-5 years of runoff data are sufficient to carry out an accurate calibration. More data are likely to improve the accuracy further, but most often only marginally.

- However, even a calibration against only one year of runoff data is most often sufficient to remove the main part of the prediction uncertainty. This is particularly true in cases where calibration is possible on longer time series from neighbouring catchments with a similar hydrological regime.

Hence, assuming limited availability of resources for collection of runoff data, a combination of long time series from a few stations and short time series from many stations will be optimal.

For simulation of other hydrological variables such as evaporation and energy flux to the atmosphere, it will be necessary to carry out calibration/validation on the same data types. In such a case, it is not sufficient to calibrate against runoff data.

In relation to global change, a coupling between Global Circulation Models (GCMs) and hydrological models is interesting. In this regard, two issues are important, namely the scale of operation and the processes and outputs being described on that scale:

Scale of operation. A GCM usually operates with a resolution which is much larger than the typical resolution of a lumped catchment model (100-5000 km^2), which again is much larger than the grid size in a distributed catchment model (0.01-25 km^2). Thus, seen from the GCM point of view, the resolution of the output from the lumped hydrological model is sufficient. On the other hand, for application of GCM output in hydrological models, a transformation to a finer resolution is required. This disaggregation must take into account, among others, the effects of the local topography.

Process descriptions and outputs. The outputs from lumped hydrological models of the NAM type are average catchment values of:

- runoff and
- evaporation.

By use of distributed, physically based models, among others, the following outputs may be generated on a grid basis:

- runoff,
- soil moisture,
- surface temperature,
- evaporation, i.e., water flux to the atmosphere,
- energy flux into the soil, and
- energy flux to the atmosphere.

A GCM generates outputs in terms of precipitation, temperaturem and radiation. A GCM, on the other hand, requires information on energy and water flux from the land surface as boundary conditions.

Thus, it is seen that some types of distributed, physically based hydrological models and GCMs can provide each other with the necessary information required for a successful coupling, whereas the traditional lumped, conceptual hydrological model cannot provide outputs with regard to energy flux to the atmosphere.

Global environmental change is an extremely complex issue, where changes may appear successively in different parts of the ecosystems. For instance, with regard to hydrology, a climate change will cause changes in vegetation characteristics: farmers may choose other crops, the time variation of root depths and leaf area index may change, etc. These changes are not straightforward to incorporate in hydrological models. Lumped, conceptual models cannot take such changes into accountat all, whereas it is possible to some extent to consider such aspects in distributed, physically based models.

7. Acknowledgements

The modeling work on the Zimbabwe catchments was carried out by my colleagues, Jesper Knudsen, Borge Storm, Merete Styczen, and Roar Jensen. During the data collection and field reconnaissance in Zimbabwe, kind help and assistance were provided by the University of Zimbabwe, National Herbarium, Department of Meteorological Services and Hydrological Branch, Ministry of Energy, and Water Resources and Development.

8. References

Abbott MB, JC Bathurst, JA Cunge, PE O'Connell, J Rasmussen (1986a) An introduction to the European Hydrological System--Systeme Hydrologique Europeen 'SHE' 1: history and philosophy of a physically based distributed modeling system, *Journal of Hydrology*, 87:45-59.

Abbott MB, JC Bathurst, JA Cunge, PE O'Connell, J. Rasmussen (1986b) An introduction to the European Hydrological System--Systeme Hydrologique Europeen 'SHE' 2: structure of a physically based distributed modeling system, *Journal of Hydrology*, 87:61-77.

DHI (1993) Validation of hydrological models, phase II, Danish Hydraulic Institute, Horsholm, Denmark.

Jain SK, B Storm, JC Bathurst, JC Refsgaard, RD Singh (1992) Application of the SHE to catchments in India, 2: field experiments and simulation studies with the SHE on the Kolar subbasin to the Narmada River, *Journal of Hydrology*, 140:25-47.

Klemes V (1985) Sensitivity of water resources systems to climate variations, WCP Report No. 98, WMO, Geneva, Switzerland.

Knudsen J, A Thomsen, JC Refsgaard (1986) WATBAL: A semi-distributed, physically based hydrological modeling system, *Nordic Hydrology*, 17:347-362.

Nash IE, IV Sutcliffe (1970) River flow forecasting through conceptual models, I, *Journal of Hydrology*, 10:282-290.

Nielsen SA, E Hansen (1973) Numerical simulation of the rainfall-runoff process on a daily basis, *Nordic Hydrology*, 4:171-190.

Refsgaard JC, SM Seth, JC Bathurst, M Erlich, B Storm, GH Jorgensen, S Chandra (1992) Application of the SHE to Catchments in India, 1: general results, *Journal of Hydrology*, 140:1-23.

WMO (1975) Intercomparison of conceptual models used in operational hydrological forecasting, WMO Operational Hydrology Report No. 7, WMO No. 429.

WMO (1985) Third Planning Meeting on World Climate Programme--Water, WCP-114, WMO/TD-No. 106, WMO and UNESCO.

WMO (1986) Intercomparison of models for snowmelt runoff, WMO Operational Hydrology Report No. 23, WMO No. 646.

Runoff and River Flow Measurements

Odon Starosolszky
Water Resources Research Centre (VITUKI)
Kvassay Jenő ut 1
H-1095 Budapest
HUNGARY

1. Trends in the Development of River Flow Measurement

The development of hydrology and water management depends to a great extent on the advances in runoff and river flow hydrometry. In several fields, however, no further improvement can be achieved without improving the reliability and accuracy of flow measurements and observations.

Due to rapid technical development, there are practically no technical limitations to real-time flow data collection, even in remote areas where satellite coverage may offer solutions for telemetering and data-collection platforms. However, it may be concluded that the impact of the new developments has not been as great as expected, due to high costs and the adverse changes in economic conditions.

During the last decade, few completely new instruments have been invented, the applicability of existing ones has been improved by refining their design and introducing more electronics with digital outputs and computer technology. The technology has been advanced appreciably by improving, in particular:

- transducers for water stage observations (submersible and nonsubmersible)
- digital recorders and data loggers
- telemetry (WMO, 1987)
- ultrasonic (acoustic) and electromagnetic velocity and flow meters
- multiparameter observing systems
- complete data acquisition systems

NATO ASI Series, Vol. I 46
Land Surface Processes in Hydrology
Trials and Tribulations of Modeling and Measuring
Edited by S. Sorooshian, H. V. Gupta, and S. C. Rodda

Because real-time observations are often essential in up-to-date water management systems, the need for them has certainly influenced the development of both sensors and transducers. However, budgetary restrictions impose limits of their application, even in new schemes. Complete modernization of the hydrometric network has been initiated only in few developed countries. Thus, a co-existence of the new and conventional technology can be characteristic in several countries only.

The following main trends in the development of hydrometric networks may be noted:

(1) replacement of manual (visual) observations by automatic or semi-automatic instruments;

(2) moves towards monitoring all hydrological elements simultaneously and with similar output signals;

(3) compatibility among different transducers and computers; and

(4) wider application of real time telemetry particularly for forecasting purposes and/or operation of water management systems (e.g., flood control, hydropower and irrigation).

In spite of the technical development, conventional methods are still widely used; e.g., the float operated water level recorders and rotating element current meters for the area/velocity method. It is not realistic to consider that the general applications of the up-to-date technology can be expected within a few years and on a global scale.

Thus, efforts should be accelerated to improve reliability and accuracy of flow measurements, even by using traditional instruments and technology. As an example, ultrasonic flow gauges still have several limitations and cannot replace area-velocity method on large shallow and wide rivers. Therefore, the methods of measurement should be refined, maintaining the traditional instruments and equipment. On the other hand, the growing demand for continuous flow measurement on tidal estuaries may enhance the expansions of ultrasonic flow gauges capable of detecting discharges during continuous variation of flow.

Appropriate technology is needed in the following fields:

- streamflow measurement on rivers of extreme water regime;
- periods when flow measurements are inapplicable (extreme high and low flows, or existence of ice cover);
- flow measurement under the effect of man-made changes (e.g., on canalized rivers);
- reduction of time of current metering due to variations in water regime during depth sounding and velocity measurement;
- replacement of variable rating curves due to changes in cross-section, slope and roughness;
- the correction of streamgauging errors due to oblique flow in the streamgauging cross-section; and
- simultaneous measurement of velocity, sediment, and other water quality parameters in order to ensure data for the calculation of transport rates.

In view of the above, it can be concluded that a development program in the field of flow measurement is necessary, including man-power training, safety regulations, and purchase of special equipment.

Runoff measurement is almost exclusively based on river flow measurement (Herschy, 1978). Thus, runoff is a catchment characteristic (e.g., an average or lump value) and not a directly measured value. Efforts to determine runoff as overland flow have been made in laboratories and runoff plots on representative or experimental areas (USDA, 1979). In this field, there appears to be a lack of standardization, with different methods being used in almost every case. It is not foreseen that an extensive direct runoff measurement program can be launched on a global scale; hence, data for any global and continental runoff will be based on streamgauging.

Measurement of flow on ephemeral streams is difficult because low discharges and high floods should all be measured, and FLASH floods dictate that hydrometric staff be present on the site, at the streamgauging measuring section. Rapid flow can also hinder the use of wading or boat technology for current metering.

These are the major reasons why so-called composite control sections are constructed and equipped (Starosolszky, 1987). The investment costs of such structures, even with simple construction technology, are considerably high. However, consistency of rating curves can only be ensured by such structures, and the shaping of the control section should offer measurements of flow over a wide range. Thus, the application of man-made structures (culverts, bridges, dams, gates, barrages) becomes necessary wherever applicable. Water intakes and outlets can also be used to measure water abstractions and discharges. Scale model calibrations may offer opportunity to determine the rating curves, but field measurements cannot be completely neglected. Therefore, the calibration of gates, hydropower, or pumping plants became a general procedure to replace traditional rating curves of two variables (stage/discharge curves). The validity and reliability of such curves are highly dependent on the hydraulic design of the structures. It is noted that several hydrological services are not familiar with the application of the different structures available to estimate flow.

The field calibration procedure needs special instruments and equipment (e.g., multicurrent metering or transportable acoustic flow meters). Hydrological service may not be equipped for such types of measurement; thus, in many cases expensive field measurements should be institutionalized.

Similar problems are observed in connection with oblique flow (e.g., either the checking of the measuring section from this point of view or the measurement in order to correct or calculate discharge values in an oblique flow field). If the direction of flow is not perpendicular to the cross-section, a direction current meter is necessary to measure the obliquity of the flow, e.g., for the simultaneous measurement of velocity in magnitude and direction. When bridges are used for flow measurement during floods in meandering rivers of wide floodplains, it is probable that flow is not parallel during low, medium, and high water levels. Hence, flow direction along the cross-section (and in the vertical) should be detected. In deep rivers, the obliquity of flow may also vary along the verticals.

Particular consideration should be devoted to the unsteady flow regime during field measurements in the vicinity of hydraulic structures. Operation of dams, gates, barrages, and power stations may cause abruptly varied flow conditions when stage varies to a lesser extent than velocities. Consequently, the velocity may vary irregularly, depending on the opening or closing of gates or turbines. Methods are available to correct flow measurement data when measurements are made during floods. However, this methodology needs revision or may be inapplicable if the abrupt changes are due to artificial conditions.

The above reasons are partially responsible for the uncertainties of flow measurement and for the scattering of the points around the stage-discharge relation.

Discharge time series can, in principle, show the effect of climate change on runoff. A detailed statistical analysis of the long time series of discharges of several rivers indicates no characteristic change (World Meteorological Organization, 1992). This undetectable change can be caused by the uncertainties in the measurements. The improvement of the reliability and accuracy of flow data may help to increase the capability of the time series to detect any man-made change. Thus, the environmental impact studies also need the improvement of river flow data which can only be achieved by improving the instruments and methods of observation.

2. Particular Problems of Flow Metering

After this general introduction of flow metering problems, some selected cases are now reviewed, e.g.:

- use of hydraulic measuring structures for flow measurement;
- impact of oblique flow;
- impact of turbulence on discharge measurement;
- fluctuation of velocity; and
- use of floats and areal photography.

Use of Hydraulic Measuring Structures

Hydraulic measuring structures (Starosolszky et al., 1972) are constructed in order to stabilize the stage-discharge relation and to pre-determine this relation without local calibration.

For the head-discharge relation to remain constant, a constant direction of approach flow is required. In most cases, flow normal to the control section is a basic requirement. Any oblique component in the approach flow may influence the calibration curve, which is usually determined on laboratory flumes on ideal prototypes.

Stability of the bed in the approach channel is also an essential requirement, because deposition or scour may influence the discharge coefficient and change the velocity distribution in, and the direction of, flow. However, due to meandering flow, which is basically three-dimensional, sediment may be deposited or removed in some parts of the approach channel.

The main problem is that the flow in natural channels is highly variable and the cross-section of the approach channel cannot be designed for all discharges. This is the main reason why measuring structures may not usually have the same accuracy at all stages and over the whole water year. Variations in turbulence of the original river bed and in the approach channel also affect deposition.

On the downstream side, especially where the submergence limit is low, scouring and deposition may change the limit of free and submerged flow, thus indirectly influencing the accuracy of flow measurement.

Oblique flow through the measuring structures may also be critical for the stability of the structure, because bank scouring may damage the complete streamgauging control section.

Impact of Oblique Flow

In general, the velocity may vary spatially across the whole flow field. Therefore, instead of the flow component being perpendicular to the cross-section, it is necessary to measure the velocity vector, its magnitude, and direction.

The two basic methods of measurement are:

(a) to measure the magnitude of the oblique velocity and its direction (in two planes, usually in the vertical and horizontal), thereby obtaining the velocity in polar coordinates; and

(b) to measure three components of velocity and to calculate the resultant velocity vector and its direction (Muszkalay and Starosolszky, 1968).

These two methods in practice mean that either a cable suspension system, with a directional current meter, needs to be deployed or that rigidly oriented meters are deployed, having component-measuring rotors. Both methods assume that the point of measurement can be determined during the observation.

The potential magnitudes of the lateral and vertical velocity components are still not fully appreciated. The illustrations given in Figures 1-6 provide examples of the problem. Figure 1 shows the variation of flow and the variation of turbulence in a small creek (Muszkalay, 1979). Figures 2 and 3 show the variation of flow in relatively large rivers. Both cross-sectional and plan presentations may be applied, depending on the purpose of the measurement (Figures 4-6).

It should be noted here that, in large rivers, cable-suspended (rather than fixed) current meters are still widely used for discharge measurement by the velocity/area method. One may imagine what differences are introduced in the discharge computation if velocity is represented by oblique vectors measured in the verticals without calculating their components normal to the cross-section.

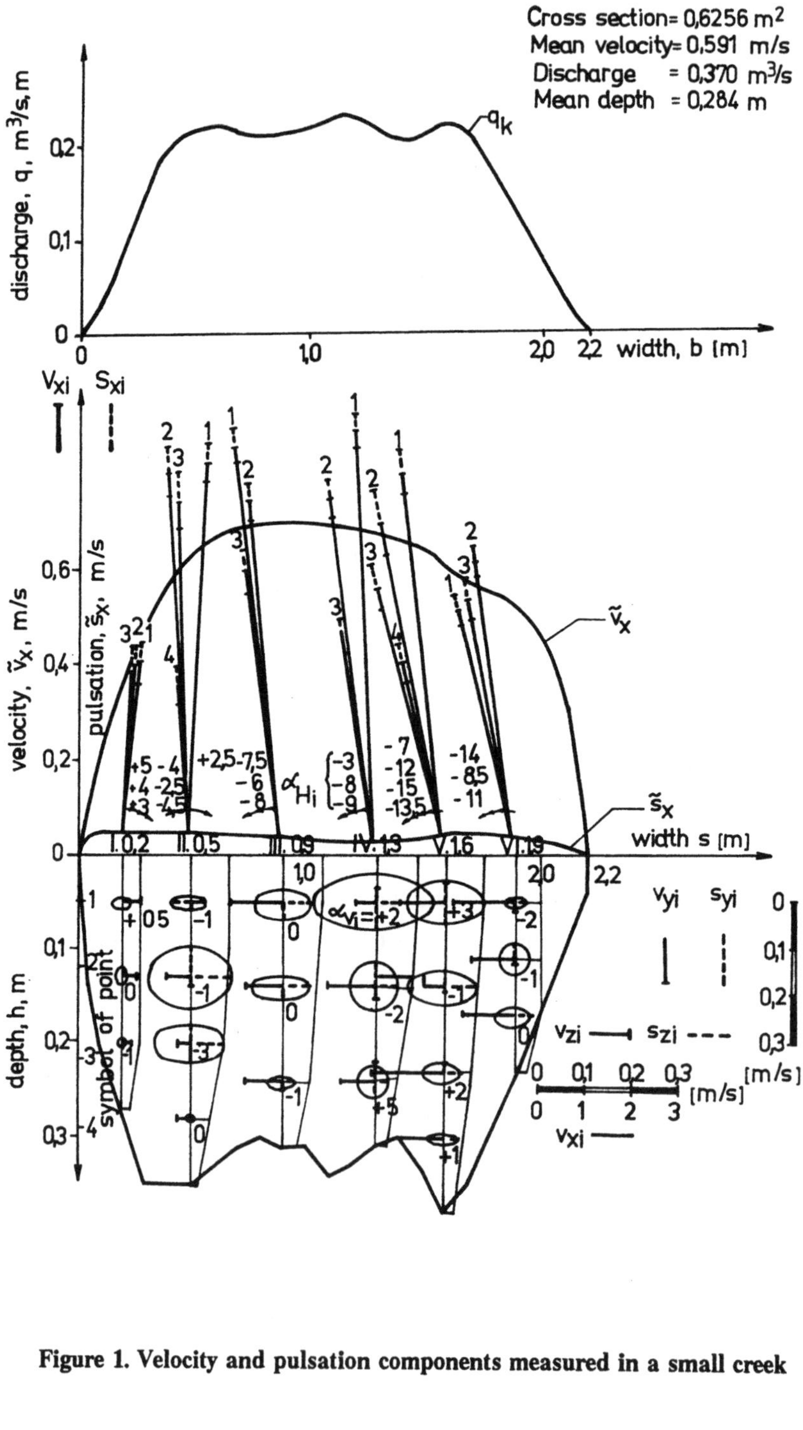

Figure 1. Velocity and pulsation components measured in a small creek

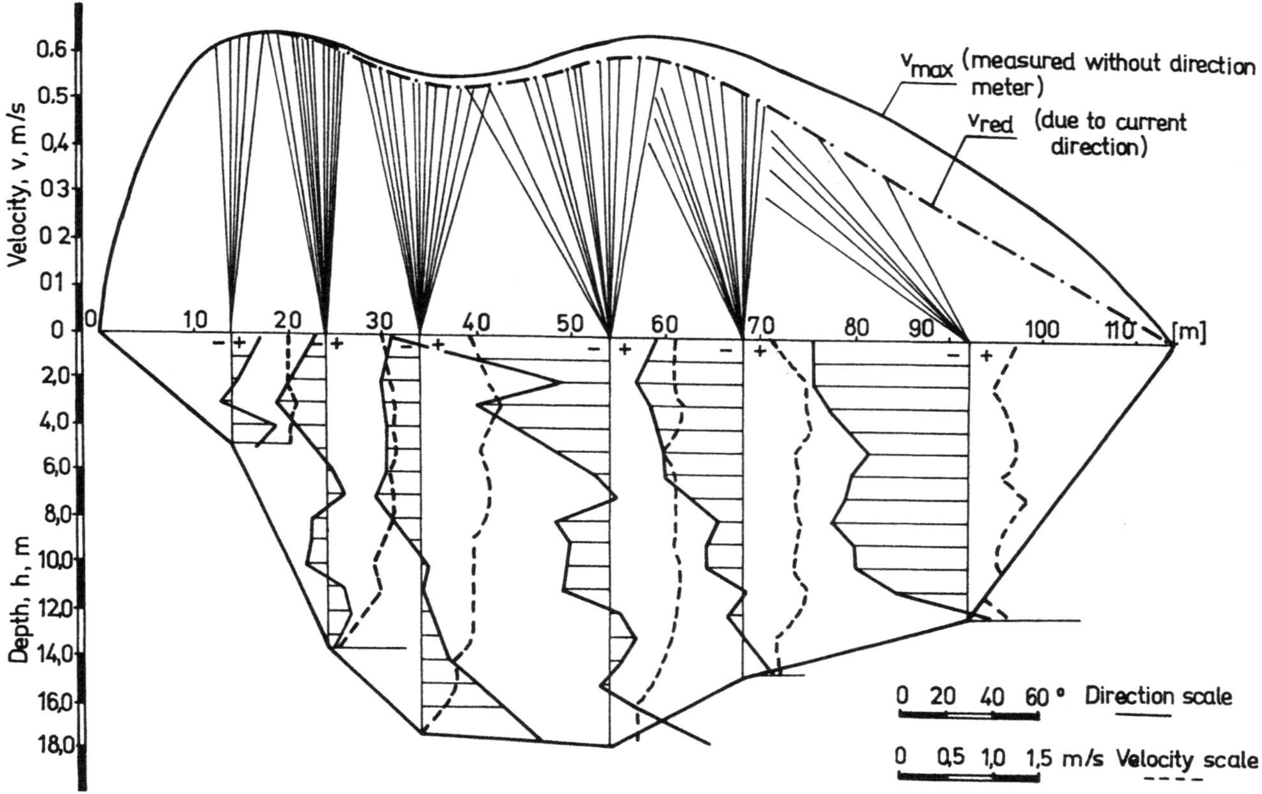

Figure 2. Distribution of the horizontal velocity components in a river cross section

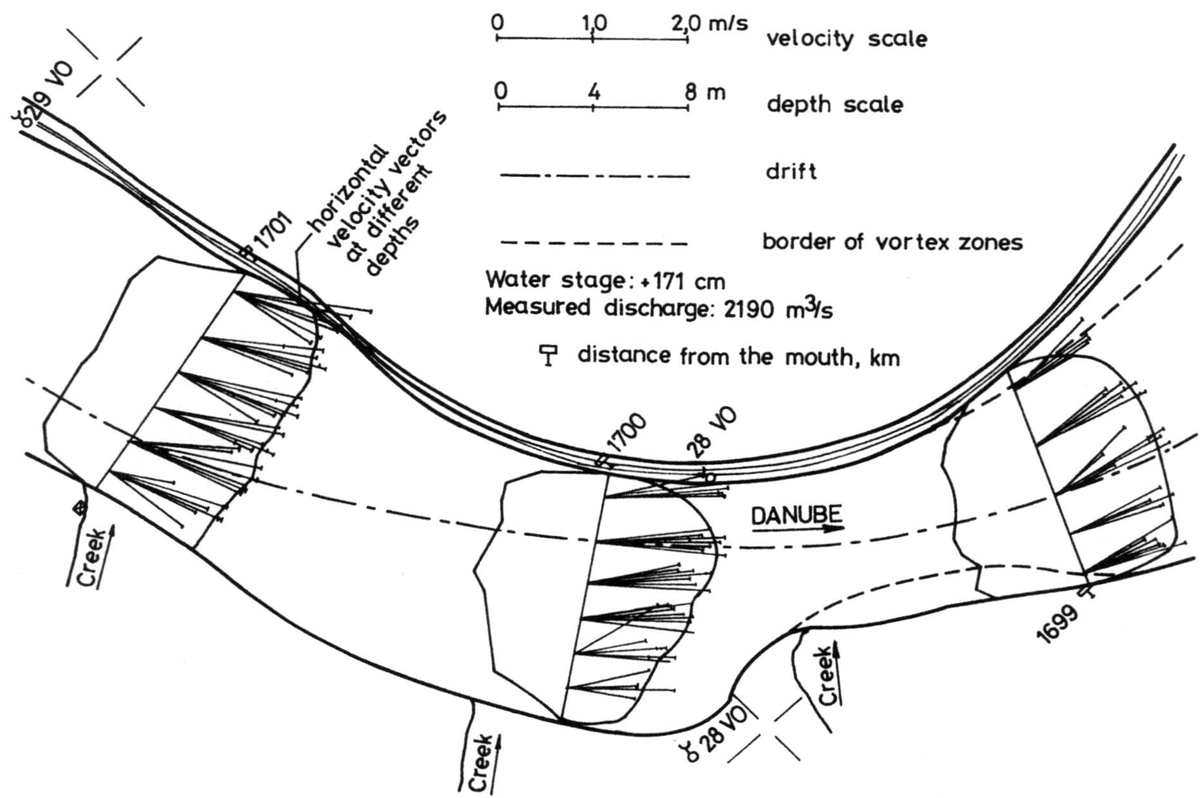

Figure 3. Distribution of velocity directions in a bend on a major stream

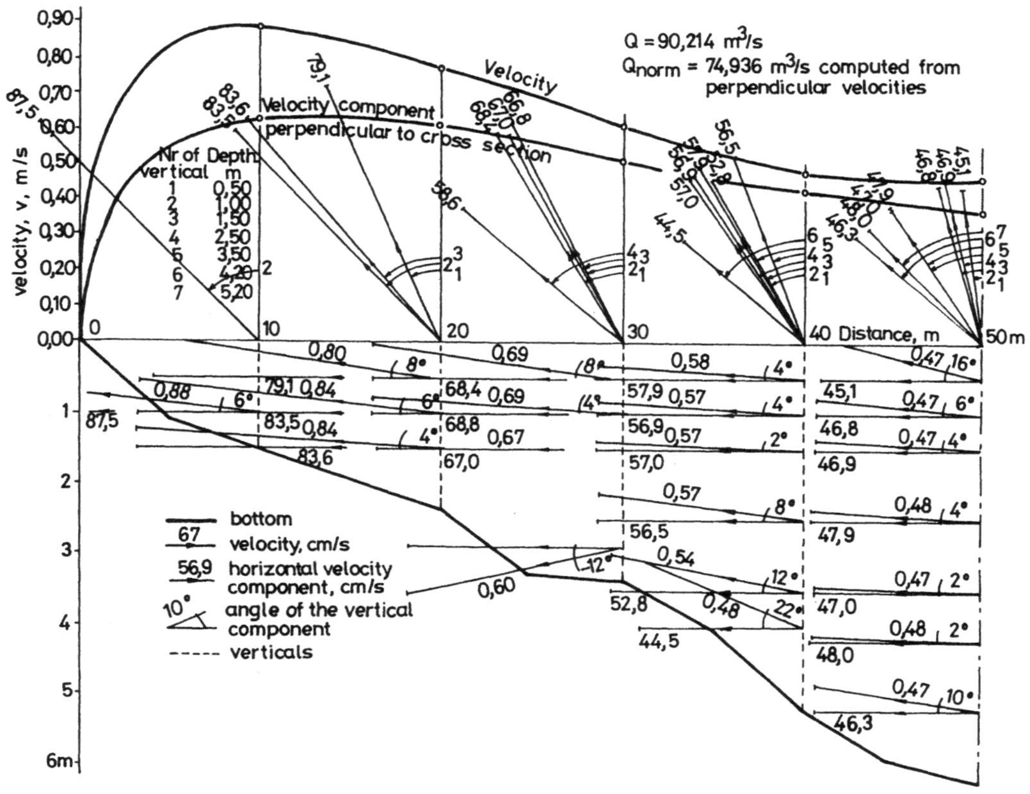

Figure 4. Comparison of one- and two-dimensional velocity distributions and computed discharges before a power station on a river

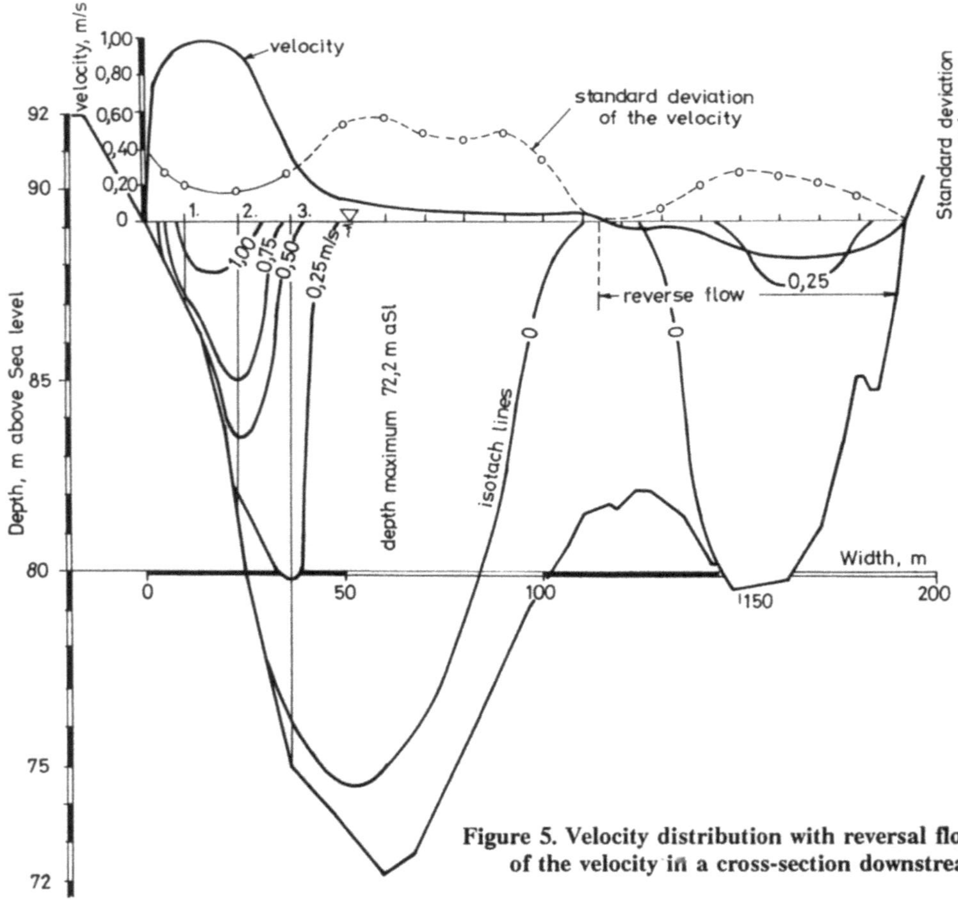

Figure 5. Velocity distribution with reversal flow and standard deviation of the velocity in a cross-section downstream of a power station

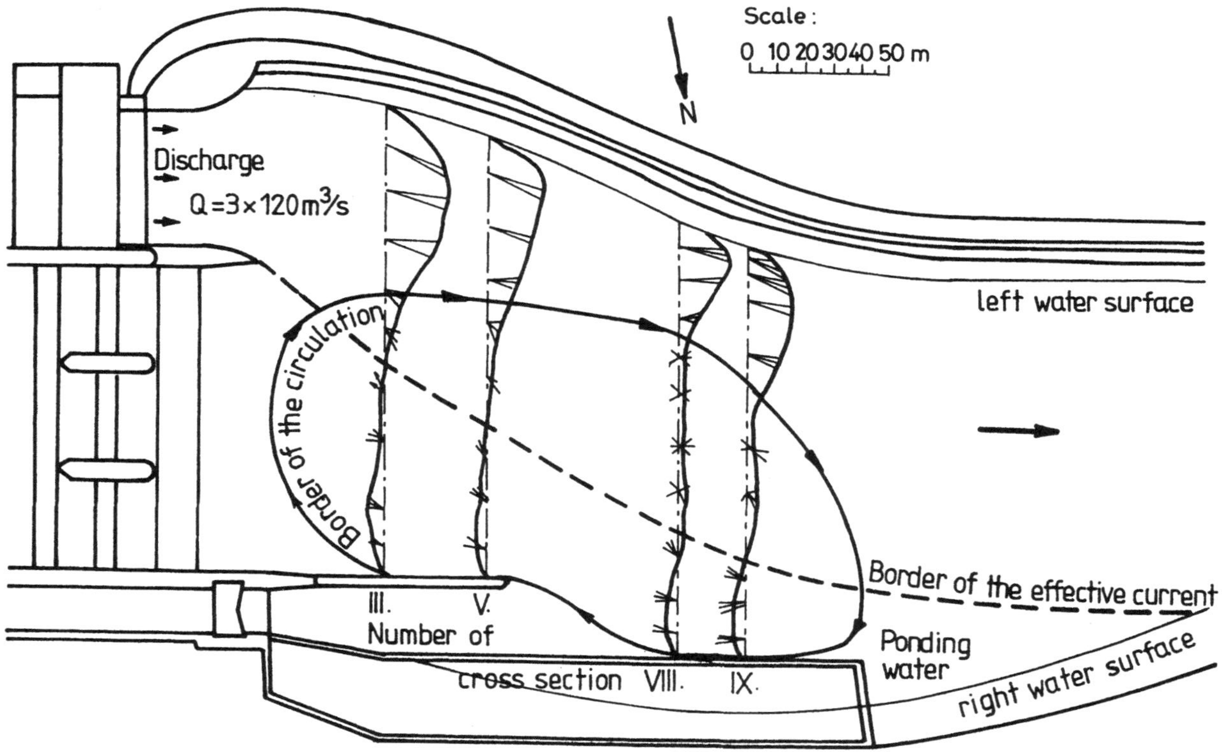

Figure 6. Distribution of the velocities with reversal flow downstream of a power station

Three-dimensional flow has a strong influence on the accuracy of discharge measurement. The variation of flow direction within the cross-section may cause a regular systematic error. On the other hand, any improper application of noncomponent meter rotors (giving only a scalar measure of current) in a fixed position may result in the measurement of oblique flow without realizing that the normal component of velocity was not being measured.

As an example, the results of velocity measurements with scalar current meters and current direction meters upstream of a low-head hydropower station (Figure 4) demonstrate the difference between the discharges measured with and without taking the strong obliqueness of flow into consideration (Muszkalay and Starosolszky, 1968; Starosolszky et al., 1972).

The effect of oblique flow on discharge measurement can be allowed for by the use of a component-measuring rotor up to an angle of 40° with the flow direction. When obliqueness is more extreme, a directional current-measuring meter is necessary for discharge measurement.

In the initial period of a streamgauging measurement, the obliqueness of flow within the cross-section at different water levels and/or discharges should be estimated. The potential error introduced by neglecting it should at least be checked.

Impact of Turbulence on Discharge Measurement

The evaluation of current meter calibrations shows that they are sensitive to turbulence. However, few efforts have been made in the study of turbulence on flow, especially in three dimensions.

To characterize flow turbulence, continuous records of velocity are necessary for which the speed of the current meter must be observed at very close intervals of time. Traditional current meters emitting a signal only every ten or 20 revolutions are completely unsuited for turbulence measurement. Current meters signaling after each revolution can be used with revolution recorders (e.g., chronographs) or by converting the speed to frequencies for recording the velocity variations.

In three-dimensional flow fields, it is essential to obtain turbulence data from velocity variations in all three components. The polar vector measuring instruments do not yield turbulence components and give only the relatively slow pulsation of the resultant velocity vector. The computed average velocity components are not necessarily in linear correlation with turbulence intensities, and the turbulence components cannot be derived by simple proportionality.

The distribution of the turbulence components in a cross-section is shown in Figure 1. The variation of the turbulence follows a probabilistic distribution.

Turbulence is highly variable with the size and features of the stream. Muszkalay (1979, 1980-1981) estimated its extent for Hungarian rivers on the basis of his measurements (Table 1). Given such information, the influence of turbulence on current meter calibration can be allowed for.

Within the framework of the WMO intercomparison project on current meters (World Meteorological Organization, 1977; Muszkalay and Starosolszky, 1981), a test was specified for detecting the effect of turbulence which can also be a sound basis for selecting the proper instruments for pulsating flow.

The systematic error of discharge measurement due to turbulence can be reduced by applying lightweight rotors (e.g., of plastic), frictionless sensing of their rotation, and improvements in counting circuits. The random error of discharge measurement due to turbulence can be reduced by increasing the length of the observation period at the points and in the number of observation points, depending on the shape of the cross-section and velocity distribution.

Table 1. Turbulent Components of the Three-Dimensional Velocity Vectors (x, y, and z) in Hungarian Rivers and Canals (after Muszkalay) (in m s^{-1}).

Name of stream	Mean velocity (m s^{-1})	Type of stream								
		Straight			Meandering			Bifurcating		
		σ_x	σ_y	σ_z	σ_x	σ_y	σ_z	σ_x	σ_y	σ_z
Danube	1.2-2.0	0.04	0.03	0.03	0.15	0.08	0.08	0.18	0.10	0.10
Tisza	0.6-1.5	0.05	0.04	0.040	0.10	0.08	0.08	0.15	0.12	0.12
Raba, Sio, Sajo	0.5-1.2	0.10	0.08	0.08	0.14	0.10	0.10	0.18	0.12	0.12
Bodrog, Zala, Zagyva	0.5-1.2	0.06	0.04	0.05	0.13	0.06	0.08	0.16	0.08	0.09
Creeks	0-0.4	0.010	0.005	0.005	0.020	0.010	0.010	0.030	0.015	0.015
	1.0-2.0	0.035	0.015	0.020	0.100	0.050	0.060	0.120	0.080	0.090
Irrigation canals	0.3-0.5	0.04	0.02	0.02	0.06	0.025	0.030	0.10	0.035	0.040
Lined channels	0.8-1.0	0.020	0.10	0.010	0.050	0.018	0.020	0.08	0.025	0.030
Drainage canals	0-0.2	0.015	0.005	0.005	0.020	0.010	0.010	0.03	0.015	0.015
Maximum measured (downstream of a barrage)								0.72	0.40	0.60

Fluctuation of Velocity

In the previous section, microturbulent effects were discussed, but under certain conditions larger-scale pulsations should also be considered. Their pulsations may be caused by man-made structures (operation of dams, reservoirs, navigation locks, power stations, gates, etc.), but may also be generated by plume-like wandering of the main flow in the channel (especially in lakes).

Muszkalay (1980-1981) observed an interesting time series of velocity in two reaches in the Danube. He measured the velocity variation at one point of each cross-section continuously and periodically (e.g., 10 minutes every hour for 24 hours). One characteristic time series is given in Figure 7. The high intensity of pulsation at this point, far downstream from any artificial effect, calls attention to the bends, where cross-flows may contribute certain fluctuations. In this example, it is impossible to ascertain the factor generating such large pulsation effects, but one may perceive the serious effect on the accuracy of flow measurement. Similar pulsations were also observed on an upper, steeper reach of the Danube, as shown in Figure 8.

This effect cannot be filtered out by instrumentation alone, but can be studied by continuously recording the mean velocity with an ultrasonic discharge recorder. This is an interesting field of further STUDY, which should result in a better understanding of discharge measurements and improve their accuracy.

Floats

Although floats are seldom used in stream gauging at present, they are useful for measuring high flows under emergency conditions, especially in the presence of floating ice and debris. Floats or buoys are useful tools for determining surface currents in lakes and reservoirs or for observing surface flow patterns recorded by surface or aerial photography (Starosolszky, 1987). Floats with torches also make night measurements possible where photography may be employed.

Floats are used in Russia for aerial surface velocity measurements. Floats are dropped from an airplane and their paths traced by aerial photography. The method has the advantage of making measurements in isolated areas which otherwise would be nearly impossible. The floats are photographed twice during their travel. As the photograph scale is known, the travel distance L during time Δt is determined. The numerical value of the surface velocity v_s is calculated from $v_s = L/\Delta t$.

The integrated-velocity method is similar to the air bubbling method. In this method, several devices, known as "hydrobomb" floats, are dropped from an airplane. Upon reaching the bed, these bombs eject a special tracer (such as oil) which follows the

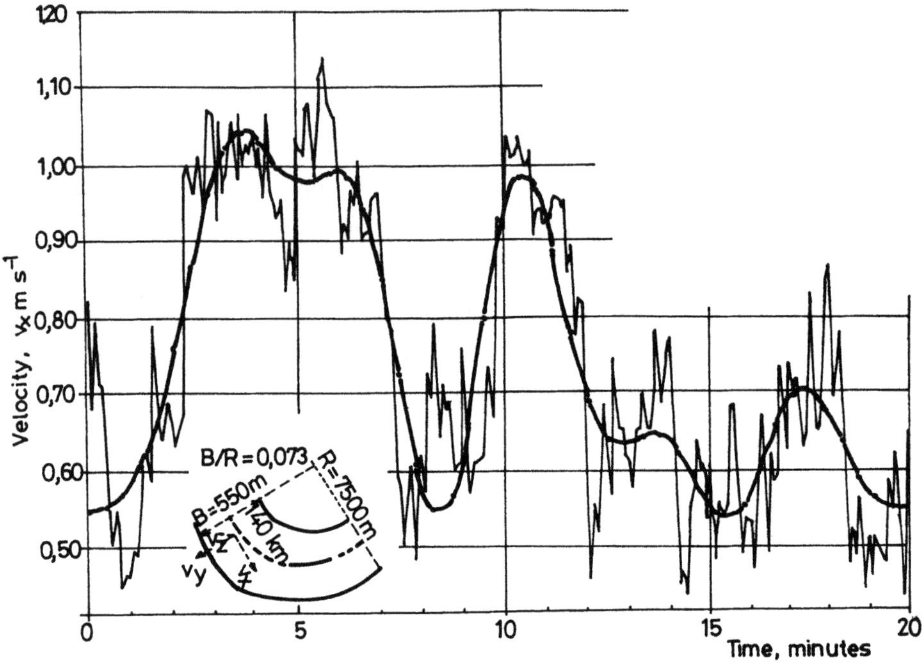

Variation of the fluctuating velocity. Danube. Cross-section: Domos, 1976.
0,5 m above the river bed

$v_x = 0{,}76$ m s^{-1}; $v_y = 0{,}045$ m s^{-1}; $v_z = 0{,}076$ m s^{-1};
$\delta_x = 0{,}194$ m s^{-1}; $\delta_y = 0{,}119$ m s^{-1}; $\delta_z = 0{,}111$ m s^{-1};
$S_x = 25{,}47$ %; $S_y = 15{,}62$ %; $S_z = 14{,}65$ %;
$S_x = 11{,}87$ % reduced to random variation

Figure 7. Large fluctuations in the measured velocities

velocity distribution both in the vertical and horizontal on the surface. Aerial photography fixes the point at which the hydrobomb float hits the surface and photographs the path which the oil traces on the surface. The uncertainty in a single determination of discharge for both methods is estimated at between 10 and 15% at the 95% confidence level. The range of measured discharges in Russia varied from 100 to 100,000 m^3/s.

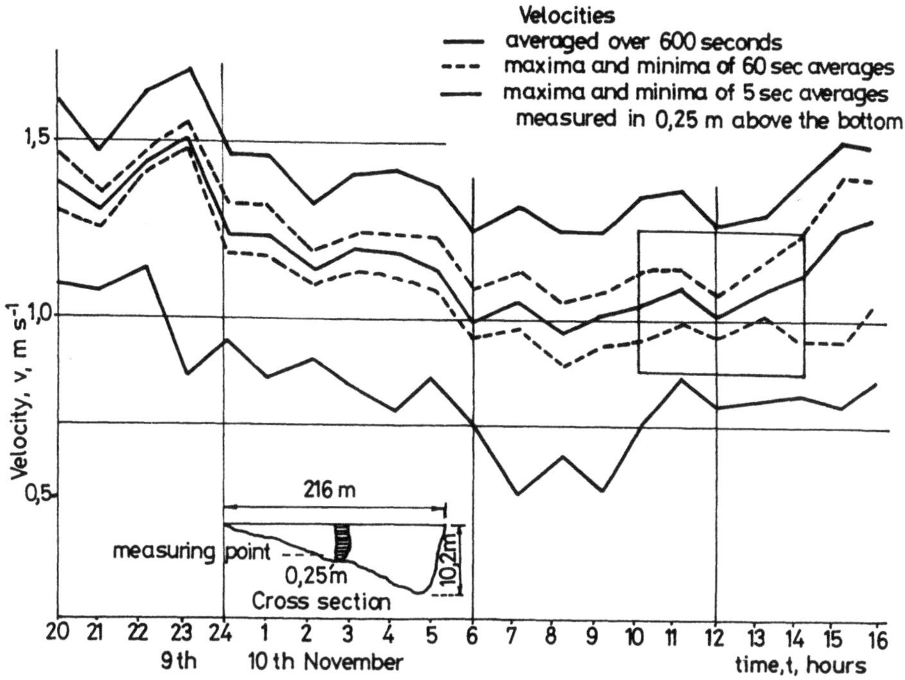

Figure 8. Time series of velocities in a large stream

Another combination of aerial photography and surface tracer floats (consisting of fine granulated white plastic foam) was successfully used in Hungary for determining the flow patterns and the discharge at a river barrage. The application of rod floats with torches has been used for determining the circulation of the thermal Lake Heviz in Hungary. These examples demonstrate the necessity of reconsideration of our standardized techniques (Framji, 1978) if the aim is to improve reliability of flow measurement.

3. The Situation of Runoff Measurement

The global situation of runoff measurement can be characterized quantitatively by the number of streamgaging stations and the measured discharge data, and qualitatively by the reliability of the measured data, depending mostly on the methodology used and which is influenced by the water regime at the measuring sites.

The available statistical data of streamflow measurements can be considered acceptable because they have been supplied through the World Meteorological Organization and are parts of the Hydroinform system. According to relatively recent surveys, the global distributions of streamflow measuring stations are shown in Table 2.

Table 2. Number of Streamflow Measuring Stations

	1977	1987	1989
ECA (Africa)	918	1694	1695
ECE (Europe + North Africa)	9549	24 228	23 946
ECLAC (Latin America)	3086	5730	5762
ESCAP (Asia and Pacific)	5923	6282	7023
ESCWA (West Asia and Arab countries)	1222	1262	1383
Totals	20 698	39 196	39 809

It can be concluded that the number of the stations may be classed as appropriate in the European Region, but scarce in Africa. The problem is more complicated when the number of actual discharge measurements is investigated. The practice of direct flow measurement is rather diverse. At certain streamgauging stations (e.g., in India or China), regular daily measurements are usual and complementary. In other countries monthly measurements are exercised. In certain countries, almost no direct measurement is carried out, and consistency of the rating curves (stage-discharge relations) is assumed.

The problem of the frequency of measurement is a qualitative factor. Random and little direct measurement cannot compete with the accuracy of regular daily measurements, particularly if there are large variations in the natural water regime.

Due to lack of financial means, the hydrology of the main rivers of the world is studied differently. Several large and medium size rivers in the developing countries have been observed with great discontinuity and unreliability. If a global hydrological observing system is planned, the primary aim should be the promotion of flow measurements to ensure continuity in data acquisition. The major aim is to maintain the observations according to acceptable requirements. The starting point can be the improvement of measurement at the downstream sections of the major rivers, where normally difficult conditions may prevail, and subsequently sections upstream can be improved if necessary. The new system would need a manual (like GEMS) which would include standardization of the flow measurements. The compilation of such a manual can be the starting point to launch the global project.

Human activities, e.g., the construction and operation of large water schemes, may cause serious conditions for the traditional direct area/velocity flow measurements, indicated above. Thus, traditional flow measurement may not supply sufficiently accurate flow and runoff data. This is a recent recognition in many situations, which was reflected by the application of new technology in the form of acoustic (ultrasonic) flow recorders.

One might assume that an acoustic flow station will provide the solution to most difficult sites but this may not always be the case. The applicability of the available ultrasonic flow meters can help in certain situations, particularly where strong daily variation in the flow can be observed. The sensitivity of the method, and its possible need for site-calibration, in alluvial rivers, where there is change in cross-sectional area and velocity distribution within the cross-section, can be a limiting factor which influences uncertainty in measurement.

Basically, the combination of daily current metering and continuous flow metering by ultrasonic flow meters can offer an appropriate although very expensive method for runoff measurement. The cost factor of flow measurement cannot be neglected and, in certain cases, cost effectiveness is the definitive factor in the selection of methods for streamflow measurement.

4. A Vision for Development

From the global point of view, the large rivers contributing to world water balance are important and need the most accurate measurements, because errors of a few percent can be equivalent to the total flow of a medium-sized river. This is the reason why first priority in an international hydrological network should be given to rivers having a mean discharge of more than 2000 m^3/s and a second priority to major rivers. The World Meteorological Organization Technical Regulation defines the terminology as follows:

"**Large River.** A river with a mean annual discharge at the mouth exceeding 2000 m^3/s or with a drainage basin exceeding 500,000 km^2.
Major River. A river with a mean annual discharge at the mouth exceeding 100 m^3/s or with a drainage basin exceeding 100,000 km^2."

For such a river, the investment cost of a permanent hydrometric station equipped with (1) a current metering/sounding team, and (2) an ultrasonic flow meter (with water level recorder) can be as much as 400,000 US dollars. The annual operation cost of such a station--varying with staff requirements-- may be in the region of 60,000-300,000 US dollars. Such a cost can be considerable for a developed country and hardly possible for a developing nation.

It is for this reason that, as a general goal, the desirable accuracy of flow measurement cannot be expected within the next decades. Ultrasonic flow measurement stations can be operated to advantage, where no change in cross-section can be expected and where the stage-discharge relation is unstable. The investment costs are high, but the operational costs are low. If manpower costs still remain low, one can conclude that conventional current metering will continue to be exercised in several countries.

For the estimation of the areal distribution of runoff at certain sites, medium- and small-sized rivers should also be gauged. In this case, the ultrasonics with certain civil engineering works may gain more consideration for the medium-sized rivers. For small rivers, the construction of artificial control sections (weirs and flumes) can be an acceptable method. These may not totally replace current-metering on these rivers, because calibration and checking of the rating may be necessary.

Current metering, however, cannot necessarily be executed by rotating current meters, e.g., ultrasonic or electromagnetic velocity meters can partially replace the conventional rotating current meters. At present, however, most of the hydrological services prefer traditional rotating current meters, particularly due to the higher prices of ultrasonic/electromagnetic meters.

Observation of stage remains necessary for all rivers. It is difficult to predict when mechanical float sensors may be replaced by electric/electronic water level (or pressure) gauges.

On the other hand, the media for collecting measured data will be more and more microprocessor solid state in the form of data loggers. The advantages of the digitized data acquisition systems are evident; therefore, only slow price-reduction may hinder the wider application of different data loggers, particularly developed and in use in the United States.

The technological advance in flow metering depends on the introduction of integrated networks. The need for different water quality parameters and the advance of different sensors (temperature, suspended solids, conductivity, pH, etc.) may offer an opportunity to use common data loggers in the same river cross-sections for water quality, stage, and/or flow. Most probably, again the financial resources are the limiting factor for such development on a global scale.

Rapidly growing water quality monitoring (or sampling) stations may support this conclusion. The need to measure transport of pollutants may also promote the joint sampling of velocities and water quality, in the same cross-sections and sampling points, as it has been advised some time ago in the GEMS/Water Manual. The dream of a global hydrological network includes such stations, preferably 'hydrological observatories', where all hydrological elements are measured continuously and in a complementary way. This is probably some time in the future, but we must accept it as a final target, at least for the so-called global stations.

5. Practical Aspects and Actions to be Executed

The Use of Up-to-Date Technology

With respect to up-to-date hydrological instrumentation, the Hungarian national hydrological service received valuable information from selected national services of developed countries. It was interesting to note that data loggers have become widely used, particularly for recording water levels. Ultrasonic flow meters are also expanding.

Considering the USGS experiences in the application of data loggers or acquisition systems, one may note that the figure for certain types may exceed 600. In Australia, eight agencies reported extensive use of data loggers produced by seven manufacturers, and some types number are 1000.

Acoustic flow meters are also used widely in the USA (like the productions of Affra Corp. or the portable velocity meters of ORE International, Inc.), in the Netherlands (Stork type), and in Switzerland.

Many multiparameter water-quality data systems, including depth or water-level sensors, are in use by the USGS (e.g., Scout 2 or Surveyor 3 systems of Hydrolab Corp).

The expansion of up-to-date technology depends highly on the available financial resources. It is interesting to note that certain companies may offer a relatively good discount (e.g., up to 40%) if a greater number of instruments is purchased (e.g., over 50-100). Hence, the larger development projects can supply less expensive instrumentation and can be economical for the participating parties.

Actions

In the action plan of the International Conference on Water and the Environment (ICWE), a chapter is devoted to a knowledge base which can be cited as follows:

"Measurement of components of the water cycle, in quantity and quality, and of other characteristics of the environment affecting water is an essential basis for undertaking effective water management. Research and analysis techniques, applied on an interdisciplinary basis, permit the understanding of these data and their application to many uses."

In the report of ICWE, paragraphs 3.9, 3.10, and 3.11 deal with the information acquisition and storage. Some important statements can be quoted, as follows:

"3.9 Reliable information on the condition and trends of a country's water resources--surface water, water in the unsaturated zone and groundwater, quantity, and quality--is required for a number of purposes, such as assessing the resource and its potential for supplying the current and foreseeable demand; protecting people and property against water-related hazards; planning, designing, and operating water projects; and monitoring the response of water bodies to anthropogenic influences, to climate variability and change, and to other environmental factors.

3.10 Integrated monitoring and information systems should be established and data collected and stored on all aspects of water resources which are required for a full comprehension of the nature of those resources and for their sustainable development. These include not only hydrological data, but also related geological, climatological, hydrobiological, and topographic data and data on soil types, land use, desertification and deforestation, as well as information on such topics as water use and reuse, sewage discharges, point and nonpoint sources of pollution, and runoff to seas and oceans. This involves the installation of observation networks and other data-gathering mechanisms designed to monitor various climatic and topographic regimes, plus the development of data storage facilities. Where water-related information at national, regional, and international levels is handled by a number of information systems, it is important that these systems be coordinated.

3.11 Recommended actions include:

(a) Installing monitoring systems designed to provide valid and comparable water-related information;
(b) Ensuring the continuous operation of such systems in support of studies requiring long-term data, such as those relating to climate change;
(c) Upgrading facilities and procedures for storing, validating, and safeguarding such data;
(d) Implementing techniques for processing such data and assimilating related information;
(e) Comparing, selecting, and applying hydrological technology appropriate to each country's needs and ensuring the transfer of appropriate technology, particularly between hydrological services."

In paragraph 3.14, the suggested action (b) is rather important for our topic:

"Recommended actions include:

(b) Assembling and disseminating basin-wide, regional, and global sets of water-related data and information for use, inter alia, in the management of resources within international river basins, and in climate change studies."

In order to achieve the goals and to make an advance before the year 2000 when national and global evaluations are again expected, urgent planning of the global hydrological network is desirable. The statement of the Ninth Session of the WMO Commission for Hydrology indicated the need for coordinated steps to identify the financial background for such an action. The approach can be itemized as follows:

(a) to agree on the principles in the selection of the hydrological observing (at least streamgaging) stations for a Global Hydrological System;
(b) a global survey on the selected rivers and stations;

(c) a tentative estimation of the costs of the development of the global network;
(d) to agree on the network based on correspondence with the national hydrological services and a conference with the proper representation of the countries;
(e) to design the stations and estimate the costs of the development of the selected stations;
(f) to ensure the necessary financial means and to plan the development project;
(g) to initiate the installation (and/or construction) of the streamgaging stations;
(h) to prepare the operation manual of the GHS and to organize training workshops for the staff at the operating agencies;
(i) to organize the regular data flow to regional and global centers (e.g., the Global Runoff Data Centre in Koblenz);
(j) to arrange the annual survey of the network and its annual report; and
(k) to have a detailed evaluation of the progress in the year 2000.

An international working group to be established, e.g., by WMO, UNESCO, and ICSU, can serve as an advisory board (or steering committee) for the execution of the project.

Most likely, the system can cover precipitation, evaporation, and soil moisture components as well, but the first priority should be given to runoff (streamgauging).

6. Conclusions

The NATO Advanced Study Workshop held in Tucson may serve as a starting point for the development of the global hydrological observing network which ensures:

(1) the adequate global coverage of the large and major rivers;
(2) the sufficient continuity in the data acquisition and processing;

(3) the desired reliability in the hydrological data, at least for key points and catchments of the world;

(4) support to the determination of the world water balance and to refine the available water resources of the main catchments; and

(5) the chance to monitor human and climate impact on the hydrological regime (at least on the streamflow) of the main rivers of the world.

NATO can be particularly instrumental in the initiation of the network over the countries of the Treaty as a pilot project for the global network, over North America and Europe. If this action can be triggered by the NATO Workshop in Tucson, it will commemoratively contribute to the development of global hydrology.

References

Framji KK (1978) International Standardisation in River Flow Measurement, Chapter 16 in: Hydrometery: Principles and Practices, John Wiley, Chichester.

Herschy RW (editor) (1978) Hydrometry: Principles and Practices, John Wiley, Chichester.

IAHS/UNESCO/WMO (1970) Proceedings of the International Symposium on Hydrometry (Koblenz, 1970), IAHS Publication No. 99.

International Conference on Water and the Environment (ICWE) (1992) The Dublin Statement and Report of the Conference.

Muszkalay L, O Starosolszky (1968) Field Investigations of Hydraulic Structures (in Hungarian), VITUKI-VIZDOK, Budapest.

Muszkalay L (1979) Three-dimensional turbulence in streams and its role in mixing, In: Proceedings of XVIII IAHR Congress (Cagliari) Paper Ba7.

Muszkalay L, O Starosolszky (1981) Intercomparison of principal hydrometric instruments In: Proceedings of XIX IAHR Congress (New Dehli), Paper A(b)5.

Starosolszky O, L Muszkalay, A Borzsonyi (1972) Discharge Measurement (in Hungarian), VITUKI. Budapest, Hungary.

Starosolszky O (1983) Trends in the development in hydrometry, *Hydrological Sciences Journal*, 28(1-3/1983).

Starosolszky O (1987) Applied Surface Hydrology Instrument and Methods of Observation and Measurements. Water Resources Publications, Littleton, Colorado.

United States Department of Agriculture (1979) Field Manual for Research in Agricultural Hydrology, Handbook No. 224, Washington, D.C.

World Health Organization (WHO) (1978) GEMS/Water Operational Guide, Published on behalf of the UNEP/WHO/UNESCO/WMO project on global water quality monitoring, WHO, Geneva, Switzerland.

World Meteorological Organization (WMO) (1975) Modern Developments in Hydrometry (Proceedings of WMO/UNESCO/IAHS/University Padua Seminar), WMO No. 427, Geneva, Switzerland.

World Meteorological Organization (WMO) (1977) Report on Intercomparison of Principal Hydrometric Instruments (first phase, 1972-1976), WMO Geneva, Switzerland.

World Meteorlogical Organization (WMO) (1980) Stream gauging manual, Vols. I and II, Operational Hydrology Report No. 13, WMO 519, Geneva, Switzerland.

World Meteorological Organization (WMO) (1987) Telemetry and Data Transmission for Hydrology, *Proceedings of the International Workshop*, Toulouse, France, March 23-27, 1987.

World Meteorological Organization (WMO) (1987) Infohydro Manual, Hydrological Information Referral Service, WMO-No. 683, Geneva, Switzerland.

World Meteorological Organization (WMO) (1992) Technical Regulations, Volume III, Hydrology, Basic Documents, No. 2, WMO No. 49.

World Meteorological Organization (WMO) (1993) Commission for Hydrology, Abridged Final Report of the Ninth Session, WMO-No 789, Geneva, Switzerland.

The Challenges We Face: Panel Discussion on Runoff

Edited by Hoshin Gupta and Soroosh Sorooshian

1. Introduction

This panel discussion was moderated by *Francois de Troch* and *Jene Michaud*. The following is a summary of the major points of discussion, based primarily on notes prepared by moderator *Jene Michaud* and written contributions by participants *Muhammud Shaikh*, *Chuck George,* and *Jim Washburne*. Their assistance in documenting the discussions is greatly appreciated.

2. What is runoff, and how accurate are our estimates?

Moderator *de Troch* initiated the panel discussion by posing the following questions:

(a) Is "runoff" identical to "river discharge"?, and

(b) What does "accuracy of runoff data" actually mean?

These questions sparked a great deal of discussion, focused primarily on the second question. *Michaud* noted that while the accuracy of runoff measurements is satisfactory for many applications, it is quite poor for circumstances such as ice cover, inundated flood plains, shifting cross-sections and roughness, backwater conditions, or unsteady flow. *Sorooshian* added that although we do typically have detailed runoff data at the catchment scale, many studies are interested in the flow extremes (e.g., flood flows) where measurement uncertainties (largely due to uncertainties in the rating curve) are the largest. Thus, what do we do to improve rating curves at high flood stage? *Starosolszky* remarked that data uncertainties arise from many different sources, that we sometimes don't know how much of the uncertainty is caused by unknowns, and how much is caused by measurement error. Nonetheless, *de Troch* suggested that we must start including metadata with each data set, including explicit estimates of known measurement uncertainties. *Gupta* suggested that a workshop recommendation be that metadata

NATO ASI Series, Vol. I 46
Land Surface Processes in Hydrology
Trials and Tribulations of Modeling and Measuring
Edited by S. Sorooshian, H. V. Gupta, and S. C. Rodda

should be included with all data sets. On the flip side, it would be helpful for modelers to estimate the degree of measurement uncertainty which would be "acceptable" for different applications.

Goodrich raised the issue that if we want to improve confidence in our estimates of runoff, we need to relate the small-scale measurements to the nested large-scale grids. For example, a proper representation of runoff-related sediment transport needs knowledge of the small-scale distribution of properties. *Michaud* commented that this need for very detailed data is necessary for process studies. *Seguin* remarked that when going from smaller to larger scales, the quantities referred to as soil moisture and evapotranspiration remain conceptually the same, and he wanted to know if runoff is similar in that way. The answer to this question is not immediately apparent. Clearly, from the point of view of large-scale water balances, it is important to understand what "runoff" is at the scale of the GCM grid. *Smith* reported that the U.S. Geological Survey has data on more than 1,000 streams in the U.S., several of which are small streams. However, data at finer than a one-day time interval are difficult to obtain, and this causes difficulties when trying to obtain accurate estimates of runoff. *Kabat* wanted to know if the contribution of the base flow component of runoff is important on the global scale, to which *Sorooshian* replied that it was. *Michaud* stated that techniques to separate the surface and base flow components of runoff do exist but are difficult to apply. *Goodrich* also pointed out that, in Arizona, 98% of all precipitation evaporates, and only 1% is stored or becomes runoff. If this evaporation figure were to change to 99%, it would represent a significant amount of the state's total water balance. The question is: are changes of a couple of percentages even observable? Clearly, much more needs to be known about the issue of runoff at the larger scales.

A complication related to the above discussion is that in some parts of the world, particularly mountain, jungle, and arctic regions, even basic data on runoff data are lacking. Some of these areas are ones expected to be most influenced by climate change. For example, the Boreas region has no measurements of small-scale basins. *Goodison* felt that rather than putting a great deal of effort into improving the accuracy of existing gaging stations, we should concentrate on new, ungaged areas in which we can start from scratch. *Rodda* suggested that the arctic and mountainous regions are expected to be most significantly affected by global climate change, and that we need to install measurement devices there. *Hirschboeck* made a plea, however, that it is critical to keep stations with long records open, because these stations provide the best available history of extreme events and are critical for the calibration of proxy data sources.

3. Are we satisfied with measurements of discharge?

Michaud conducted a survey among the participants to determine what the general perception of the community was of its satisfaction with existing hydrologic flux data (see Chapter 5), and reported that the workshop participants felt most satisfied with the quality and availability of runoff data in comparison with data on other hydrologic fluxes such as precipitation, soil moisture, or evapotranspiration. It was generally agreed that existing discharge measurement stations will continue to be important for operational and scientific purposes, including calibration and validation of hydrologic models. Detailed discharge measurements, including measurements at small time steps, will continue to be of value for hydrologic process studies. *Michaud* and *Goodrich* pointed out, however, that "nested" data are essential for the study of scale issues. In particular, because runoff is an "integrating" variable (i.e., it provides a spatially and temporally integrated measure of variability in factors such as precipitation, temperature, soil moisture, and watershed characteristics), it is an ideal parameter for validation of monitoring. Therefore, it is important that a global hydrologic observing network be established for environmental monitoring. Long-term continuous records are particularly valuable, both for documenting global change and for studying extreme events. *Goodrich* indicated that improvements in runoff data would mainly be in terms of measurements being made at locations where they are not presently available.

4. How satisfied are we with our runoff models?

There are a wide variety of runoff models available, and we are gradually gaining experience with them. *Michaud* pointed out that the choice of model depends on, and will continue to depend on, various factors such as the modeling purpose, data constraints, scale, and climate. *Famiglietti* noted that we should generally use the simplest models which account for the above-mentioned constraints. However, there continues to be a controversy over the use of lumped versus distributed models. Support for use of distributed models appears to be based largely on a desire to take advantage of the increasing availability of distributed data (both ground and remotely sensed) and the interest in scale issues and modeling of large land surfaces. However, *Goodrich* pointed out that there is much debate over the appropriate degree of process or geometric complexity and concerns about parameter estimation and validation (establishing model believability). *Gupta* mentioned that the only real way to use complex distributed models is to set up the model using prior information and to improve the parameterization as the model outputs are compared to field data. He felt that this was an entirely legitimate approach to employ.

As might be expected, we have had the greatest success in modeling runoff phenomena for watersheds for which observation data are available. *Michaud* noted that we have been less successful in modeling runoff in ungaged watersheds or modeling runoff under watershed/climate conditions that are not found in the calibration data. There is still uncertainty regarding the accuracy of model outputs when used to evaluate impacts of environmental changes.

5. Challenges for the hydrological community

The moderators *Jene Michaud* and *Francois de Troch* summarized the major conclusions of this panel discussion as follows:

(a) It is important to keep in mind the following facts:

- River discharge is not necessarily the same as "runoff".
- Baseflow is a component of both the hydrologic cycle and river flow.
- Anthropogenic effects are often present in discharge data.
- Sediment loads and water quality are important aspects of runoff.

(b) While satisfaction with runoff data is generally high, there are conditions under which discharge cannot be accurately measured, and there are regions of the world that lack basic discharge data.

(c) A global hydrologic observing network, including long-term stations and observations from data-poor regions, is important for environmental monitoring.

(d) Runoff modeling continues to be an active research area. Current issues include the appropriate degree of process or geometric complexity and the effects of scale and climate. There is a continuing controversy over lumped versus distributed models, fueled by the availability of distributed data on the one hand, and by parameter estimation and validation concerns on the other. There is increasing evidence that simple models can be as accurate as complex models.

(e) At the GCM scale, a comparison between runoff and the difference between precipitation and evaporation is strongly recommended.

SUBJECT INDEX

The ASI Series Books Published as a Result of Activities of the Special Programme on Global Environmental Change

This book contains the proceedings of a NATO Advanced Research Workshop held within the activities of the NATO Special Programme on Global Environmental Change, which started in 1991 under the auspices of the NATO Science Committee.

The volumes published as a result of the activities of the Special Programme are:

Vol. 1: **Global Environmental Change.** Edited by R. W. Corell and P. A. Anderson. 1991.
Vol. 2: **The Last Deglaciation: Absolute and Radiocarbon Chronologies.** Edited by E. Bard and W. S. Broecker. 1992.
Vol. 3: **Start of a Glacial.** Edited by G. J. Kukla and E. Went. 1992.
Vol. 4: **Interactions of C, N, P and S Biogeochemical Cycles and Global Change.** Edited by R. Wollast, F. T. Mackenzie and L. Chou. 1993.
Vol. 5: **Energy and Water Cycles in the Climate System.** Edited by E. Raschke and D. Jacob. 1993.
Vol. 6: **Prediction of Interannual Climate Variations.** Edited by J. Shukla. 1993.
Vol. 7: **The Tropospheric Chemistry of Ozone in the Polar Regions.** Edited by H. Niki and K. H. Becker. 1993.
Vol. 8: **The Role of the Stratosphere in Global Change.** Edited by M.-L. Chanin. 1993.
Vol. 9: **High Spectral Resolution Infrared Remote Sensing for Earth's Weather and Climate Studies.** Edited by A. Chedin, M.T. Chahine and N.A. Scott. 1993.
Vol. 10: **Towards a Model of Ocean Biogeochemical Processes.** Edited by G. T. Evans and M.J.R. Fasham. 1993.
Vol. 11: **Modelling Oceanic Climate Interactions.** Edited by J. Willebrand and D.L.T. Anderson. 1993.
Vol. 12: **Ice in the Climate System.** Edited by W. Richard Peltier. 1993.
Vol. 13: **Atmospheric Methane: Sources, Sinks, and Role in Global Change.** Edited by M. A. K. Khalil. 1993.
Vol. 14: **The Role of Regional Organizations in the Context of Climate Change.** Edited by M. H. Glantz. 1993.
Vol. 15: **The Global Carbon Cycle.** Edited by M. Heimann. 1993.
Vol. 16: **Interacting Stresses on Plants in a Changing Climate.** Edited by M. B. Jackson and C. R. Black. 1993.
Vol. 17: **Carbon Cycling in the Glacial Ocean: Constraints on the Ocean's Role in Global Change.** Edited by R. Zahn, T. F. Pedersen, M. A. Kaminski and L. Labeyrie. 1994.
Vol. 18: **Stratospheric Ozone Depletion/UV-B Radiation in the Biosphere.** Edited by R. H. Biggs and M. E. B. Joyner. 1994.
Vol. 19: **Data Assimilation: Tools for Modelling the Ocean in a Global Change Perspective.** Edited by P. O. Brasseur and J. Nihoul. 1994.

Vol. 20: **Biodiversity, Temperate Ecosystems, and Global Change.**
Edited by T. J. B. Boyle and C. E. B. Boyle. 1994.

Vol. 21: **Low-Temperature Chemistry of the Atmosphere.**
Edited by G. K. Moortgat, A. J. Barnes, G. Le Bras and J. R. Sodeau. 1994.

Vol. 22: **Long-Term Climatic Variations – Data and Modelling.**
Edited by J.-C. Duplessy and M.-T. Spyridakis. 1994.

Vol. 23: **Soil Responses to Climate Change.**
Edited by M. D. A. Rounsevell and P. J. Loveland. 1994.

Vol. 24: **Remote Sensing and Global Climate Change.**
Edited by R. A. Vaughan and A. P. Cracknell. 1994.

Vol. 25: **The Solar Engine and Its Influence on Terrestrial Atmosphere and Climate.**
Edited by E. Nesme-Ribes. 1994.

Vol. 26: **Global Precipitations and Climate Change.**
Edited by M. Desbois and F. Désalmand. 1994.

Vol. 27: **Cenozoic Plants and Climates of the Arctic.**
Edited by M. C. Boulter and H. C. Fisher. 1994.

Vol. 28: **Evaluating and Monitoring the Health of Large-Scale Ecosystems.**
Edited by D. J. Rapport, C. L. Gaudet and P. Calow. 1995.

Vol. 29: **Global Environmental Change Science: Education and Training.**
Edited by D. J. Waddington. 1995.

Vol. 30: **Ice Core Studies of Global Biogeochemical Cycles.**
Edited by R. J. Delmas. 1995.

Vol. 31: **The Role of Water and the Hydrological Cycle in Global Change.**
Edited by H. R. Oliver and S. A. Oliver. 1995.

Vol. 32: **Atmospheric Ozone as a Climate Gas.**
Edited by W.-C. Wang and I. S. A. Isaksen. 1995.

Vol. 33: **Carbon Sequestration in the Biosphere.**
Edited by M. A. Beran. 1995.

Vol. 34: **Climate Sensitivity to Radiative Perturbations: Physical Mechanisms and Their Validation.**
Edited by H. Le Treut. 1996

Vol. 35: **Clouds, Chemistry and Climate.**
Edited by P. J. Crutzen and V. Ramanathan. 1996.

Vol. 36: **Diachronic Climatic Impacts on Water Resources.**
Edited by A. N. Angelakis and A. S. Issar. 1996.

Vol. 37: **Climate Change and World Food Security.**
Edited by T. E. Downing. 1996.

Vol. 38: **Evaluation of Soil Organic Matter Models.**
Edited by D. S. Powlson, P. Smith and J. U. Smith. 1996.

Vol. 39: **Microbiology of Atmospheric Trace Gases.**
Edited by J. C. Murrell and D. P. Kelly. 1996.

Vol. 40: **Forest Ecosystems, Forest Management and the Global Carbon Cycle.**
Edited by M. J. Apps and D. T. Price. 1996.

Vol. 41: **Climate Variations and Forcing Mechanisms of the Last 2000 Years.**
Edited by P. D. Jones, R. S. Bradley and J. Jouzel. 1996.

Vol. 42: **The Mount Pinatubo Eruption: Effects on the Atmosphere and Climate.**
Edited by G. Fiocco, D. Fuà and G. Visconti. 1996.

Vol. 43: **Chemical Exchange Between the Atmosphere and Polar Snow.**
Edited by E. W. Wolff and R. C. Bales. 1996.

Vol. 44: **Decadal Climate Variability: Dynamics and Predictability.** Edited by D. L.T. Anderson and J. Willebrand. 1996.

Vol. 45: **Radiation and Water in the Climate System: Remote Measurements.** Edited by E. Raschke. 1996.

Vol. 46: **Land Surface Processes in Hydrology: Trials and Tribulations of Modeling and Measuring.** Edited by S. Sorooshian, H. V. Gupta, and J. C. Rodda. 1997.

Vol. 47: **Past and Future Rapid Environmental Changes: The Spatial and Evolutionary Responses of Terrestrial Biota.** Edited by B. Huntley, W. Cramer, A.V. Morgan, H. C. Prentice, and J. R. M. Allen. 1997.

Vol. 48: **The Mathematics of Models for Climatology and Environment.** Edited by J. I. Díaz. 1997.

Vol. 49: **Third Millennium BC Climate Change and Old World Collapse.** Edited by H. N. Dalfes, G. Kukla, and R. Weiss. 1997.

Vol. 50: **Gravity Wave Processes: Their Parameterization in Global Climate Models.** Edited by K. Hamilton. 1997.

Springer and the environment

At Springer we firmly believe that an international science publisher has a special obligation to the environment, and our corporate policies consistently reflect this conviction.

We also expect our business partners – paper mills, printers, packaging manufacturers, etc. – to commit themselves to using materials and production processes that do not harm the environment. The paper in this book is made from low- or no-chlorine pulp and is acid free, in conformance with international standards for paper permanency.

Springer

MIX
Papier aus verantwortungsvollen Quellen
Paper from responsible sources
FSC® C105338

If you have any concerns about our products,
you can contact us on
ProductSafety@springernature.com

In case Publisher is established outside the EU,
the EU authorized representative is:
Springer Nature Customer Service Center GmbH
Europaplatz 3, 69115 Heidelberg, Germany

Printed by Libri Plureos GmbH
in Hamburg, Germany